Energy and Technical Building Systems—Scientific and Technological Advances

Energy and Technical Building Systems—Scientific and Technological Advances

Special Issue Editors

Jarek Kurnitski
Andrea Ferrantelli

MDPI • Basel • Beijing • Wuhan • Barcelona • Belgrade

Special Issue Editors
Jarek Kurnitski
Tallinn University of Technology
Estonia

Andrea Ferrantelli
Tallinn University of Technology
Estonia

Editorial Office
MDPI
St. Alban-Anlage 66
4052 Basel, Switzerland

This is a reprint of articles from the Special Issue published online in the open access journal *Energies* (ISSN 1996-1073) in 2019 (available at: https://www.mdpi.com/journal/energies/special_issues/energy_and_built_environment).

For citation purposes, cite each article independently as indicated on the article page online and as indicated below:

LastName, A.A.; LastName, B.B.; LastName, C.C. Article Title. *Journal Name* **Year**, *Article Number*, Page Range.

ISBN 978-3-03928-178-7 (Pbk)
ISBN 978-3-03928-179-4 (PDF)

© 2020 by the authors. Articles in this book are Open Access and distributed under the Creative Commons Attribution (CC BY) license, which allows users to download, copy and build upon published articles, as long as the author and publisher are properly credited, which ensures maximum dissemination and a wider impact of our publications.

The book as a whole is distributed by MDPI under the terms and conditions of the Creative Commons license CC BY-NC-ND.

Contents

About the Special Issue Editors vii

Preface to "Energy and Technical Building Systems—Scientific and Technological Advances" ix

Kristina Mjörnell, Dennis Johansson and Hans Bagge
The Effect of High Occupancy Density on IAQ, Moisture Conditions and Energy Use in Apartments
Reprinted from: *Energies* **2019**, *12*, 4454, doi:10.3390/en12234454 1

Andrea Ferrantelli, Jevgeni Fadejev and Jarek Kurnitski
Energy Pile Field Simulation in Large Buildings: Validation of Surface Boundary Assumptions
Reprinted from: *Energies* **2019**, *12*, 770, doi:10.3390/en12050770 12

Mehdi Taebnia, Sander Toomla, Lauri Leppä and Jarek Kurnitski
Air Distribution and Air Handling Unit Configuration Effects on Energy Performance in an Air-Heated Ice Rink Arena
Reprinted from: *Energies* **2019**, *12*, 693, doi:10.3390/en12040693 32

Janne Hirvonen, Juha Jokisalo, Juhani Heljo and Risto Kosonen
Towards the EU Emission Targets of 2050: Cost-Effective Emission Reduction in Finnish Detached Houses
Reprinted from: *Energies* **2019**, *12*, 4395, doi:10.3390/en12224395 53

John Clauß, Sebastian Stinner, Christian Solli, Karen Byskov Lindberg, Henrik Madsen and Laurent Georges
Evaluation Method for the Hourly Average $CO_{2eq.}$ Intensity of the Electricity Mix and Its Application to the Demand Response of Residential Heating
Reprinted from: *Energies* , , 1345, doi:10.3390/en12071345 82

Jan Vanus, Ojan Majidzadeh Gorjani and Petr Bilik
Novel Proposal for Prediction of CO_2 Course and Occupancy Recognition in Intelligent Buildings within IoT
Reprinted from: *Energies* , *12*, 4541, doi:10.3390/en12234541 107

Simplice Igor Noubissie Tientcheu, Shyama P. Chowdhury and Thomas O. Olwal
Intelligent Energy Management Strategy for Automated Office Buildings
Reprinted from: *Energies* **2019**, *12*, 4326, doi:10.3390/en12224326 132

József Menyhárt and Ferenc Kalmár
Investigation of Thermal Comfort Responses with Fuzzy Logic
Reprinted from: *Energies* **2019**, *12*, 1792, doi:10.3390/en12091792 159

Ming Hu and Mitchell Pavao-Zuckerman
Literature Review of Net Zero and Resilience Research of the Urban Environment: A Citation Analysis Using Big Data
Reprinted from: *Energies* **2019**, , 1539, doi:10.3390/en12081539 172

Han Chang and In-Hee Lee
Environmental and Efficiency Analysis of Simulated Application of the Solid Oxide Fuel Cell Co-Generation System in a Dormitory Building
Reprinted from: *Energies* **2019**, *12*, 3893, doi:10.3390/en12203893 188

About the Special Issue Editors

Jarek Kurnitski is a professor at Tallinn University of Technology and Aalto University, as well as a vice-president of REHVA, Federation of European Heating and Air-Conditioning Associations, a non-profit organization representing more than 120,000 HVAC engineers and energy experts. He is the leader of the Estonian Center of Excellence in Research ZEBE, Zero Energy and Resource Efficient Smart Buildings, and the leader of the Nearly Zero Energy Buildings (NZEB) research group, which today operates at both universities. He is internationally known for the preparation of technical definitions for nearly zero energy buildings through many activities in the REHVA Technology and Research Committee and contributions to European standards. Recently he chaired a task force preparing a European residential ventilation guidebook. He has been deeply involved in the work to improve the energy efficiency of the built environment in Estonia and Finland with major contributions to the development of dynamic simulation-based energy calculation frames for present energy performance regulations.

Andrea Ferrantelli is a postdoctoral researcher at Tallinn University of Technology. He obtained his MSc in theoretical physics at Turin University (Italy) and his PhD in theoretical particle cosmology at the University of Helsinki (Finland). He is interested in the physical modelling of HVAC and in energy efficient buildings.

Preface to "Energy and Technical Building Systems—Scientific and Technological Advances"

Future buildings require not only energy efficiency but also proper building automation and control system functionalities in order to respond to the needs of occupants and energy grids. These development paths require a focus on occupant needs, such as good indoor climate, easy operability, and monitoring. Another area to be tackled is energy flexibility, which is needed to make buildings responsive to the price signals of electricity grids with increasing amounts of fluctuating renewable energy generation installed both in central grids and at building sites. This Special Issue is dedicated to HVAC systems, load shifting, indoor climate, energy, and ventilation performance analyses in buildings. All these topics are important for improving the energy performance of new and renovated buildings within the roadmap of low energy and nearly zero energy buildings (NZEB). To improve energy performance and, at the same time, occupant comfort and wellbeing, new technical solutions are required. The research in this Special Issue provides the evidence and experience of how such new technical solutions have worked in practice in new or renovated buildings, also showing potential problems and how the solutions should be further developed. Energy performance and indoor climate improvements are also a challenge for calculation methods. More detailed approaches are needed in order to be able to correctly design and size dedicated systems, and to be capable for accurate quantification of energy savings. To avoid common performance gaps between calculated and measured performance, occupant behavior and building operation must be adequately addressed. This demonstrates the challenge of the type of highly performing buildings, comfortable buildings with adequate indoor climate, and easy and cheap operation and maintenance, expected by end customers. Occupancy patterns and recognition, intelligent building management, demand response and performance of heating, and cooling and ventilation systems are some common keywords in the articles of this Special Issue contributing to the future of reliable, high performing buildings.

Jarek Kurnitski, Andrea Ferrantelli
Special Issue Editors

Article

The Effect of High Occupancy Density on IAQ, Moisture Conditions and Energy Use in Apartments

Kristina Mjörnell [1,2,*], Dennis Johansson [3] and Hans Bagge [2]

1. RISE Research Institutes of Sweden, 412 58 Gothenburg, Sweden
2. Division of Building Physics, Lund University, 221 00 Lund, Sweden; hans.bagge@byggtek.lth.se
3. Division of Building Services, Lund University, 221 00 Lund, Sweden; dennis.johansson@hvac.lth.se
* Correspondence: kristina.mjornell@ri.se; Tel.: +46-730-88-57-45

Received: 31 October 2019; Accepted: 20 November 2019; Published: 22 November 2019

Abstract: Apartments built in Sweden during the record years 1961–1975 with the aim to remedy the housing shortage and abolish poor standards, were designed for a normal-sized family of 2–4 persons. The mechanical ventilation system, if existing, was primarily designed to ensure an air exchange in the apartment according to Swedish building regulations. During the last few years, the number of overcrowded apartments has increased due to housing shortage in general but also due to migration. Another aspect is that the ventilation in many apartments built during the record years is already insufficient at normal occupant load. The question is how doubling or tripling the number of occupants and thus, the moisture load will affect the risk of bad air quality and moisture damage. To find out, simulations were made to estimate whether it is possible to obtain sufficient air quality and low risk of moisture damage by only increasing the ventilation rates in existing systems or introducing new ventilation systems with and without heat recovery and what the consequence would be in terms of the additional energy demand. Measurements from earlier studies of CO_2 and moisture supply in Swedish apartment buildings were used as input data.

Keywords: occupancy density; moisture conditions; energy use; indoor air quality; ventilation rate

1. Introduction

1.1. Housing Shortage Results in Overcrowded Apartments

The National Board of Housing, Building and Planning in Sweden has estimated a need for 600,000 new apartments by the year 2025 [1]. Shortage of housing and divergence in supply in relation to demand, together with an increasing population, has led to high-occupancy-density housing in many areas. The percentage of persons living in a dwelling with more than one person per bedroom, which is defined as "overcrowded" according to norm 3 [2], has increased from 15 to 17 percent from 2008 to 2018. Norm 3 defines overcrowding as less than one room for each occupant or two partners, except for kitchen, bathroom and living room. Among people born abroad the number is 38 percent in 2018 [3]. Many Swedish municipalities have major challenges to offer housing to both Swedes and migrants. The law states that migrants have the right to arrange their own accommodation with friends or relatives which, in some neighborhoods, has led to overcrowding in some residential areas. The average occupancy level in rental apartments in Sweden is 33 m^2 per person [4], which corresponds to 0.03 person/m^2.

1.2. Effects of High Occupancy Density on Indoor Quality and Demand for Ventilation

The buildings, where high occupancy density is common, were built during the "record years" (1961–1975), which means that the design, construction and heating-and ventilation technology are not adapted to the high occupancy density. These buildings, if not recently renovated, are in need of renovation since most materials and components are promptly approaching their technical end of life. Almost 70% of the houses from this era are heated by waterborne radiators connected to district heating and ventilated by natural ventilation (45%), exhaust ventilation (22%) or supply and exhaust ventilation (8%) [5]. A well-functioning ventilation system in buildings is a prerequisite both for achieving good indoor air quality and thermal comfort. However, in order to save energy, the ventilation rates are often decreased to the lowest possible levels, which, in some cases, has led to the risk of CO_2 concentrations exceeding the maximum National permissible level and formaldehyde concentrations close to the value recommended by World Health Organization (WHO) [6]. In addition, risk of mold growth in the building's envelope, which may cause indoor environment problems must be avoided. At the same time, the energy demand of the building depends on the building envelope and ventilation system. Former studies have shown that the users have great influence on the indoor environment, energy use and moisture safety in homes, while a recent study showed that the variation in the users' behavior is large, [7–9]. Overcrowding, in the form of high personal load and related activities such as cooking, washing and shower affect the building's constructions, materials and systems. If, for example, furniture or textiles are placed against poorly insulated external walls, the inside surface of the wall becomes cold, and in combination with high moisture load, this may result in high levels of relative humidity, condensation and risk of microbial growth. In this case, a higher ventilation rate limits the moisture load on the building's structural parts and avoids high concentrations of indoor pollutants. Even though more people are contributing to an internal heating load, an increased ventilation rate without heat recovery will result in higher energy demand. To force the ventilation to higher levels than the system is designed for is probably possible with existing fans, but it would probably cause noise and draft indoors. According to the Swedish Building regulations, the exhaust air flow should at least equal 0.35 liter per second per square meter of living space. It is allowed to reduce the exhaust air flow to a minimum of 0.10 liter per second and square meters of living space when no one is home, to save energy. The air in the home must not cause unpleasant odors or adverse health effects. To achieve this, a higher outdoor air flow is sometimes required than the lowest permissible [10]. The building owner must also carry out an Obligatory Ventilation Control, OVK, in all buildings on a regular basis. Its purpose is to show that the indoor climate is good and that the ventilation systems are functioning [10]. The control must be carried out by a certified inspector. The controller must also provide suggestions on how to reduce energy consumption for ventilation without this giving rise to a worse indoor environment [11]. The Swedish Building Regulations (BBR) regulate the design of building in terms of function and technical systems but do not regulate the number of people using it. Since 1974, the Swedish norm (norm 3) for apartments is one room per person or couple, except for kitchen, bathroom and living room [12].

1.3. Effects of High Occupancy Density on General Wellbeing

International research studies on overcrowding in housing show that children living in such conditions more often have problems associated with mental and physical health. Overcrowded housing also has higher demands on ventilation and hygiene, as respiratory diseases and gastric disease otherwise spread more easily. Research has also shown that the relationship between parents and children is often subjected to strains when the number of residents per room increases—with conflicts as well as lack of personal integrity and peace as a result [13,14]. Not enough beds in the apartment and no room for storage of mattresses, makes it difficult to keep the apartment clean, which affects the indoor air quality and may cause respiratory diseases. Kitchens and bathrooms are exposed to high moisture loads, which can lead to moisture problems, fungi and mold. An extensive relocation of furniture, for example, to making temporary beds, also lead to wear of surface layers, floors, lining and sockets. Overall, high occupancy levels result in higher consumptions of water and energy [15].

1.4. The Effect of Occupancy Density on Energy Use

It can be noted that the use of energy for heating, electricity and water is unreasonably high in some apartments compared to the average household, which could be due to high occupancy, [16]. On the other hand, comparing the use per person, the numbers are more reasonable as when many people live in the same apartment it is obvious that they consume more electricity and water than an apartment with less occupants. Johansson [16] compared energy use per resident to the energy use in kWh/m^2 (heated floor area) and showed that energy use per capita is almost the same for different building types. Even though the multifamily buildings often have poor energy performance with a high total energy use, the energy use per capita is not so high since the floor area per capita is lower for multifamily buildings than for 1–2 family buildings [16]. The proportionally lower energy use per person in apartments with high occupancy level is probably because the occupants contributes to internal heat load and that activities such as cooking are done simultaneously for the entire household. A Finnish study has showed similar results, measuring an increase in energy use in school buildings when intensifying the use of facilities, and suggests that alternative indicators such as energy use adjusted for occupancy or energy use adjusted for usage and space efficiency, should be used when evaluating energy efficiency in buildings [17].

1.5. The Effect of Occupancy Density on Indoor Environment and Moisture supply

Previously, in 1993, Markus et al. stated that overcrowding and inadequate ventilation increase interior moisture load [18]. Moisture supply, defined as the difference in absolute vapor content indoors and outdoors, is the result of moisture-generating activities, which are typically related to the occupants of the building. Examples of such activities are cooking, showering and watering of plants but also, moisture produced by the human body. The relation between moisture supply and occupancy is important to understand in order to make appropriate predictions on ventilation need to assure a good indoor air and to assure the moisture safety of the building. Unfortunately, there are few studies based on measured detailed data during longer periods of time of those two parameters. Møller and de Place Hansen [19] analyzed the influence of occupancy on moisture supply in Danish single-family houses and found a slight indication of increased moisture supply for low-social-status areas and increased moisture supply for households with high occupancies in terms of number of square meters per occupant and total hours spent at home. Moisture supply, however, decreases somewhat with increasing average age, which could be due to less moisture producing activities such as cooking and showering. Moisture supply varied for different types of rooms; bathrooms had the highest moisture supply, followed by basements and living rooms [19].

1.6. Aim of the Study

There is a lack of research dealing with the effects of overcrowding in apartments, and also, how to design and manage apartments as a function of occupant density. The hypothesis is that a higher occupancy level results in a higher moisture supply that needs to be reduced by increased ventilation to avoid the negative effects on indoor air quality and mold growth. The increased demand for ventilation will, in turn, affect the energy use, depending on the type of ventilation system and degree of heat recovery. As a starting point, this study aims to describe the problem with increased occupancy density and overcrowded apartments, estimating the relation between occupancy and moisture supply based on theoretical assumptions and measurements, and use this relation to show the effect of occupancy level on energy demand for heating and ventilation due to the need of increased ventilation airflow.

2. Materials and Methods

2.1. General Outline of the Methods Used

The analyses of the relationship between occupancy and moisture supply is based on measurements in apartments carried out in previous studies. The analyses of the energy demand for ventilation to assure to not exceed a certain moisture supply is made by simulations for different occupancy levels, climatic conditions, building envelope and ventilation systems. Moisture supply, v_{ms}, is the difference between the indoor vapor content and the outdoor vapor content. The moisture supply is the potential that drives the moisture transport, for example, through an exterior wall, and is, therefore, a crucial parameter, which is used in this initial study for determining the effect of moisture on ventilation and energy demand. The relationship between moisture supply and occupancy was determined with a top-down approach from estimations of occupancy level, based on measurements of CO_2, and the moisture supply, to describe the relationship between the two parameters using statistical analysis of the measured data. The alternative approach would have been a bottom-up approach describing every persons' activities in the household, and how they relate to generate vapor. There are numerous types of activities, all more or less difficult to describe, and there are more or less unknown relationships between the activities and the people performing them, meaning that estimating moisture supply based on occupancy activity schedules introduces many uncertainties. Therefore, the top-down approach was used as a first step since it gives the actual input data of interest directly and will be based on actual measurements. To be able to maintain a certain allowable moisture supply, the ventilation rate must be adjusted, which will influence the energy use of the building. For a given occupancy level, and the generated moisture supply, based on the given relationship estimated from measurement, the needed ventilation airflow and thus, the annual energy demand was determined from simulations of a certain building in a certain outdoor climate. The assumption is that no windows are opened, which is the worst-case scenario, but valid for approximately 30% of the apartments in Swedish multifamily houses [5].

2.2. Measurements of Carbon Dioxide, Temperature and Moisture Supply

In a former study, moisture supply and occupancy have been measured at the building level during more than one and a half years in four multi-family buildings, located in the south of Sweden, latitude N55°34' comprising a total of 72 apartments and an apartment floor area of 4920 m². The buildings used mechanical-exhaust-air ventilation with an air flow of 0.35 l/(s·m²$_{floor}$). Exhaust air was taken out from bathrooms, kitchens, and fresh outdoor air from inlets located in bedrooms and living rooms. Measurement equipment was placed outdoors, and in each buildings' central exhaust air duct. In this way, the measurements covered a number of apartments that were included at a reasonable cost, compared to measuring in each individual apartment. On the other hand, it is not possible to find distributions between apartments inside a certain building or distributions between different rooms in an apartment. The measured parameters were temperatures, Onset, specified error ±0.35 °C, relative humidity, Onset, specified error ±2.5% absolutely, and carbon dioxide, SenseAir, specified error ±30 ppm ±3% of reading with a repeatability error of ±20 ppm ±1% of reading. Measurements were performed with readings every 30 min to make it possible to obtain daily and weekly time distributions. Moisture supply was calculated based on these parameters [7].

2.3. Calculations of Occupancy Level based on CO_2 Measurements

Occupancy levels were calculated based on the measured CO_2 concentrations, indoors and outdoors, and the ventilation airflow. The ventilation airflow was assumed to be constant and the ventilation rate was obtained from measurements made by the operation team during commissioning of the building. The production of carbon dioxide can be described by Equation (1), where C_p is the carbon dioxide production in l/s, C_{in} is the carbon dioxide volume concentration in the exhaust air, and C_{out} is the carbon dioxide volume concentration in the outdoor air and q is the ventilation airflow of the building in l/s including leakage. It is assumed that there is no buffering of CO_2. The effect of buffering and time lags will be a matter for future analysis. If a single person produces a concentration of c_p l/s carbon dioxide indoors, Equation (2) gives the equivalent number of persons, n, in the building. C_p can be corrected for considering other producing or reducing sources in the building, using the constant q. This gives the equilibrium state of occupancy, as there is a transient build up and a transient decay time. Due to the linearity, the average will be described by Equation (2).

$$C_p = (C_{in} - C_{out}) \cdot q \tag{1}$$

$$n = \frac{C_p}{c_p} \tag{2}$$

$$O = \frac{n}{A_{floor}} \tag{3}$$

The metabolic rate varies with activity level, age, sex and weight, and, therefore, an average c_p must be determined. Reference values are presented in the literature [20] as values per body weight for adults at the different activity levels sleeping, 0.17 l/(h·kg), sitting, 0.26 l/(h·kg) and standing, 0.30 l/(h·kg). Regarding children, the carbon dioxide generation was analyzed by [21] based on reference values from [22], where children produce 1.74 times as much CO_2 as adults per kg body mass. If it is assumed that there are equal numbers of men and women, the average weight of adult Swedes is 74.5 kg according to Statistics Sweden. In this study, it was assumed that 8 h were spent sleeping, 4.5 h sitting and 1.5 h standing. Children were assumed to weigh, on average, half as much as adults, and 60% of the population were assumed to be adults. Taking all the above into account, the average c_p became 0.00417 l/s. Swedes are assumed to spend 14 h a day in their homes according to [23], while those measurements gave actual numbers. Equation (3) gives the occupancy, O, as used in this paper, where A_{floor} is the apartment floor area.

2.4. Relation between Occupancy and Moisture Supply

Linear regression analysis was carried out to describe the moisture supply as a function of occupancy level according to Equation (4), where A and B are constants and O the occupancy in persons/m². Figure 1 shows that the lower the time resolution is, the higher the coefficient of determination, R^2, becomes. Time resolution is the monitoring interval of the measurements. This is expected since fewer data will vary less. It is also shown that the influence on the coefficient of performance from time offset and reasonably higher resolution is rather low, indicating that the correlation is more dependent on the average occupancy than time lag aspects of it. Figures 2 and 3 show the moisture supply over time and as function of O, respectively. Negative moisture supply occurs when outdoor climate changed.

$$v_{ms} = A + B \cdot O \tag{4}$$

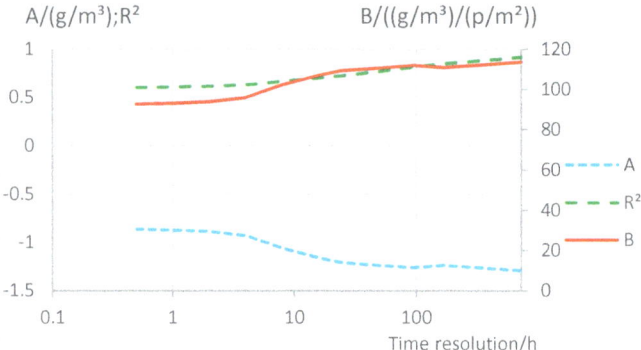

Figure 1. The resulting correlation parameters according to Equation (4) as a function of time resolution for measured data.

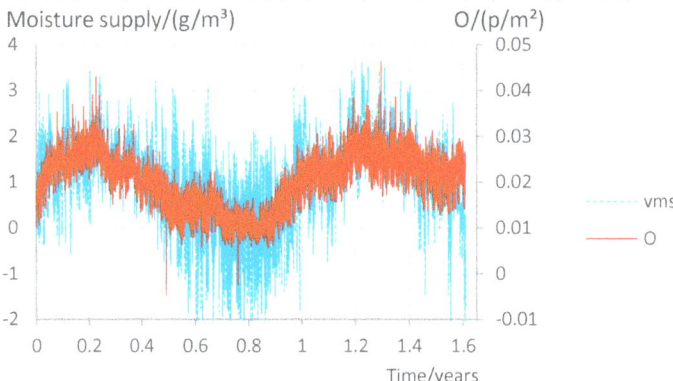

Figure 2. The yearly variation of measured moisture supply (v_{ms}) and occupancy level (O), estimated from CO_2 measurements in exhaust ventilation from apartment buildings.

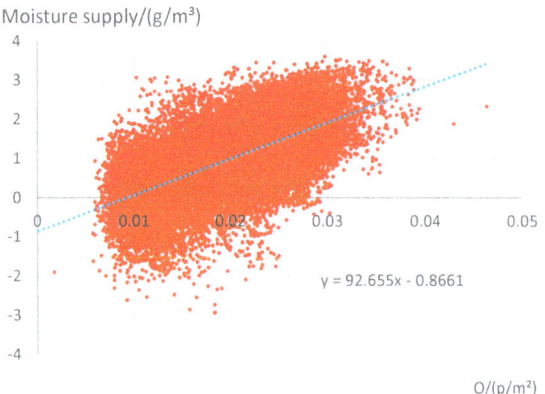

Figure 3. The linear regression curve describing the relation between moisture supply and occupancy level.

2.5. Calculations of Energy Demand for Heating and Ventilation

Based on the linear regression, the relation between occupancy levels and moisture supply is estimated. In the calculation, the ventilation airflow is varied in order to keep the moisture supply constant if the occupancy level deviates. The annual heating energy demand including the heating of the ventilation airflow is calculated. A power-balance model according to Equation (5) was developed in Visual Basic for Application based on the size of the measured building, taking into account, heat transmission of the building envelope, air leakage, ventilation and internal heat gains that were assumed to be constant over time. P_{heat} is the heating power, or hourly heating energy, which must be supplied by the heating system, per floor area and q is the ventilation airflow per floor area. A heat loss factor of the building is, for the purpose of this study, defined as heat loss as a result of transmission and leakage calculated per envelope area and per temperature difference. Equation (6) describes the heat loss factor, Q, with U as the heat transmission coefficient, q_{leak} the average leakage flow, ρ the density of air (1.2 kg/m³) and c the specific heat capacity of air (1000 J/(kg·K)) and A_{env} the total envelope area.

$$P_{heat} = max(Q \cdot (t_{in} - t_{out}) - P_{gain}; 0) + max(q \cdot \rho \cdot c \cdot (t_{in} - t_{hr}); 0) \tag{5}$$

$$Q = \frac{(U + q_{leak} \cdot \rho \cdot c) \cdot A_{env}}{A_{floor}} \tag{6}$$

$$t_{hr} = \eta \cdot (t_{in} - t_{out}) + t_{out} \tag{7}$$

The calculation was carried out with three different heat loss factors, (Q = 0.2, 0.6 and 1 W/(m²$_{floor}$·K), respectively, and a nominal airflow, q, according to the Swedish building code [2] of 0.35 l/(s·m²) at the average occupancy level in the measured cases of 0.0204 persons/m²$_{floor}$. The simulation was performed for every hour over one year. The annual energy use is the sum of each hour's energy use. The temperature efficiency of the heat recovery, η, of the ventilation air handling unit was either zero, representing no heat recovery, or 0.75, representing a typical heat recovery unit during operation. Equation (7) gives the temperature after the heat recovery, t_{hr}. A heat gain, P_{gain}, of 100 W/person together with a heat gain of 3.4 W/m²$_{floor}$ from household electricity heating [24] was assumed. The outdoor climate data, t_{out}, was retrieved from Meteonorm (Meteotest, 2016) for the cities of Malmö, south Sweden, latitude N55°36′ and Kiruna, latitude N67°51′. The floor area, A_{floor}, of the simulated and measured building was 4920 m², and the building envelope area, A_{env}, was 4788 m². The indoor temperature, t_{in} was assumed to be 22 °C.

3. Results

The energy demand for space heating and ventilation air heating were calculated for two fictive buildings with Malmö and Kiruna outdoor climates, respectively, see Figures 4 and 5. This was based on the value of a measurement resolution of 24 h according to Figure 1, which means that A = 109, B = 1.21 and coefficient of determination = R^2 = 0.72. It is shown that the energy demand for heating increases when occupancy levels increase if the ventilation airflow is adjusted to assure the same moisture content in the air. Of course, a more insulated envelope with lower Q also decreases the energy demand. It can also be seen that the benefits from a better insulated envelope are less at high occupancy levels because the internal heat loads decrease the number of heating degree hours. In Figure 6, heating for the ventilation air is shown. If the ventilation airflow is not increased, the energy used for heating is lower with higher occupancy due to more internal heat gains. On the other hand, this would result in more need for cooling or risk of overheating indoors.

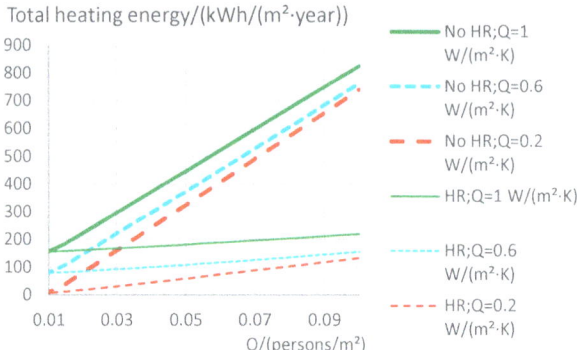

Figure 4. Calculated total heating demand as a function of average occupancy level, O, and heat loss factor, Q, for the building with outdoor climate from Kiruna. 'HR' means heat recovery in the ventilation system.

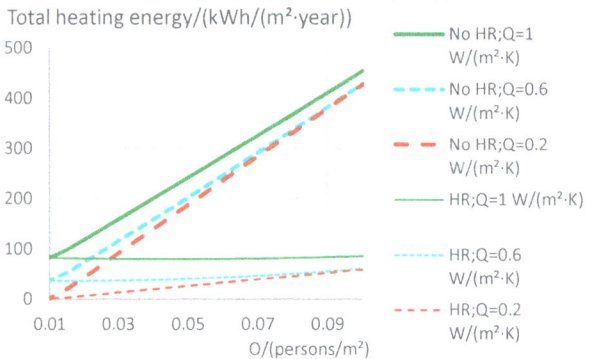

Figure 5. Calculated total heating demand as a function of average occupancy level, O, and heat loss factor, Q, for the building with outdoor climate from Malmö. 'HR' means heat recovery.

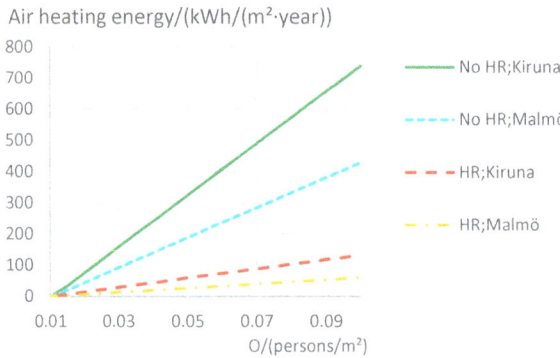

Figure 6. Calculated heating demand for the ventilation air flow as a function of average occupancy level, O, with outdoor climate from Kiruna and Malmö with and without heat recovery (HR). The heat-loss factor is 0.1 W/(m²·K).

4. Discussion

In this study, the occupancy level was estimated based on measured concentration of CO_2 because the concentration is relatively proportional to the number of people staying in the building. The moisture supply, on the other hand, depends on the number of people but also, on which activities are carried out, especially cooking and showering, which generate water vapor. It would, of course, be interesting to know how many people are staying in the apartment to be able to directly see correlation with moisture supply but there is a lack of such measurements. In the data which forms the basis of this study, there is a time lag based on the fact that the measurements were made on exhaust air. This air has been transported a certain distance and time after being in the rooms, but both parameters (CO_2 concentration and moisture content) were measured in the same way and, therefore, have the same time lag, and the linear relationships between the parameters mean that the average over many hours is still correct. The moderate coefficient of determination between moisture supply and occupancy, 0.72 in the simulated case, confirms that there are many aspects of the occupants' behavior that influence the moisture supply and the relationship between the number of occupants and moisture supply is, therefore, quite weak. This relationship should be studied more carefully in the future since it will be crucial for designing proper ventilation in apartments with high occupancy. It was also assumed that the level of presence in apartments was the same over the year, which can be overestimated for the summertime, since people are spending more time outdoors during summer. Generally, in Sweden, the society has the responsibility to ensure a sound and healthy housing market not based on overcrowding parts of the housing stock. However, the current situation with housing shortage cannot be solved in a quick way, and high occupancy levels will be a fact in parts of the housing stock during many years to come. Therefore, it is important to increase knowledge about measures to assure an acceptable indoor environment for the occupants and minimal risk of damage to the buildings in the concerned areas. The impact of high moisture content in the indoor air involves an increased risk for condensation at windowpanes, mold growth on cold indoor surfaces, moisture transfer and condensation in colder parts of the building envelop, etc. This must be studied more carefully, and measures have to be taken in exposed buildings. A possible measure could be to install exhaust and supply air ventilation with capability to force ventilation or even demand controlled ventilation, with heat recovery, especially for cold climates. A recent project in Sweden reported an approximate investment cost of 20,000 €/apartment for installing such a system in a residential building with eight apartments [25]. The cost per apartment should decrease if the same central air handling unit can serve more apartments. These efforts should be compared in relation to the consequences of not doing anything, which would involve risk for negative effects on occupants' health and for damages of the buildings in the form of mold growth, etc. It is obviously much more cost effective to replace the ventilation system and allow a higher occupancy level than to build more new homes. However, it must be borne in mind that there are more negative consequences of high occupancy density than just poor air quality and the risk of moisture damage, namely the difficulty in keeping the dwelling clean as well as providing a silent and suitable environment for children to do their homework and avoiding conflicts in the household.

This study is a first attempt to elaborate on the possibilities of increasing ventilation rate in order to improve the indoor environment in apartments with high occupancy. The relation between moisture supply and occupancy calculated with consideration of CO_2 concentrations is evaluated from measurements in apartments with normal occupancy rates and extrapolated to high occupancy rates, which might be a weakness since some activities generating moisture supply such as cooking, washing, laundry, plants, etc., might not be proportional to the number of occupants. The theoretical ventilation rates that are needed to keep the moisture supply down to acceptable levels were calculated using a simplified model of the building, which is a limitation. Based on the required ventilation rates, the total heating demand, as well as the total air heating demand, were estimated. Even though the results are not precise, they clearly reflect the increased demand of ventilation and consequently, an increased heating demand at increased occupancy levels, in order to provide a decent indoor environment.

5. Conclusions

This study has shown that increased occupancy density leads to an increased moisture supply, depending on the number of people staying in the apartment. To ensure good air quality and to keep the moisture supply at levels not exceeding levels involving a risk of damage to the building in the form of condensation and mold growth with negative consequences on the residents' health, the building must be ventilated with a certain exchange of air. It is possible to increase the airflow in apartments with high occupancy density so that the moisture supply does not exceed the critical levels to risk moisture damage. Simulations were made to estimate whether it is possible to obtain sufficiently good air quality and low risk of moisture damage by only increasing the ventilation rates in existing systems or introducing new ventilation systems with and without heat recovery. However, there will be an increased energy demand for heating of the air due to increased airflow, but the energy demand can be limited if heat-recovery is installed. As an example, the energy demand for heating of the ventilation airflow in an apartment of 70 m^2 will be almost twice as high, approximately, 370 to 700 kWh/m^2 per year for the Kiruna climate and 200 to 400 kWh/m^2 per year for the Malmö climate, if the occupancy level increases from three to six persons. However, the energy demand can be limited to about 50 KWh/m^2 per year for Kiruna and even less for Malmö, if a ventilation system with heat recovery on the exhaust air is installed.

Author Contributions: Conceptualization, K.M., D.J. and H.B.; methodology, K.M., D.J. and H.B.; software, D.J.; validation, D.J. and H.B.; formal analysis, D.J.; investigation, K.M., D.J. and H.B.; resources, K.M., D.J. and H.B.; data curation, D.J. and H.B.; writing—original draft preparation, K.M., D.J. and H.B.; writing—review and editing, K.M. and D.J.; visualization, D.J.; project administration, K.M. and H.B.; funding acquisition, K.M. and H.B.

Funding: This research was funded by Swedish Research Council Formas grant 2013-1804 SIRen national research environment on sustainable integrated renovation and by the Swedish Energy Agency, program E2B2 grant 41819-1, NOVA.

Conflicts of Interest: The authors declare no conflict of interest.

References

1. Boverket, Swedish Board of Housing Building and Planning. *Calculation of the Need of Housing Until 2025*; Swedish Board of Housing Building and Planning: Karlskrona, Sweden, 2017.
2. Boverket. *Trångboddheten i Storstadsregionerna*; Report 2016:28; Boverket: Karlskrona, Sweden, 2016; ISBN 978-91-7563-419-7.
3. SCB Statistics. Undersökningarna av Levnadsförhållanden (ULF), Statistics Sweden, Stockholm. Available online: http://www.scb.se (accessed on 13 September 2019).
4. Swedish Tenant Association. 2018. Available online: https://hurvibor.se/hur-vi-bor/bostadsyta/ (accessed on 21 November 2019).
5. Boverket. *BETSI Study—Buildings' Energy Use, Technical Status and Indoor Environment*; Boverket: Karlskrona, Sweden, 2009.
6. Dovjak, M.; Slobodnik, J.; Krainer, A. Deteriorated Indoor Environmental Quality as a Collateral Damage of Present Day Extensive Renovations. Strojniški vestnik. *J. Mech. Eng.* **2019**, *65*, 31–40. [CrossRef]
7. Bagge, H. *Building Performance—Methods for Improved Prediction and Verification of Energy Use and Indoor Climate, Building Physics LTH*; Lund University: Lund, Sweden, 2011.
8. Bagge, H.; Johansson, D.; Lindstrii, L. Measured indoor hygrothermal conditions and occupancy levels in an arctic Swedish multi-family building. *HVAC R Res. J.* **2014**, *20*, 376–383. [CrossRef]
9. Fransson, V.; Bagge, H.; Johansson, D. Impact of variations in residential use of household electricity on the energy and power demand for space heating—Variations from measurements in 1000 apartment. *Appl. Energy* **2019**, *254*, 113599. [CrossRef]
10. Boverket. Boverket's Building Regulations. Available online: https://www.boverket.se/en/start/building-in-sweden/swedish-market/laws-and-regulations/national-regulations/building-regulations/ (accessed on 17 November 2019).

11. Boverket. OVK Obligatory Ventilation Control. Available online: https://www.boverket.se/en/start/building-in-sweden/swedish-market/laws-and-regulations/national-regulations/obligatory-ventilation-control/ (accessed on 17 November 2019).
12. SCB Statistics. Available online: https://www.scb.se/hitta-statistik/artiklar/var-femte-person-fodd-utanfor-europa-ar-trangbodd/ (accessed on 13 September 2019).
13. Krieger, J.; Higgings, D.L. Housing and Health: Time Again for Public Health Action. *Am. J. Public Health* **2002**, *92*, 758–768. [CrossRef] [PubMed]
14. Bornehag, C.G.; Blomquist, G.; Gyntelberg, F.; Järvholm, B.; Malmberg, P.; Nordvall, L.; Sundell, J.; Pershagen, G.; Nielsen, A. Dampness in buildings and Health. Nordic interdisciplinary review of te scientific evidence on associations between exposure to "dampness" and health effects. *Indoor Air* **2001**, *11*, 72–86. [CrossRef] [PubMed]
15. Evidens. Trångboddhet i Sverige—Beskrivning av Nuläget och Diskussion om Effekter, Stockholm i december 2016. Available online: http://www.mynewsdesk.com/se/hsb_riksforbund/documents/traangboddhet-i-sverige-beskrivning-av-nulaeget-och-diskussion-om-effekter-64435 (accessed on 21 November 2019).
16. Johansson, T. Performance Visualization of Urban Systems. Ph.D. Thesis, Luleå University of Technology, Luleå, Sweden, 2017.
17. Sekki, T. Evaluation of Energy Efficiency in Educational Buildings. Ph.D. Thesis, Aalto University, Helsinki, Finland, 2017.
18. Markus, T.A. Cold, condensation and housing poverty. In *Unhealthy Housing Research Remedies and Reform*; Burrage, R., Ormandy, D., Eds.; Spon Press: New York, NY, USA, 1993; pp. 141–167.
19. Møller, E.B.; de Place Hansen, E. Moisture Supply in Danish single-family houses—The influence of occupant behavior and type of room. Proceeding of the 11th Nordic Symposium on Building Physics, NSB2017, Trondheim, Norway, 11–14 June 2017.
20. Swegon. *Technical Documentation on Ventilation Systems*; Swegon: Tomelilla, Sweden, 2006.
21. Coley, D.A.; Beisteiner, A. Carbon dioxide levels and Ventilation Rates in Schools. *Int. J. Vent.* **2003**, *1*, 45–52. [CrossRef]
22. Ruch, T.C.; Patton, H.D. *Physiology and Biophysics*; W B Saunders Company: Philadelphia, PA, USA, 1965.
23. Levin, P. Anvisningar för val av Brukarindata för Beräkningar av specifik Energianvändning i Bostäder, Final Report 2009-01-20. Available online: http://www.sbuf.se (accessed on 21 November 2019).
24. Bagge, H.; Lindstrii, L.; Johansson, D. *Brukarrelaterad Energianvändning. Resultat från Mätningar i 1300 Lägenheter*; FoU-Väst Rapport 1240; Sveriges Byggindustrier: Stockholm, Sweden, 2012.
25. Kristoffersson, J.; Bagge, H.; Abdul Hamid, A.; Johansson, D.; Almgren, M.; Persson, M.-L. *Användning av Värmeåtervinning i Miljonprogrammet*; Report 2017:17; Swedish Energy Agency: Frösön, Sweden, 2017.

 © 2019 by the authors. Licensee MDPI, Basel, Switzerland. This article is an open access article distributed under the terms and conditions of the Creative Commons Attribution (CC BY) license (http://creativecommons.org/licenses/by/4.0/).

Article

Energy Pile Field Simulation in Large Buildings: Validation of Surface Boundary Assumptions

Andrea Ferrantelli [1,*], **Jevgeni Fadejev** [1,2] **and Jarek Kurnitski** [1,2]

[1] Department of Civil Engineering and Architecture, Tallinn University of Technology, Ehitajate tee 5, 19086 Tallinn, Estonia; jevgeni.fadejev@taltech.ee (J.F.); jarek.kurnitski@taltech.ee (J.K.)
[2] Department of Civil Engineering, Aalto University, P.O. Box 12100, 00076 Aalto, Finland
* Correspondence: andrea.ferrantelli@taltech.ee

Received: 21 January 2019; Accepted: 20 February 2019; Published: 26 February 2019

Abstract: As the energy efficiency demands for future buildings become increasingly stringent, preliminary assessments of energy consumption are mandatory. These are possible only through numerical simulations, whose reliability crucially depends on boundary conditions. We therefore investigate their role in numerical estimates for the usage of geothermal energy, performing annual simulations of transient heat transfer for a building employing a geothermal heat pump plant and energy piles. Starting from actual measurements, we solve the heat equations in 2D and 3D using COMSOL Multiphysics and IDA-ICE, discovering a negligible impact of the multiregional ground surface boundary conditions. Moreover, we verify that the thermal mass of the soil medium induces a small vertical temperature gradient on the piles surface. We also find a roughly constant temperature on each horizontal cross-section, with nearly identical average values when either integrated over the full plane or evaluated at one single point. Calculating the yearly heating need for an entire building, we then show that the chosen upper boundary condition affects the energy balance dramatically. Using directly the pipes' outlet temperature induces a 54% overestimation of the heat flux, while the exact ground surface temperature above the piles reduces the error to 0.03%.

Keywords: energy piles; validation; floor slab heat loss; energy; computer simulations

1. Introduction

According to the European Parliament directive 2010/31/EU [1], each and every new construction should be nearly zero-energy buildings (nZEB) by the end of 2020. Such a requirement clearly demands an extensive use of renewable sources.

A recent review [2] on the utilization of geothermal energy [3,4] revealed that, by 2015, 49 countries invested over 20 billion USD in geothermal plants, which resulted in energy savings of ca 52.5 million tonnes of equivalent oil. This prevented respectively 46 and 148 million tonnes of carbon and CO_2 from being released into the atmosphere every year. Furthermore, the installed ground-source heat pump capacity grew 1.51 times from 2010 to 2015.

Ground-source heat pumps utilize ground heat exchangers (GHE) [5–7] to exploit geothermal energy. In buildings with pile foundations, installation of heat exchange piping into such piles enables them to perform as a GHE; the resulting systems are known as geothermal energy piles [8,9]. Their immediate advantage is that installation of heat exchange piping into a foundation pile is much cheaper than drilling a new borehole, therefore energy piles tend to be a very cost effective GHE solution.

Heat exchange processes occurring in boreholes are object of continuous studies from many different sides; a more refined design can reduce installation costs and improve heat transfer and heat pump efficiency (see for instance the reviews [10,11] and references quoted therein). Specifically, the physical processes involved can be addressed at various levels: from heat transfer into the soil to heat transport within the heat exchanger and the absorber pipes, investigating the role of design geometry, quantifying thermal interactions in multiple borehole fields etc. [11–13]. Classical analytical models include line-source [14] and cylindrical-source [15] solutions, which tend however to make too strong approximations, such as infinitely extended line sources and a constant heat flux inside an infinite and homogeneous medium. These could be extended in several ways though [13,16,17], from analytical *g-functions* [18] to superposition of thermal response functions (the so-called Hellström approach) [19,20]. Unfortunately, unavoidable approximations (such as constant wall temperatures) still limit the predictive power of these models [11,13].

Numerical computations can instead relax many assumptions that are generally needed to solve analytical models [21,22]. The geometry in particular can be more complex, accounting for finite lengths, thermal interaction of the heat exchangers, multilayer soil [23] and inclusion of the ground water flow [24,25]. A large number of different software types is being used in these investigations. Finite difference methods (FDM) used by TRNSYS [18,26] and IDA-ICE [27–30] are widely adopted, due to their ability to account for variable ground surface temperatures [31,32]. In some recent approaches, these can even be combined with probabilistic analysis [33].

The Finite Element Method (FEM) [34–36] is extensively used as well, for instance with ABAQUS [37–39] or COMSOL Multiphysics [40–43]. Several sensitivity studies assess design issues such as the role of heat exchanger configurations with a given energy pile, or the spacing and configuration of an energy pile group [44] and its vertical displacement [45]. FEM modelling is also used for developing analytical GHE heat transfer models [12,46], and Computational Fluid Dynamics (CFD) allows e.g., comparison of different types of ground heat exchangers [47]. Finally, thermomechanical effects that are induced on the heat exchanger constitute an emergent and promising field of investigation [48,49]. Experimental research is very active as well, on each and every level, providing empirical support and validation to both analytical and numerical models [11,50].

Despite a number of encouraging results, even in the numerical modelling of GHE a number of issues still need to be addressed [11,13,51]. In this paper we are mostly concerned with the assessment of variable ground surface temperatures. In the case of energy piles, this is generally described as the surface temperature of the building floor slab, which is assumed to coincide with the indoor air temperature. The soil region surrounding the building floor slab is exposed to outdoor air temperature, which unfortunately cannot be modelled in TRNSYS nor IDA-ICE due to lack of implementation [13,28]. Hypothetically, heat transfer in the soil region surrounding the building should affect the soil temperature development underneath, altering the heat transfer between the energy piles and, ultimately, their thermal yield. However, according to the present literature, the transient soil temperature profile still needs to be quantified.

Clearly, since the floor heat loss and temperature profile right under the concrete slab (or floor) are crucial for the correct implementation of energy pile analyses, one should pay particular attention to the upper boundary condition. One first objective of this study was indeed to quantify how the soil temperature and the energy piles thermal yield are affected by heat transfer processes in the soil region surrounding the building. This was accomplished numerically with COMSOL Multiphysics [34], which as remarked above, uses a finite element method (FEM).

We believe that reducing the gap between simulations and reality is essential: as it was proven in [52], energy consumption estimates from building simulations can differ by 30% when compared to measured values. Accordingly, to provide a phenomenological foundation for our analysis, we validate this specific FEM numerical model against measured data. These were taken from an undisturbed reference energy pile of an office building called Innova 2, located in Jyväskylä, Finland. Such validation is then used in simulations of heat transfer for a multi-pile system under a building, to investigate how the heat transfer processes depend on different temperature boundary conditions (b.c.) at the surface (multiregional or floor slab). We find that assuming a simple floor slab gives virtually no different result than a more complex multiregional b.c. proposed in [13], which can be safely disregarded in practical studies.

Another known bottleneck of energy piles modelling in IDA-ICE, prior to the November 2017 release, is the lack of calculated ground surface temperature as disturbed by the operation of an energy piles plant [28]. Such temperature constitutes the boundary condition in the floor slab model of a zone located directly above the energy piles. Since operation of the heat pump in heating mode cools the soil, heat losses in the floor slab increase the building heating need. To account for this phenomenon, a rough estimate of the disturbed ground surface temperature was applied in [28] (in this paper we will refer to this earlier method as "IDA-ICE outlet"). However, up to now the accuracy of such rough estimation was still unknown.

In this work we attempt to fill this gap by means of analogous calculations performed with a new version of IDA-ICE and with COMSOL. We define a borehole model for 20 energy piles, and compare the effect of two distinct upper boundary conditions: (i) energy piles outlet temperature—"IDA-ICE outlet", and (ii) ground surface temperature—"IDA-ICE slab". We find that the heat flux through the floor and the yearly energy demand computed according to (i) are overestimated by 54% and 5% respectively, in comparison with (ii), which is adopted both in COMSOL and in the updated version of IDA-ICE. Furthermore, when using the ground surface temperature as in case (ii), differences in the thermal yield between the two software are found to be quantitatively negligible.

The present paper is organized as follows: In Section 2 we define a 3D COMSOL model for a single energy pile, and validate it against measured data. In Section 3 a 2D reduction of the previous model performs ground surface boundary analysis for multiple energy piles. In Section 4 we validate an IDA-ICE energy piles model against a 3D COMSOL computation extended to multiple piles, and finally in Section 5 we compare the effect of different boundary conditions on a calculation of yearly energy demand. Our findings are then exposed in the Section 6.

2. Validation of a COMSOL Model for a Single Pile against Measured Data

In this section we discuss the validation of an FEM-based numerical simulation, performed with the software COMSOL Multiphysics. This preliminary was necessary as we use the same setup in the energy pile heat transfer analysis discussed in the rest of the paper, Sections 3–5.

2.1. Method

The first building with pile foundation used as GHE in Finland is an office building called Innova 2, built in summer 2012 in Jyväskylä. The geothermal heat pump plant is equipped with energy meters and two piles of foundation, with temperature sensors placed along the depth, Figures 1 and 2 (see [9] for a thorough description of the piles' construction, in combination with heat exchangers and heat pumps).

Additionally, a reference energy pile located near the building measures the undisturbed soil temperature with 11 sensors installed along its depth as well, as in Table 1. In our COMSOL simulation, we modelled precisely this isolated energy pile, together with its surrounding soil layers. The temperatures calculated in the model were logged from the location of each sensor of the reference pile. Depth, density ρ and thermal conductivity λ of each layer were measured on site [53].

As illustrated in Figure 3, the reference pile was modelled per measurements as a 22 m-long concrete cylinder, with $\lambda = 1.8$ W/mK, $\rho = 2400$ kg/m^3 and $c_p = 900$ kJ/kgK, and diameter 170 mm. It was embedded in a 10 m × 10 m, 26.7 m deep multilayer block with material properties listed in Table 2.

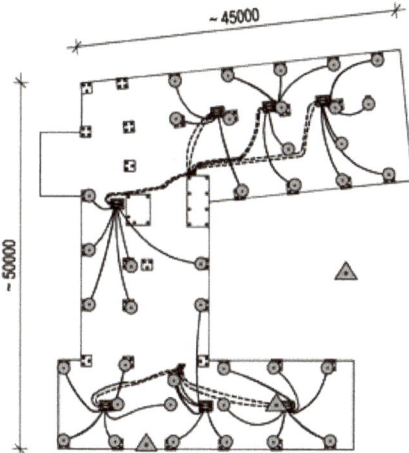

Figure 1. Energy piles (circles) and monitoring layout (undisturbed *T* monitored at the isolated triangle) [53].

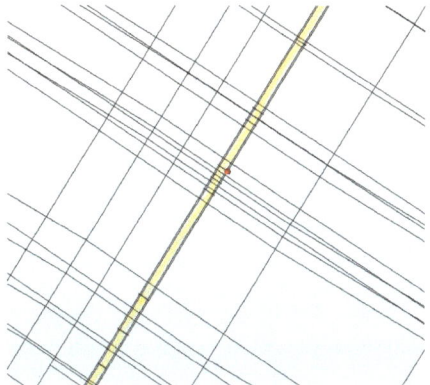

Figure 2. Sketch of sensor placement in soil, on the surface of each pile.

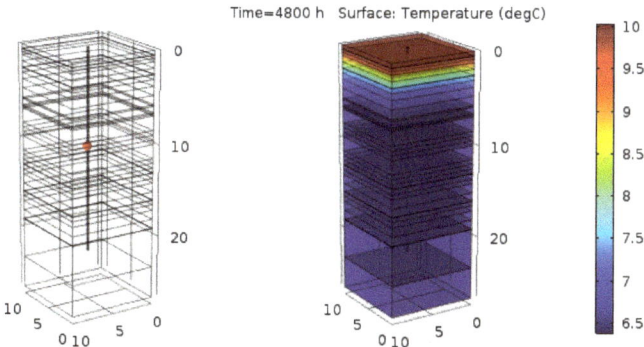

Figure 3. COMSOL simulation for the Innova office building reference pile. (**Left**) the pile geometry (sensor T31 in red). (**Right**) the result after t = 4800 h.

Table 1. Temperature sensors location for the reference energy pile [53].

Temperature Sensor	Depth, m
Ground surface	0
Pile top	−0.5
T28	−0.5
T29	−2.5
T30	−4.5
T31	−6.5
T32	−8.5
T33	−10.5
T34	−12.5
T35	−14.5
T36	−16.5

Table 2. Soil layer properties for the single pile simulation [53].

Layer nr	Depth, m	ρ, t/m^3	c_p, kJ/kgK	λ, W/mK
1	3.73	1.4	1.8	0.87
2	5.67	1.72	1.82	1.24
3	5.84	1.66	1.78	1.08
4	6.5	1.80	1.71	1.25
5	6.67	1.83	1.72	1.39
6	6.84	1.91	1.57	1.42
7	12.9	2.03	1.40	1.89
8	12.91	2.01	1.39	1.81
9	15.90	2.06	2.32	1.92
10	15.91	2.05	2.33	1.91
11	19	1.99	2.39	1.53
12	19.01	1.95	2.41	1.5
13	23.3	2.28	2.10	2.52
14	26.7	2.21	2.16	2.44

The specific heat c_p was obtained instead by combining dry specific heat values from [54,55] with the measured humidity content of each soil layer, again obtained from [53], assuming a value 4.2 kJ/kgK for the water specific heat (we will comment on the effect of this uncertainty on the results in the next section).

The version of COMSOL used was 5.3a, with the module Heat Transfer in Solids: the FEM method solves the time-dependent heat conduction equation in three dimensions with no heat source, namely

$$\rho C_p \frac{\partial T}{\partial t} - \nabla \cdot k \nabla T = 0, \tag{1}$$

where $\mathbf{q} = -k\nabla T$ is the heat flux through each layer of the medium. According to the large simulation scale, the mesh was defined as follows (a few tests performed with finer meshes showed a negligible impact of the resolution on the temperature profiles): maximum and minimum element sizes respectively 2.67 m and 0.481 m, maximum element grow rate 1.5, curvature factor 0.6 and resolution of narrow regions 0.5.

The actual pile temperatures were measured with sensors placed in soil on the edge of the structure, Figure 2, at depths listed in Table 1. The temperatures of each soil layer at $t = 0$ were defined as initial conditions in COMSOL according to the measured data. The upper boundary condition, namely the temperature at the ground level, consisted of measured data of a sensor located on top of the pile for the period 7 March 2014–2 October 2014 (i.e., for 4800 h). The transient study was performed with a constant time step $\Delta t = 1$ h.

Even though the above setup is rather simple and the physical phenomenon investigated is straightforward, involving a relatively small amount of degrees of freedom, reproducing the measurements accurately was not trivial. This is due to the inherent inhomogeneity of the soil layers, whose composition and thermophysical properties might well be approximated by Table 2 *globally*, but on a smaller scale this lack of accuracy becomes relevant.

For instance, the computed specific heat for Layer 1 was 2.5 kJ/kgK, however this value returned erroneous initial temperatures for the layer's sensors T28 and T29 (see Table 1). In the simulation we therefore fine-tuned the specific heat to 1.8 kJ/kgK, which gives correct initial T for T28 (0.5 m curves in Figure 4). This c_p value is also consistent with Layer 2 right below, as one would more naturally expect. We will comment more on this in the Results, Section 2.2.

The initial temperature of each pile section was the same of the surrounding soil layer, while the measured surface T was interpolated as in Figure 4 and used as boundary condition. The main difficulty in validating the simulation consisted of matching the initial conditions also for the other sensors, not only for T28. This occurs since several layers (for which the software demands uniform initial conditions) contain sensors with different initial temperatures.

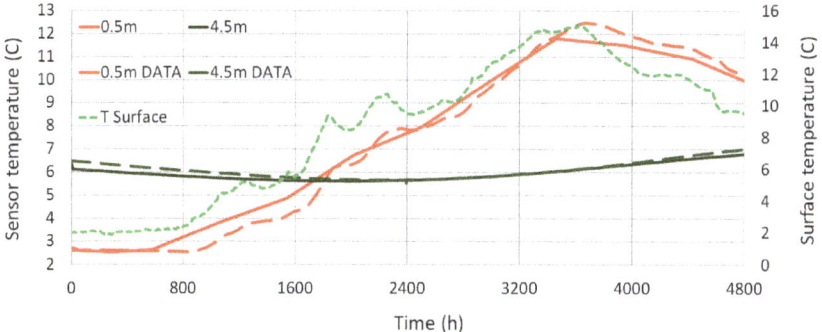

Figure 4. Measured (dashed) vs. COMSOL simulated (solid) temperatures for a single pile, sensors at 0.5 m and 4.5 m. Surface temperature in small dashes.

To increase accuracy, we accordingly refined the soil stratification around the pile. As illustrated in Figure 3, the original 14 layers were split when necessary into a total of 36 layers, to allow for a finer initial temperature profile. This means that 36 different initial temperatures were set in the model, still keeping the thermophysical properties in Table 2 unaltered. Avoiding sharp differences at the interface of contiguous layers, we were indeed aiming to a more physical initial temperature profile in the soil.

2.2. Results

The plot presented in Figure 4 compares the temperature profile for the sensors at 0.5 m, 2.5 m and 4.5 m as computed by COMSOL with the data measurements. We find a very good agreement for all sensors. In Figure 5 the temperature profiles at 4.5 m (with error well below 5% for almost all time steps), at 10.5 m and at 16.5 m show a remarkable agreement, considering that the specific heat was unknown and had to be computed. The difference data-simulation is very minimal, never larger than 0.2 °C~3%, and is clearly a reflection of the uncertainty in the thermal mass.

These results thus seem to be convincing; more accurate soil properties around the sensors would provide with an even more precise validation. In any case, we can conclude that our COMSOL setup is reliable enough for our purpose, and constitutes a solid foundation for the set of simulations described in the next sections.

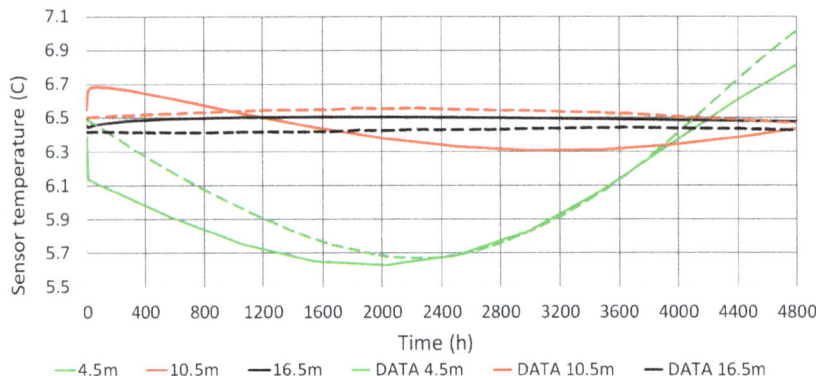

Figure 5. Data (dashed) vs. COMSOL T (solid) for sensors at 4.5 m, 10.5 m and 16.5 m depth.

3. Ground Surface Boundary Analysis in COMSOL

After validating the simulation setup, we are now going to use an analogous COMSOL model to study a 2D heat transfer analysis in energy piles with two different boundary conditions at the surface: case (a) single uniform indoor floor, and case (b) outdoor soil/indoor floor slab. The goal of this second case study was assessing the importance of the multiregional boundary condition proposed in [13].

3.1. Method

The calculation at hand consists of a 2D dimensional reduction of the 3D model addressed in Section 2, extended to a multiple piles layout. The 2D COMSOL model in Figures 6 and 7 studies the heat transfer processes occurring between five 15 m-long energy piles, placed under the building, and the surrounding homogeneous soil.

We modelled a 20 m-large floor slab, consisting of two layers: a lower 20 cm-thick EPS layer with $k = 0.034$ W/mK, $\rho = 20$ kg/m^3 and $c_p = 750$ kJ/kgK, and an upper 10 cm-thick concrete slab with $k = 1.8$ W/mK, $\rho = 2400$ kg/m^3 and $c_p = 880$ kJ/kgK. Each pile was implemented as a 15 m-long concrete grout with diameter 115 mm, surrounding a U-pipe of external diameter 25 mm (their modelling is discussed more into detail in the next Section 3.1). The piles spacing was 4.5 m and they were buried into a soil medium with $k = 1.1$ W/mK, $\rho = 1800$ kg/m^3 and $c_p = 1800$ kJ/kgK.

The two cases investigated correspond to two different upper boundary conditions: (a) floor only, set at 20 °C, namely the average annual indoor air temperature of a commercial hall-type building; (b) floor + soil, where the soil extends for 5 m further from each floor edge (see Figure 8). The soil surface was set at $T = 5.67$ °C, which corresponds to the average annual outdoor air temperature in Southern Finland.

The initial T values were the following: soil layer and grout 5.67 °C, U-pipes 0 °C, upper concrete floor layer 20 °C. The U-pipes were always kept at constant $T = 0$ °C (corresponds to constant heat pump operation), the floor at 20 °C for both (a) and (b) and the soil surface for case (b) at 5.67 °C. A 2D heat conduction module was used, defined by an equation analogous to (1) that naturally takes into account the mutual thermal interaction of adjacent piles. The mesh was normally sized, tetrahedral and physics-controlled, finer at the soil/pile interface and coarser near the boundaries (Figures 6 and 7), with minimum element size 9 mm and maximum 2 m. The simulation was carried out for 2400 h, with time interval $\Delta t = 1$ h.

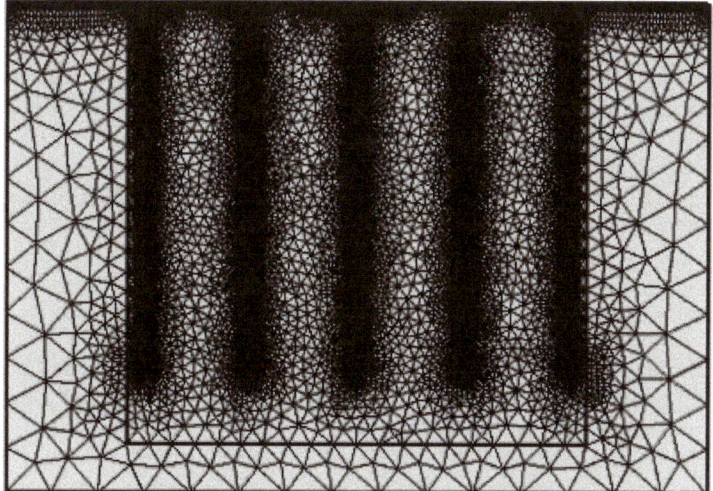

Figure 6. COMSOL mesh for case (a), only floor.

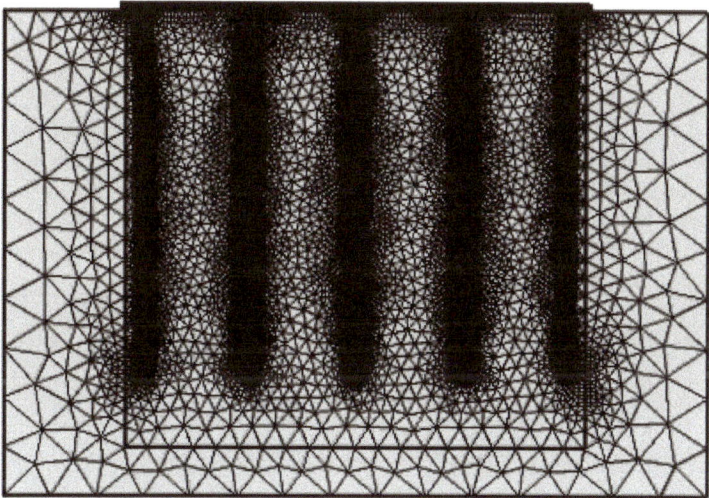

Figure 7. COMSOL mesh for case (b), floor and soil.

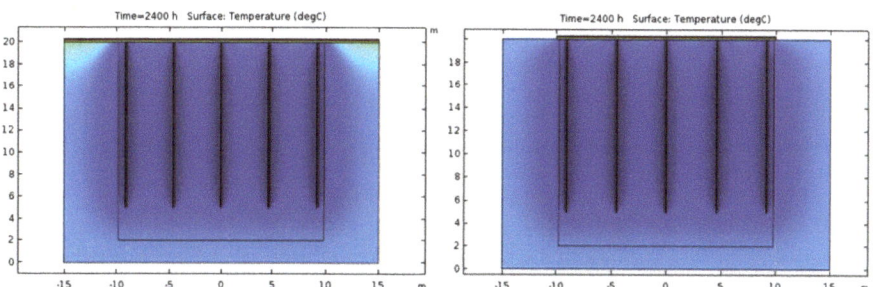

Figure 8. 2D COMSOL result after 2400 hrs for (**left**) (a) only floor and (**right**) (b) floor + soil as upper boundary.

Brine Flow Modelling

In standard constructions, the U-pipes inside the energy piles are usually made of high-density polyethylene (HDPE), with a brine fluid (mostly a water/ethanol mixture) flowing inside [53].

Since the simulations here performed are characterized by large geometry and time scales ($\Delta t = 1$ h and $t_{tot} = 2400$ h), it is legitimate to be concerned with the computational problems related to the inclusion of a fluid dynamics module. For the same reasons though, simulating the fluid flow in the pipes should not be necessary, since high-resolution microphysical processes (inducing fluctuations, turbulences and local irregularities in the brine flow and temperature) should not be relevant.

This reasoning seems to find support in the literature: several examples (see e.g., [56,57]) can be found illustrating how models of transient fluid transport inside the tubes would be justified only for much smaller time scales. In particular, Ref. [56] shows that a steady-state, a transient and a semi transient model converge already after ~3 h.

There are different ways to simplify the implementation of fluid flow, in order to reduce the computational time; for example, in [25,43] the convective heat transfer associated with the fluid flow was simulated by an equivalent solid with the same thermophysical properties of the actual circulation fluid. In order to quantify the error induced by neglecting the fluid flow, we therefore considered the system in Figure 6 only with heat conduction, and modelled the U pipes as made of concrete at constant temperature 0 °C. The surrounding grout was at 5.67 °C at $t = 0$, and then subject to heat transfer for 2400 h. On the other hand, we created another simulation based on the exact same setup, but this time with U-pipes made of water that was flowing at 0 °C, with inlet velocity $V_{in} = 0.45$ m/s and ignoring the pipe thickness.

The result is plotted in Figure 9, where we compare the average temperature in the soil area surrounding the piles, which is active for heat extraction and is highlighted in Figure 8. One can see that the difference is negligible, confirming our assumption at the beginning of this section that we can safely ignore the fluid flow. We accordingly model the U-pipes as made of concrete with constant T, both in the 2D computations and in the full 3D simulations performed in Sections 4 and 5. Let us remark however that the above discussion pertains only COMSOL; IDA-ICE, on the other hand, considers the fluid turbulence already by default when computing the convection heat transfer coefficient h.

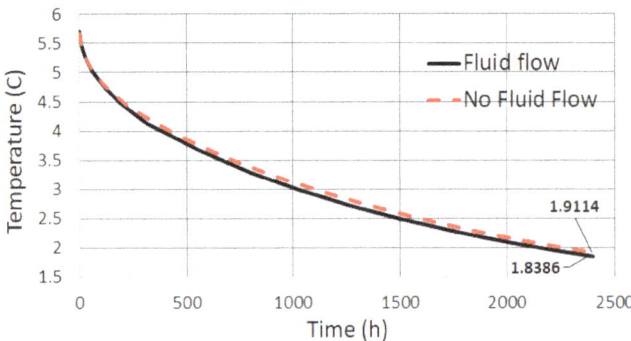

Figure 9. Average temperature in the highlighted area of Figure 8 with fluid flow modelling (solid) compared with no fluid modelling (dashed).

3.2. Results

In this section we compare and discuss the thermal profiles calculated by COMSOL in the two cases corresponding to two different boundary conditions at surface. To quantify differences in the soil region that is active for heat extraction, we computed the average temperature for the rectangular region highlighted in Figure 8 for both (a) and (b). This extends for 1.5 m from the most external piles on both sides, and for 1m from the piles bottom. Thermal insulation effects at the boundaries (at 0 m, −5 m and 15 m in the plot), which are embedded in COMSOL, should not alter the result as they are washed out by the large amount of soil between boundaries and the region's edges.

The result is plotted in Figure 10. It is rather evident that, even after so many hours of heat pump operation at full load, no appreciable difference can be discerned, only ~0.015 °C at the most. We thus conclude that when interested in average temperatures and energy balance, we can freely choose either boundary conditions without compromising our results.

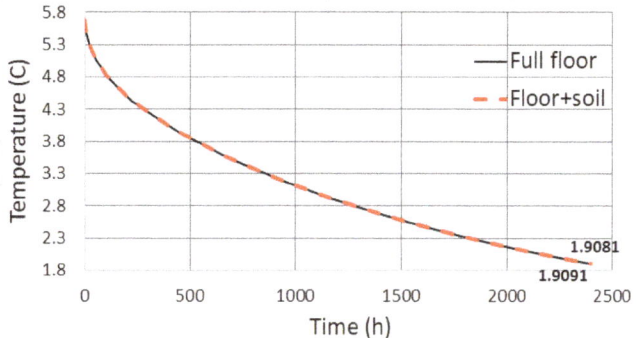

Figure 10. Average temperature in the highlighted slab portion, (a) floor (dashed) vs. (b) floor + soil (solid). T [°C] values for the last point in bold.

4. IDA-ICE Borehole Model Validation against COMSOL

In this third study we consider a 3D IDA-ICE finite difference borehole model, together with the COMSOL FEM counterpart. The latter is a direct 3D extension of the 2D model discussed in Section 3. Here we compute the outlet temperatures at different depths on the edge of 20 energy piles, which are buried in soil under a multilayer floor. Results from the IDA-ICE model are compared with those obtained from COMSOL, in order to estimate the accuracy of the IDA-ICE calculation.

4.1. Method

A 3D FEM COMSOL extension of the 2D model discussed in Section 3 is here performed, with 20 identical energy piles that are 15 m-long and with 4.5 m spacing; the same geometry is implemented both in COMSOL (Figure 11) and in IDA-ICE, with a 3D finite element (COMSOL) or finite difference (IDA-ICE) borehole model that accounts for the mutual thermal interaction of adjacent piles.

The piles were buried in a 25 m × 25 m × 20 m soil medium (layout listed in Table 3), with the same multilayer floor (concrete slab over an EPS layer) addressed in Section 3 as the upper boundary condition. In IDA-ICE this is set by default because, as remarked earlier, the option to define multiple ground surface boundary regions is not available. Temperature loggers were set on the centre pile #10, on the centre-edge pile #12 and on the edge pile #20, to quantify the impact of surrounding piles on the temperature fields. Temperatures were logged at 1.5 m, 3 m, 6 m and 12 m depth, with loggers located on the pile edge in contact with soil.

In the COMSOL simulation piles, floor and soil were modelled exactly as in Section 3.1, with the same material properties. The concrete floor was kept at 20 °C at all times, and the U-pipes were kept at steady $T = 0$ °C (corresponding to a constant heat pump operation). In order to decrease the simulation time, no fluid flow was modelled, as explained in Section 3.1; in IDA-ICE instead, the fluid at constant $T = 0$ °C entered the energy piles at a very high flow rate. Simulations in COMSOL and IDA-ICE ran for 2400 h, with time interval $\Delta t = 1$ h.

Table 3. Spatial arrangement (view from above) of the energy piles for both IDA-ICE and COMSOL.

17	18	19	20
13	14	15	16
9	10	11	12
5	6	7	8
1	2	3	4

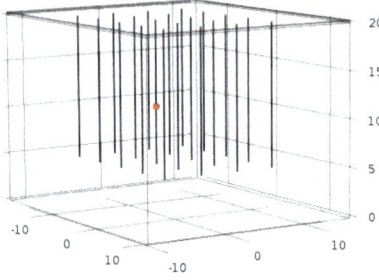 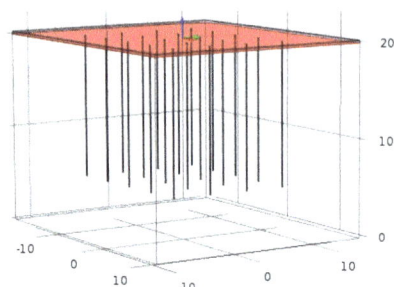

Figure 11. Floor slab with 20 energy piles in COMSOL. Data point at 9 m shown as a red dot.

4.2. Results

Looking at Figures 12–14, one can clearly see that, as expected, only the central pile #10 is slightly colder than the other two. We do not really see a major impact of the specific placement. Furthermore, predictions of COMSOL and IDA-ICE are extremely close, on the average of the order 0.05 °C. Regardless of the software used, the temperature difference is overall very small, accounting for the soil medium and piles temperature homogeneity. Even a very large depth difference shows little discrepancy, e.g., 1.5 m and 12 m give $\Delta T < 0.4$ °C. Specifically, notice how for every pile, the 12 m curves overlap exactly with those at 6 m. Surface effects are indeed strongly hindered by the large thermal mass of the soil medium. Figure 15 also illustrates that, for different boundary conditions at the surface (full floor vs. floor + soil), COMSOL shows a similar $\Delta T \sim 0.4$ °C. This effect is compensated by averaging over the large area that is active for heat extraction, as we discussed in Section 3.

Surface effects should however be relevant for small depths. We thus consider a point in between the central piles, at 0.5 m under the floor, and compare its temperature profile to that of another point right above, very close to the floor, at 3 cm depth. As illustrated in Figure 16, after about 200 h (i.e., after the FEM simulation is stabilised) there is a constant $\Delta T \sim 0.8$ °C. This means that, differently from what we have seen above for points sitting deeper underground, when approaching the surface T differences are relevant and cannot be ignored.

More importantly though, Figure 16 also shows that the integrated average temperature on a plane located at 3 cm under the floor, pictured in Figure 11, always coincides with T as computed at a central point at the same depth. This is a natural consequence of the simulation's geometrical and physical symmetry, showing that for a similar setup one can use averages instead of point values. This result might be valuable for practitioners and on-site applications.

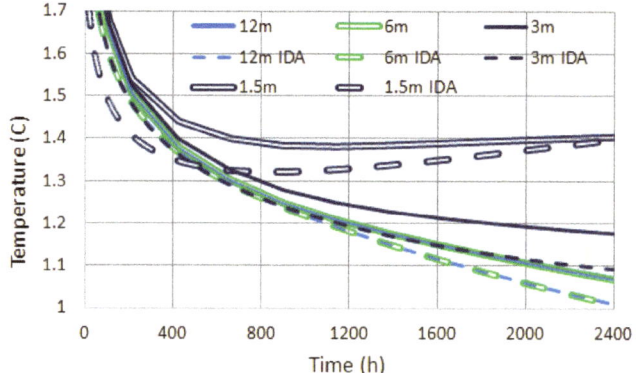

Figure 12. Pile #10.

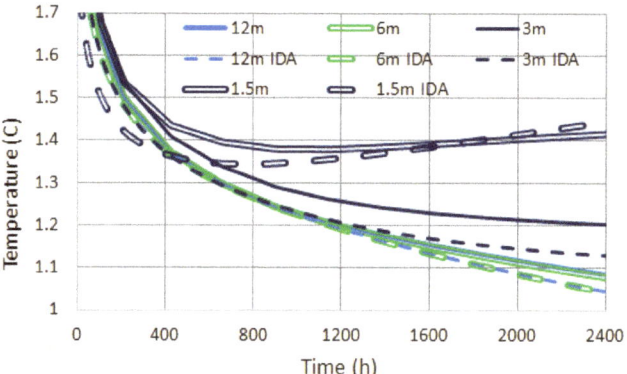

Figure 13. Pile #12.

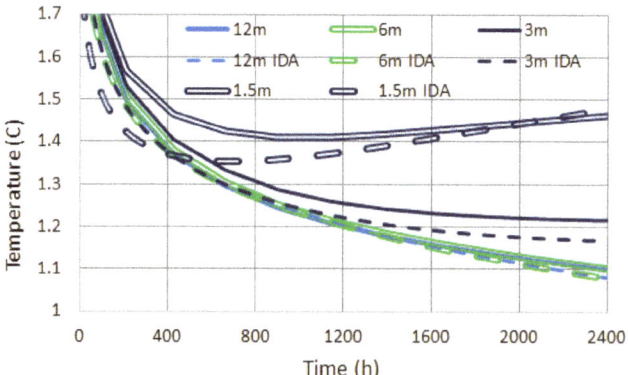

Figure 14. Pile #20.

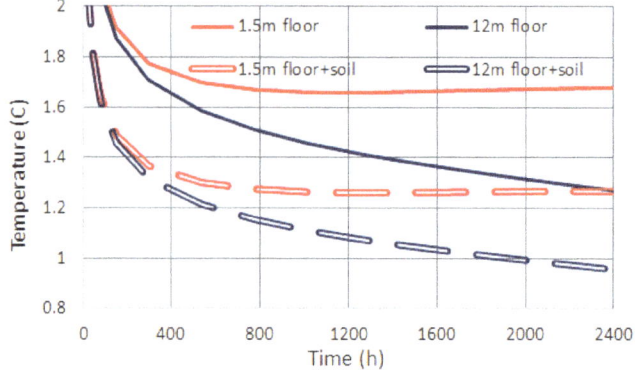

Figure 15. COMSOL comparison of different boundary conditions.

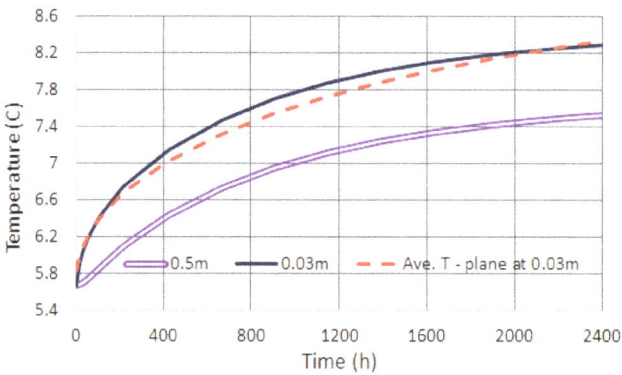

Figure 16. Temperature at mid points and on a plane parallel to the floor.

5. Impact of Boundary Conditions on Energy Efficiency Calculations

In this final case, we apply to a 3D COMSOL FEM model the data obtained in [28] by modelling the energy piles with IDA-ICE, in order to calculate the temperature underneath the floor slab (also called "ground surface temperature"). The result is then compared to the rough estimation made in [28] (the "outlet" solution here), and to an analogous IDA-ICE calculation using a new software version updated in November 2017 (the "slab" solution here). We find that the new implementation "slab" shows a remarkable improvement over the "outlet" method in terms of energy consumption assessment.

5.1. Method

Geothermal energy piles with heat pump in a whole building were modelled in IDA-ICE following the design proposed in [28] and addressing the same commercial hall-type building (Figure 17). The total number of energy piles was 192, the initial soil temperature 5.67 °C and the piles were 15 m-long. The soil properties were $k = 1.1$ W/mK, $\rho = 1800$ kg/m^3 and $c_p = 1800$ kJ/kgK.

In the present study, we performed an annual IDA-ICE simulation according to this design. We obtained inlet and outlet energy piles temperatures with hourly resolution, which we used to calculate the average energy piles fluid temperature. These were then implemented in COMSOL as boundary conditions, to avoid including a fluid dynamics module for the reasons explained in Section 3.1. Our COMSOL simulation accordingly involves only heat conduction, but uses the IDA-ICE values to increase the precision.

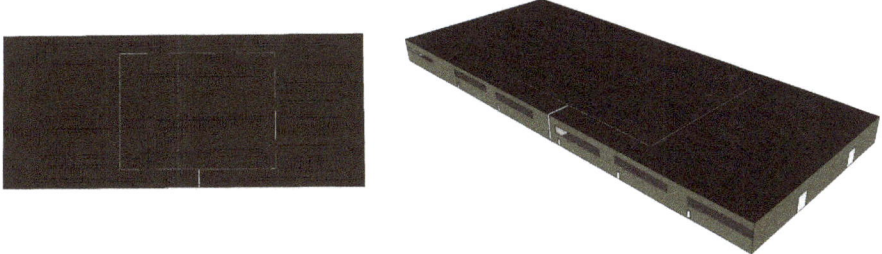

Figure 17. Building model with a separated zone (centre) above energy piles modelled in IDA-ICE.

To compute the yearly energy demand for this building, the model was divided into two zones—one with energy piles and one without. The building floor slab above the energy piles accounts for ca. 33% of the total floor slab area. In the original "outlet" approach presented in [28], a variable temperature underneath this surface is roughly estimated as the energy piles outlet temperature, with connection scheme presented in Figure 18. This was necessary since before November 2017 it was not possible for IDA-ICE to calculate the exact ground surface temperature above the piles, which we accomplish instead with the new solution called "slab".

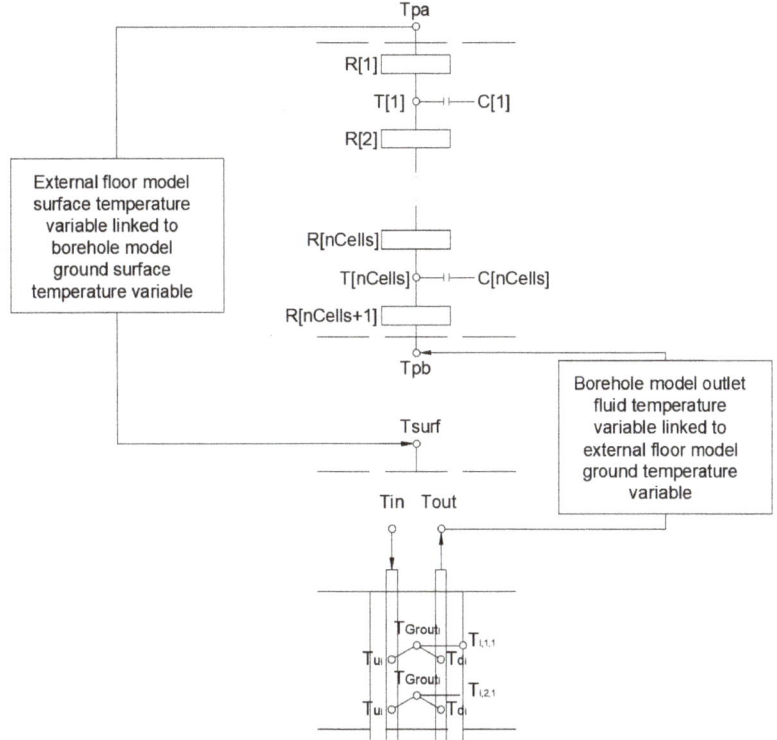

Figure 18. Connection scheme for "slab" (**left**) and "outlet" (**right**) IDA-ICE borehole models.

The annual simulation with hourly resolution performed in COMSOL returned the average ground surface temperature T_s beneath the floor slab, then compared against the same T_s calculated by the new "slab" borehole model. Finally, we quantify the discrepancy between COMSOL and the two IDA-ICE models "outlet" and "slab" when computing the yearly energy need for the above building.

5.2. Results

Figure 19 compares the COMSOL-calculated ground surface temperature against the two temperatures estimated in IDA-ICE. In the previous IDA-ICE "outlet" approach [28], the results were obtained with an outdated version of the software, where the borehole model uses the energy piles outlet temperature as a rough estimate for the ground surface temperature.

The newer IDA-ICE "slab" estimates use instead the latest version of borehole model, which is now capable of estimating the ground surface temperature. One can observe a significant difference of 5.8 °C between the new and old version over the simulated annual period. Remarkably, the new result of IDA-ICE borehole model differs by only 0.6 °C from COMSOL.

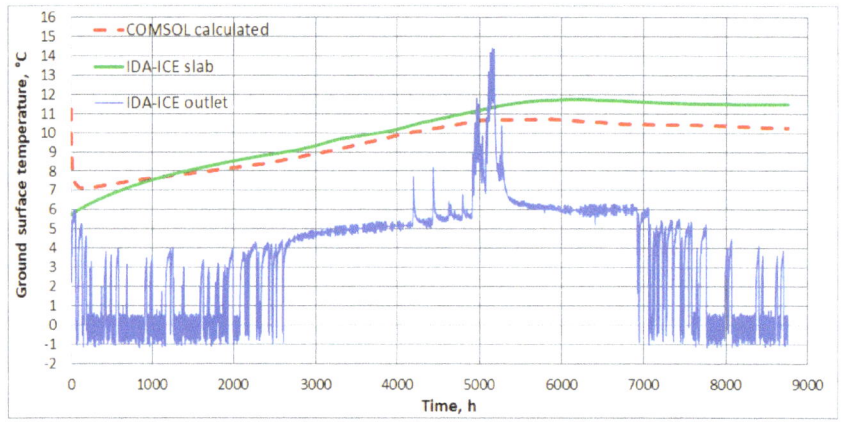

Figure 19. Estimated IDA-ICE ground surface temperature vs. calculated in COMSOL.

Table 4 shows how improper modelling of the ground surface temperature affects the floor slab heat flux over the energy piles, and accordingly the overall annual heat demand of the building. From the perspective of annual heat flux difference, simply applying the outlet temperature of energy piles as ground surface boundary condition in "outlet" produces a difference in ca. 54%, which induces nearly a 5% overestimation of building annual heating need. In contrast, the new version of borehole model "slab" performs in a very good agreement with COMSOL, within an acceptable heat flux and heating need difference of 0.03% and 0.2% respectively.

Table 4. Validation results for the IDA-ICE estimated ground surface temperature.

Case	Floor Slab Heat Flux, kWh/a	Heating Need, kWh/a	Heat Flux, % Difference	Heating Need, % Difference
COMSOL	24066	142900	-	-
IDA-ICE slab	24073	142580	0.03%	0.2%
IDA-ICE outlet	37127	150196	54%	5%

The new IDA-ICE implementation of a borehole model is therefore free from previous heating consumption overestimations, providing a reliable tool for investigations of building heating needs.

6. Conclusions

In this paper we have investigated into detail geothermal plants modelling, by comparing different boundary conditions and their impact on numerical studies of heat transfer and energy performance.

First we considered the finite element method (FEM) software COMSOL Multiphysics, and validated a 3D model of transient heat transfer against experimental data. We found an excellent agreement, despite uncertainties in the measurements and estimated soil properties.

Next we simulated transient heat transfer in a multi-pile 2D reduction of the previous simulation, and addressed the effect of different upper boundary conditions, namely either (i) a unique floor multilayer at 20 °C or (ii) a floor at 20 °C and soil at ∼6 °C. We found that for practical purposes, the average temperatures and energy balance for a yearly calculation are unaffected by the specific boundary condition. This holds by virtue of extremely small temperature differences in the active heat extraction region: modelling of energy piles can thus be performed using borehole models with variable ground surface temperature, which can be set as the indoor air temperature above the energy piles area. The implementation of multiregional surface boundary conditions proposed in [13] can be accordingly neglected in future studies.

As a side result, we discussed the role of fluid flow modelling inside the pipes. We confirm earlier findings that for the length and time scales of interest, one can safely ignore the fluid flow and model the U-pipes as made of concrete with constant T, both in 2D and 3D simulations.

A third computation was then performed with both COMSOL and IDA-ICE, addressing transient heat transfer for 2400 h. The simulations consisted of a 3D model with 20 energy piles embedded in a soil medium, under a uniform multilayer floor, revealing that temperature predictions of COMSOL and IDA-ICE are extremely close, on the average of the order 0.05 °C. In particular, the piles temperatures are not affected by the specific placement, as they are approximately constant for a given depth. The large thermal mass of piles and medium also provides a very small vertical gradient, e.g., the difference between 1.5 m and 12 m depth gives only $\Delta T < 0.4$ °C. Moreover, for every pile the 12 m curves overlap exactly with those at 6 m.

Figure 15 illustrates a comparable $\Delta T \sim 0.4$ °C also for different boundary conditions at the surface (full floor vs. floor + soil) in COMSOL. This effect is compensated by averaging over the large area that is active for heat extraction, as we discussed in Section 3. However, close to the surface the different boundary conditions (full floor vs. floor + soil) do give a different temperature. Interestingly, we also found that the integrated average T on a horizontal plane at height h is nearly identical to the one at a point laying on it. This result could be useful for practical calculations of energy balance over the same time scale.

Finally, we performed a yearly simulation to assess the heating need of a commercial building employing a geothermal plant. Assuming the COMSOL FEM calculation validated in Section 1 as our benchmark, two IDA-ICE FDM simulations illustrated the impact of different upper boundary conditions on the final energy consumption estimate. We found that in IDA-ICE, adopting the energy piles outlet temperature as a rough estimate for the ground surface temperature overestimates the heat flux in the floor slab by 54% and the heating need by 5%. On the other hand, using the ground surface temperature produced a consistent result with COMSOL, overestimating the heat flux and heating need by only 0.03% and 0.2% respectively.

One might now wonder about the long-term effects of the heat discharge. This has been studied into details in [28] for a heat pump plant with energy piles serving a one storey commercial hall building. Long-term simulations encompassing 20 years of usage showed a consistent reduction in performance of the energy piles regarding heat extraction. It was found indeed that, over the years, the heat extracted from the energy piles exceeds the amount which is loaded during the cooling season. Evidently, this implies a need for thermal storage, at least for the case of commercial hall-type buildings.

In conclusion, this paper provides a through comparison of software widely used in simulations of heating systems adopting borehole and energy piles. In particular, focusing on the role of upper boundary conditions, we showed that multiregional b.c. are fairly equivalent to a single surface with uniform temperature. Moreover, we proved that using the pipes outlet temperature induces a severe overestimation of the heat flux through the heated floor and of the energy demand for the whole building.

All these results are relevant especially for practical purposes, in the development of geothermal energy methods in compliance with international Standards and towards sustainability. In this perspective, they set the foundation for a number of necessary developments and improvements: for instance, longer term simulations (such as 10–20 years) could investigate the effect of a reverse operation mode in summer, for discharging the ground to limit the thermal drift observed in [28]. Other possibilities could be the extension of our methodology to other types of climates, and the assessment of thermomechanical effects on the piles' structure.

Author Contributions: Conceptualization, A.F., J.F. and J.K.; methodology, A.F. and J.F.; software, A.F. and J.F.; validation, A.F.; investigation, A.F. and J.F.; writing—original draft preparation, A.F. and J.F.; writing—review and editing, A.F.; supervision, J.K.; funding acquisition, J.K.

Funding: The research was supported by the Estonian Research Council with Institutional research funding grant IUT1-15. The authors are also grateful to the Estonian Centre of Excellence in Zero Energy and Resource Efficient Smart Buildings and Districts, ZEBE, grant 2014-2020.4.01.15-0016 funded by the European Regional Development Fund.

Conflicts of Interest: The authors declare no conflict of interest.

Abbreviations

The following abbreviations have been used in this manuscript:

GHE   Ground Heat Exchangers
FEM   Finite Element Method
FDM   Finite Difference Method
b.c.   Boundary Conditions

Nomenclature

ρ Density [kg/m^3]
c_p Specific heat at constant pressure [kJ/(kgK)]
T Temperature [°C]
λ Thermal conductivity [W/(mK)]
Δt Time step [h]

References

1. European Parliament. Directive 2010/31/EU of the European Parliament and of the Council of 19 May 2010 on the energy performance of buildings. *Off. J. Eur. Union* **2010**, *53*, 13–35.
2. Lund, J.W.; Boyd, T.L. Direct utilization of geothermal energy 2015 worldwide review. *Geothermics* **2016**, *60*, 66–93. [CrossRef]
3. Agemar, T.; Weber, J.; Moeck, I.S. Assessment and Public Reporting of Geothermal Resources in Germany: Review and Outlook. *Energies* **2018**, *11*, 332. [CrossRef]
4. Park, K.S.; Kim, S. Utilising Unused Energy Resources for Sustainable Heating and Cooling System in Buildings: A Case Study of Geothermal Energy and Water Sources in a University. *Energies* **2018**, *11*, 1836. [CrossRef]
5. Aydin, M.; Sisman, A. Experimental and computational investigation of multi U-tube boreholes. *Appl. Energy* **2015**, *145*, 163–171. [CrossRef]
6. Bose, J.E.; Smith, M.; Spitler, J. Advances in ground source heat pump systems-an international overview. In Proceedings of the 7th IEA Heat Pump Conference, Beijing, China, 19–22 May 2002; Volume 324, pp. 313–324.
7. Spitler, J. Editorial: Ground-Source Heat Pump System Research—Past, Present, and Future. *HVAC R Res.* **2005**, *11*, 165–167. [CrossRef]
8. Brandl, H. Energy foundations and other thermo-active ground structures. *Geotechnique* **2006**, *56*, 81–122. [CrossRef]
9. Lautkankare, R.; Sarola, V.; Kanerva-Lehto, H. Energy piles in underpinning projects—Through holes in load transfer structures. *DFI J. J. Deep Found. Inst.* **2014**, *8*, 3–14. [CrossRef]

10. Faizal, M.; Bouazza, A.; Singh, R.M. Heat transfer enhancement of geothermal energy piles. *Renew. Sustain. Energy Rev.* **2016**, *57*, 16–33. [CrossRef]
11. Bourne-Webb, P.; Burlon, S.; Javed, S.; Kuerten, S.; Loveridge, F. Analysis and design methods for energy geostructures. *Renew. Sustain. Energy Rev.* **2016**, *65*, 402–419. [CrossRef]
12. Li, M.; Lai, A.C. Review of analytical models for heat transfer by vertical ground heat exchangers (GHEs): A perspective of time and space scales. *Appl. Energy* **2015**, *151*, 178–191. [CrossRef]
13. Fadejev, J.; Simson, R.; Kurnitski, J.; Haghighat, F. A review on energy piles design, sizing and modelling. *Energy* **2017**, *122*, 390–407. [CrossRef]
14. Ingersoll, L.R.; Zobel, O.J.; Ingersoll, A.C. Heat conduction with engineering, geological and other applications. *Q. J. R. Meteorol. Soc.* **1955**, *81*, 647–648. [CrossRef]
15. Goldenberg, H. A problem in radial heat flow. *Br. J. Appl. Phys.* **1951**, *2*, 233. [CrossRef]
16. Hu, P.; Zha, J.; Lei, F.; Zhu, N.; Wu, T. A composite cylindrical model and its application in analysis of thermal response and performance for energy pile. *Energy Build.* **2014**, *84*, 324–332. [CrossRef]
17. Abdelaziz, S.L.; Ozudogru, T.Y. Selection of the design temperature change for energy piles. *Appl. Therm. Eng.* **2016**, *107*, 1036–1045. [CrossRef]
18. Hellström, G. Ground Heat Storage: Thermal Analyses of Duct Storage Systems. Ph.D. Thesis, School of Mathematical Physics, Lund University, Lund, Sweden, 1991.
19. Claesson, J.; Hellström, G. Analytical studies of the influence of regional groundwater flow on the performance of borehole heat exchangers. In Proceedings of the 8th International Conference on Thermal Energy Storage, Terrastock 2000, Stuttgart, Germany, 28 August–1 September 2000; pp. 195–200.
20. Bozis, D.; Papakostas, K.; Kyriakis, N. On the evaluation of design parameters effects on the heat transfer efficiency of energy piles. *Energy Build.* **2011**, *43*, 1020–1029. [CrossRef]
21. Rosa, M.D.; Ruiz-Calvo, F.; Corberán, J.M.; Montagud, C.; Tagliafico, L.A. Borehole modelling: A comparison between a steady-state model and a novel dynamic model in a real ON/OFF GSHP operation. *J. Phys. Conf. Ser.* **2014**, *547*, 012008. [CrossRef]
22. Kwong Lee, C.; Nam Lam, H. Thermal Response Test Analysis for an Energy Pile in Ground-Source Heat Pump Systems. In *Progress in Sustainable Energy Technologies: Generating Renewable Energy*; Dincer, I., Midilli, A., Kucuk, H., Eds.; Springer: Cham, Switzerland, 2014; pp. 605–615.
23. Li, Z.; Zheng, M. Development of a numerical model for the simulation of vertical U-tube ground heat exchangers. *Appl. Therm. Eng.* **2009**, *29*, 920–924. [CrossRef]
24. Nam, Y.; Ooka, R.; Hwang, S. Development of a numerical model to predict heat exchange rates for a ground-source heat pump system. *Energy Build.* **2008**, *40*, 2133–2140. [CrossRef]
25. Lazzari, S.; Priarone, A.; Zanchini, E. Long-term performance of BHE (borehole heat exchanger) fields with negligible groundwater movement. *Energy* **2010**, *35*, 4966–4974. [CrossRef]
26. Diersch, H.J.; Bauer, D.; Heidemann, W.; Rühaak, W.; Schätzl, P. Finite element modeling of borehole heat exchanger systems: Part 2. Numerical simulation. *Comput. Geosci.* **2011**, *37*, 1136–1147. [CrossRef]
27. Salvalai, G. Implementation and validation of simplified heat pump model in IDA-ICE energy simulation environment. *Energy Build.* **2012**, *49*, 132–141. [CrossRef]
28. Fadejev, J.; Kurnitski, J. Geothermal energy piles and boreholes design with heat pump in a whole building simulation software. *Energy Build.* **2015**, *106*, 23–34. [CrossRef]
29. Østergaard, D.; Svendsen, S. Space heating with ultra-low-temperature district heating—A case study of four single-family houses from the 1980s. *Energy Procedia* **2017**, *116*, 226–235. [CrossRef]
30. Schweiger, G.; Heimrath, R.; Falay, B.; O'Donovan, K.; Nageler, P.; Pertschy, R.; Engel, G.; Streicher, W.; Leusbrock, I. District energy systems: Modelling paradigms and general-purpose tools. *Energy* **2018**, *164*, 1326–1340. [CrossRef]
31. Lee, C.; Lam, H. A simplified model of energy pile for ground-source heat pump systems. *Energy* **2013**, *55*, 838–845. [CrossRef]
32. Mehrizi, A.A.; Porkhial, S.; Bezyan, B.; Lotfizadeh, H. Energy pile foundation simulation for different configurations of ground source heat exchanger. *Int. Commun. Heat Mass Transf.* **2016**, *70*, 105–114. [CrossRef]
33. Xiao, J.; Luo, Z.; Martin, J.R.; Gong, W.; Wang, L. Probabilistic geotechnical analysis of energy piles in granular soils. *Eng. Geol.* **2016**, *209*, 119–127. [CrossRef]

34. Pryor, R.W. *Multiphysics Modeling Using COMSOL: A First Principles Approach*, 1st ed.; Jones and Bartlett Publishers, Inc.: Burlington, MA, USA, 2009.
35. Dupray, F.; Laloui, L.; Kazangba, A. Numerical analysis of seasonal heat storage in an energy pile foundation. *Comput. Geotech.* **2014**, *55*, 67–77. [CrossRef]
36. Diersch, H.J.; Bauer, D.; Heidemann, W.; Rühaak, W.; Schätzl, P. Finite element modeling of borehole heat exchanger systems: Part 1. Fundamentals. *Comput. Geosci.* **2011**, *37*, 1122–1135. [CrossRef]
37. Park, H.; Lee, S.R.; Yoon, S.; Choi, J.C. Evaluation of thermal response and performance of PHC energy pile: Field experiments and numerical simulation. *Appl. Energy* **2013**, *103*, 12–24. [CrossRef]
38. Cecinato, F.; Loveridge, F.A. Influences on the thermal efficiency of energy piles. *Energy* **2015**, *82*, 1021–1033. [CrossRef]
39. Ng, C.; Ma, Q.; Gunawan, A. Horizontal stress change of energy piles subjected to thermal cycles in sand. *Comput. Geotech.* **2016**, *78*, 54–61. [CrossRef]
40. Go, G.H.; Lee, S.R.; Yoon, S.; byul Kang, H. Design of spiral coil PHC energy pile considering effective borehole thermal resistance and groundwater advection effects. *Appl. Energy* **2014**, *125*, 165–178. [CrossRef]
41. Loveridge, F.; Powrie, W. 2D thermal resistance of pile heat exchangers. *Geothermics* **2014**, *50*, 122–135. [CrossRef]
42. Alberdi-Pagola, M.; Poulsen, S.E.; Jensen, R.L.; Madsen, S. Thermal design method for multiple precast energy piles. *Geothermics* **2019**, *78*, 201–210. [CrossRef]
43. Batini, N.; Loria, A.F.R.; Conti, P.; Testi, D.; Grassi, W.; Laloui, L. Energy and geotechnical behaviour of energy piles for different design solutions. *Appl. Therm. Eng.* **2015**, *86*, 199–213. [CrossRef]
44. Caulk, R.; Ghazanfari, E.; McCartney, J.S. Parameterization of a calibrated geothermal energy pile model. *Geomech. Energy Environ.* **2016**, *5*, 1–15. [CrossRef]
45. Loria, A.F.R.; Vadrot, A.; Laloui, L. Analysis of the vertical displacement of energy pile groups. *Geomech. Energy Environ.* **2018**, *16*, 1–14. [CrossRef]
46. Li, M.; Lai, A.C. Heat-source solutions to heat conduction in anisotropic media with application to pile and borehole ground heat exchangers. *Appl. Energy* **2012**, *96*, 451–458. [CrossRef]
47. Park, S.; Lee, D.; Lee, S.; Chauchois, A.; Choi, H. Experimental and numerical analysis on thermal performance of large-diameter cast-in-place energy pile constructed in soft ground. *Energy* **2017**, *118*, 297–311. [CrossRef]
48. McCartney, J.S.; Murphy, K.D. Investigation of potential dragdown/uplift effects on energy piles. *Geomech. Energy Environ.* **2017**, *10*, 21–28. [CrossRef]
49. Sung, C.; Park, S.; Lee, S.; Oh, K.; Choi, H. Thermo-mechanical behavior of cast-in-place energy piles. *Energy* **2018**, *161*, 920–938. [CrossRef]
50. Singh, R.; Bouazza, A.; Wang, B. Near-field ground thermal response to heating of a geothermal energy pile: Observations from a field test. *Soils Found.* **2015**, *55*, 1412–1426. [CrossRef]
51. Suryatriyastuti, M.; Mroueh, H.; Burlon, S. Understanding the temperature-induced mechanical behaviour of energy pile foundations. *Renew. Sustain. Energy Rev.* **2012**, *16*, 3344–3354. [CrossRef]
52. Eguaras-Martínez, M.; Vidaurre-Arbizu, M.; Martín-Gómez, C. Simulation and evaluation of Building Information Modeling in a real pilot site. *Appl. Energy* **2014**, *114*, 475–484. [CrossRef]
53. Uotinen, V.M.; Repo, T.; Vesamäki, H. Energy piles—Ground energy system integrated to the steel foundation piles. In Proceedings of the 16th Nordic Geotechnical Meeting (NGM 2012), Copenhagen, Denmark, 9–12 May 2012; pp. 837–844.
54. Jõeleht, A.; Kukkonen, I.T. Physical properties of Vendian to Devonian sedimentary rocks in Estonia. *GFF* **2002**, *124*, 65–72. [CrossRef]
55. Kukkonen, I.; Lindberg, A. *Thermal Properties of Rocks at the Investigation Sites: Measured and Calculated Thermal Conductivity, Specific Heat Capacity and Thermal Diffusivity*; Working Report; Ilmo Kukkonen Antero Lindberg: Helsinki, Finland, 1998.

56. Bauer, D.; Heidemann, W.; Diersch, H.J. Transient 3D analysis of borehole heat exchanger modeling. *Geothermics* **2011**, *40*, 250–260. [CrossRef]
57. Al-Khoury, R.; Kölbel, T.; Schramedei, R. Efficient numerical modeling of borehole heat exchangers. *Comput. Geosci.* **2010**, *36*, 1301–1315. [CrossRef]

© 2019 by the authors. Licensee MDPI, Basel, Switzerland. This article is an open access article distributed under the terms and conditions of the Creative Commons Attribution (CC BY) license (http://creativecommons.org/licenses/by/4.0/).

Article

Air Distribution and Air Handling Unit Configuration Effects on Energy Performance in an Air-Heated Ice Rink Arena

Mehdi Taebnia [1,*], Sander Toomla [2], Lauri Leppä [3] and Jarek Kurnitski [1,4]

1. Aalto University, Department of Civil Engineering, P.O. Box 12100, 00076 Aalto, Finland; jarek.kurnitski@aalto.fi or jarek.kurnitski@taltech.ee
2. Granlund Consulting Oy, Malminkaari 21, PL 59, 00701 Helsinki, Finland; sander.toomla@granlund.fi
3. Leanheat Oy, Hiomotie 10, FI-00380 Helsinki, Finland; lauri.leppa@leanheat.fi
4. Department of Civil Engineering and Architecture, Tallinn University of Technology, Ehitajate tee 5, 19086 Tallinn, Estonia
* Correspondence: mehdi.taebnia@aalto.fi

Received: 14 January 2019; Accepted: 15 February 2019; Published: 21 February 2019

Abstract: Indoor ice rink arenas are among the foremost consumers of energy within building sector due to their exclusive indoor conditions. A single ice rink arena may consume energy of up to 3500 MWh annually, indicating the potential for energy saving. The cooling effect of the ice pad, which is the main source for heat loss, causes a vertical indoor air temperature gradient. The objective of the present study is twofold: (i) to study vertical temperature stratification of indoor air, and how it impacts on heat load toward the ice pad; (ii) to investigate the energy performance of air handling units (AHU), as well as the effects of various AHU layouts on ice rinks' energy consumption. To this end, six AHU configurations with different air-distribution solutions are presented, based on existing arenas in Finland. The results of the study verify that cooling energy demand can significantly be reduced by 38 percent if indoor temperature gradient approaches 1 °C/m. This is implemented through air distribution solutions. Moreover, the cooling energy demand for dehumidification is decreased to 59.5 percent through precisely planning the AHU layout, particularly at the cooling coil and heat recovery sections. The study reveals that a more customized air distribution results in less stratified indoor air temperature.

Keywords: ice rinks; air distribution solutions; indoor air temperature gradient; air handling unit configuration; building energy efficiency; building performance simulation; energy and HVAC-systems in buildings

1. Introduction

The reduction of energy use in buildings is a strategic research challenge, due to the significant contribution of the building sector in CO_2 emissions. The reduction of energy use and the improvement in energy efficiency is strongly linked to the operations and performance of passive and active systems in buildings [1]. The potential for the reduction of energy demand has to be evaluated through the prioritizing solutions based on their energy efficiency [2]. Specifically, indoor ice arenas among the building sector are an enormous consumer of energy, due to their unique indoor conditions. The yearly energy consumption of a standard single ice rink arena is typically estimated to be between 1000–1500 MWh [3,4]. However, the range of individually measured energy consumptions is even larger, within 500–3500 MWh/year, which provides a great potential for energy savings [5]. The ice pad refrigeration and hall space heating are two major contributors to the energy use of the ice rinks. By default, to maintain a steady-state condition, the heat removed from the hall primarily by the refrigeration machinery needs to be roughly matched with the heat supplied into and generated inside the hall. In the case of an air-heated arena, the vast majority of the heat balance is maintained through a heated supply of air.

Generally, ventilation efficiency in similar sports halls such as swimming pools could potentially be improved by various alternative air distribution concepts [6]. However, the unique indoor conditions of an ice rink arena proposes challenges to energy-efficient heating and ventilation. Due to the cooling effect of the ice pad, a vertical temperature gradient inside the hall space is unavoidably formed. This, accompanied with the fact that the recreational activities practiced on the ice pad require a free height of approximately five meters, makes space heating of the rink difficult. In fact, in order to maintain a set temperature at an occupational height above the ice pad, the temperature of the supply air entering the hall at a height below the ceiling has to exceed the occupational set point temperature by a large amount.

Several past studies have focused on reducing the heat load towards the ice pad, and thereby reducing the refrigeration unit's electricity consumption [7–9]. Simultaneously, numerous efforts have gone into modeling the air distribution inside the hall space in experimental, zonal model, or Computational Fluid Dynamics (CFD) form [10–14]. As a result, we have a fairly comprehensive understanding of the temperature and moisture profiles inside the hall, as well as the factors affecting the heat load.

The vertical distribution of temperature in various ice rinks has been measured in current and previous studies [12,14–17] and their outcome as temperature gradient curves is used in this paper, while similar ice rinks as case study arenas have been measured.

However, the actual role of the air handling unit (AHU), along with its components and control strategies, has only been briefly investigated in two prior publications. Seghouani [8] modeled the AHU of a simulated ice arena hall space as a two-speed system, either low or high, which is increased to high-speed mode only during ice pad resurfacing, to evacuate the combustion gases of the resurfacing vehicle, with no air recirculation or no extract air heat recovery [8]. Piché [18] continued Seghouani's research by adding two possible modifications to the AHU modeled earlier: an alternative pre-heated fresh air source, or an air-to-air heat exchanger, both of which utilized the refrigeration unit's condenser heat. While the later study obtained significant results regarding the AHU's energy demand compared to the prior one, neither implementations represented a typical, modern, real-life indoor ice rink AHU solution. This means that previous studies about AHUs are outdated, and they do not represent a modern AHU layout. Thus, the energy performance of modern AHUs, equipped with full variable-air-volume (VAV) control, a heat exchanger (HX) for extract air heat recovery, and the possibility for extract air recirculation, in the context of demanding indoor ice rink conditions, should be further investigated.

The objective is to determine (quantify) the impacts of indoor temperature stratification, as well as AHU layouts, on energy consumption, while two commonly used AHU configurations at different temperature gradients are applied. To study the two focus features, the AHU design, and the air stratification intensity, we present six simulation setups, which are based on existing ice rink arenas in Finland.

The AHU for the hall space of an air-heated ice arena usually has three main objectives. Firstly, as with any ventilation system, it should provide adequate fresh air into the space, to maintain satisfactory indoor air conditions. Secondly, in this case, it is solely responsible for supplying the space with enough heat. Thirdly, in case, no external dehumidification equipment is present, and the AHU is equipped with a condensing dehumidifier, and it is thus responsible for maintaining the moisture content under a specific set point inside the hall. The indoor air recirculation is implemented by the maximum possible rate at any moment, for energy conservation. Each AHU has its own theoretical energy demand, depending on its main objective. To maximize the AHU's energy efficiency, it should be demand-controlled, based on CO_2, temperature, and humidity set point levels, depending on their measured values. If either of the measured values in a particular moment exceeds the acceptable range, then that parameter's control signal prevails to other signals. In the case of simultaneous exceeding of set points, the automation system reacts simultaneously so that each parameter can react independently, by sending its control signal to the associated section of AHU to that parameter. Two air-handling layouts have been used as the simulation model in this study. A section of the air-handling units (AHU1.1) and (AHU2.1) are shown in Figure 1, and their specifications are described in the following paragraphs.

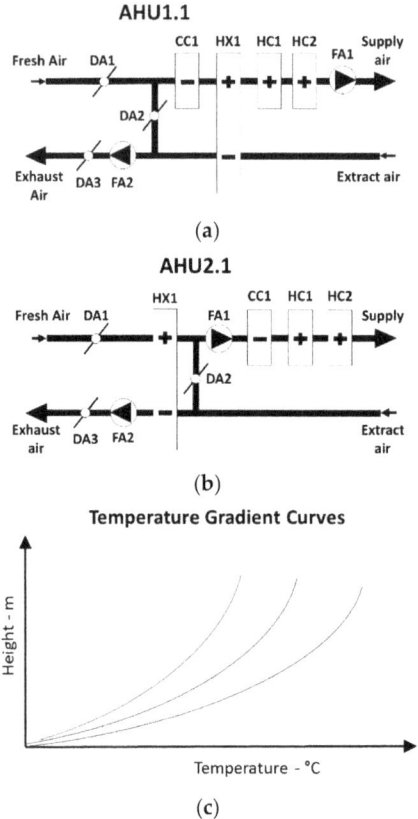

Figure 1. AHU layouts and temperature gradient curves. (a) Schematic view of AHU1.1; (b) Schematic view of AHU 2.1; (c) Estimated models for temperature gradients.

The air handling unit 1.1 (AHU1.1), depicted in Figure 1a, is fully automated, equipped with extract air recirculation, a cooling coil (CC) acting as a condensing dehumidifier, a rotary heat exchanger (HX) for extract air heat recovery, with an assumed efficiency of 85%, and two heating coils (HC). HC1 utilizes condenser heat from the refrigeration plant, while HC2 is connected to the district heating system. HC2 acts as a backup heat device in case the refrigeration unit is not operating or is not producing enough condenser heat. In the simulation, HC1 is not modeled. Both the supply and exhaust fans are fully VAV-compatible up to 4 m^3/s, and their speeds are individually controlled. The exhaust fan is placed outside the recirculation loop, making it possible to recirculate air utilizing the supply fan only. The whole unit is demand-controlled based on temperature, humidity or CO_2-level measurements from the ice rink.

AHU2.1, is in many aspects very similar to AHU1.1, except for one key difference. The rotary heat exchanger with an assumed efficiency of 85% is placed outside the recirculation loop, as presented in Figure 1b, leaving it completely unavailable for recirculation mode. The supply and extract fans are demand-controlled in the same fashion as AHU1.1 and rated up to 4 m^3/s. Supply air is cooled and dehumidified with a condensing dehumidifier. It is then heated with two heating coils. The HC1 utilizes condenser heat and the HC2 district heat, similar to AHU 1.

In this study, we concentrate first on the AHU design and its control approach, by presenting two AHU layouts that only differ from the position of their heat exchangers. Second, we study the temperature stratification of indoor air and its effects on energy consumption in a simplified way. We also study how various air distribution designs relate to a temperature gradient. The air stratification intensity of the cases is based on real measured data in three ice rink arenas, similar to a previous study [14]. Overall, six cases are presented for the simulations, two AHU layouts, and three temperature gradients. The results of on-site measurements can only verify three of these cases, since each ice rink is equipped with only one of the AHUs.

There are three ice rink arenas, each with a demand-controlled AHU equipped with a condensing dehumidifier. However, their final implementations regarding components and control strategies differ from each other. In this publication, six simulation models are presented for the ice rinks, similar to the real-case study rinks, and their measured data have been used to verify the simulation results. The heating and cooling energy demands for each AHU, along with the indoor air conditions, as well as their temperature stratifications, are also presented.

2. Methods

2.1. Buildings and Air Handling Units

The three indoor air distribution models selected for use in this study are presented in Figure 2. The reasons for selecting these particular models is firstly, because they are existing ice arenas in Finland, and second, because both the required measurements for this study (temperature gradient and energy consumption measurements) were implemented there. This means that each of the indoor air distribution models corresponds to one of the measured temperature gradients. Therefore, these selected air distribution models are were the case study models.

The rather simple air distribution system corresponding to the measured temperature gradient of 2.1 is depicted as the hall space cross-section in Figure 2a. The supply air terminals of this system are located below the ceiling level, and their air jets blow horizontally to the opposite directions. The extract air terminal is located close to one end along the space.

The air distribution system corresponding to the temperature gradient of 1.6 consists of multiple supply air terminals located above the spectator balcony, angled towards the ice pad. A cross section of the hall space is depicted in Figure 2b. Supply air jets are located along the length of the hall, while extract air is drawn from terminals located near the end alongside the hall. In the vertical direction, both the supply and extract terminals are close to the ceiling level.

The air distribution system corresponding to a temperature gradient of 1.5 is unlike the other presented systems. Non-heated supply air enters the hall space from terminals connected to small holes drilled to the sideboards of the rink. The idea is to ventilate the occupational zone above the rink without compromising the quality of the ice pad with heated air. Heated supply air is distributed at an angle towards the stands, while the extract air terminal is located towards the end of the hall below the ceiling. The system is presented in Figure 2c.

To verify the simulation results we used the experimental data from real ice arenas. It is important to present the unique features of each arena. This generates errors that might favor some outcomes. If the results do not make sense without modification, we can then modify the simulations based on the unique features of the ice arenas, which have been experimentally measured. We would need to show that the differences in the simulations are also seen in the experimental measurements.

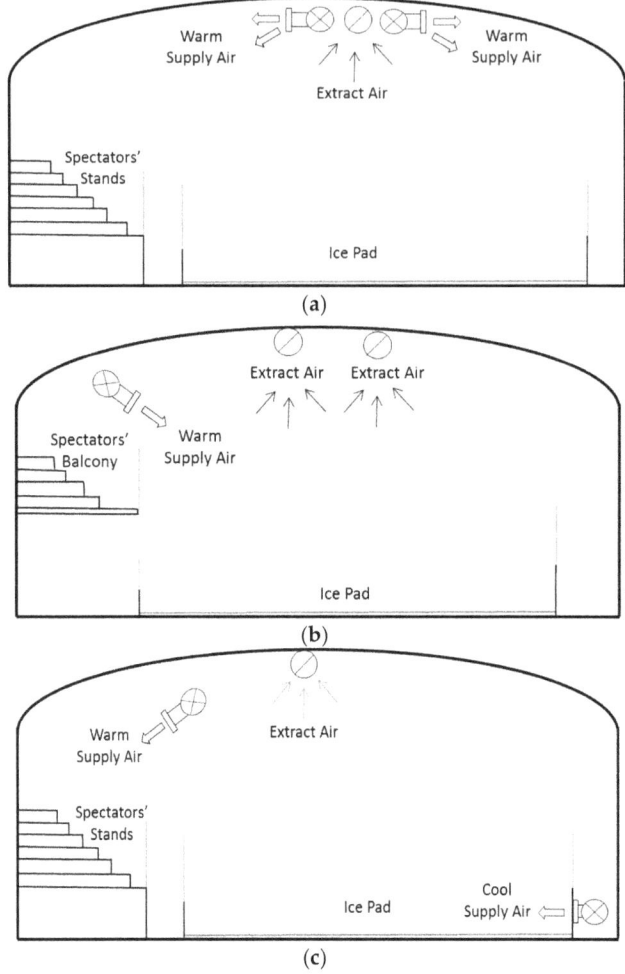

Figure 2. Air distributions corresponding to the measured temperature gradients: (**a**) A; (**b**) B; (**c**) C.

In order to compare the AHUs' performance against outdoor and indoor conditions, a series of measurements were performed. A Temperature and relative humidity (T/RH)-logger, shielded with direct insolation, was used to track the temperature and relative humidity of the outdoor air in close proximity of the studied building. Inside the hall space, T/RH/CO_2-loggers measured the temperature, relative humidity, and CO_2 level of the indoor air, both from the skater's occupational zone above the ice rink, and from the stands. Due to practical reasons, the logger measuring the skating zone was placed just outside the rink at a height of 2 to 2.5 m above the rink, depending on the case.

The case study for this publication included four similar single ice rink indoor arenas, built between 2003 and 2015, located in the southern parts of Finland. Their ice pad sizes ranged from 1456 m² to 1566 m² (56 ... 58 × 26 ... 27 m²), with the arena hall volumes falling between 13,000 m³ and 16,000 m³. The smallest arena had an elevated spectator balcony with the capacity for 60 standing spectators, while the others had stands rated for 500 to 750 seated spectators. Other spaces in the studied ice rink facilities were not considered in this publication. The descriptions for each measured AHU and their air distribution systems are as follows.

The air distribution system corresponding to AHU 1.1 is a combination of the air distribution system as shown in Figure 2b, and the AHU1.1 which is depicted in the following Figure 3.

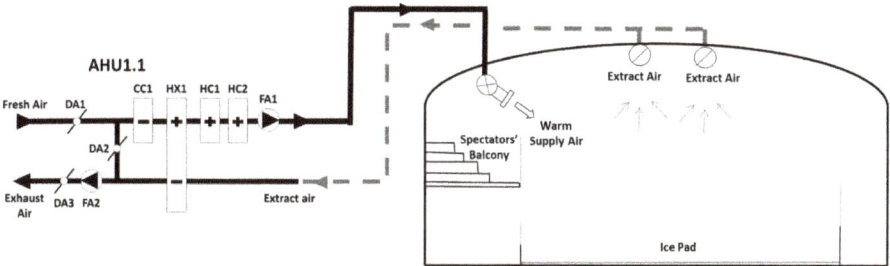

Figure 3. Schematic view of AHU1.1 and its corresponding air distribution.

AHU1.2, presented in Figure 4, is similar to AHU1.1, which is presented earlier with some modifications. The demand control strategy, based on the temperature, humidity or CO_2 level, is the same as the order of components in the supply side of the unit (recirculation to cooling coil, heat exchanger, and to the heating coils). The core differences are:

- Supply air is split into a heated and non-heated flow. The HC for the heated supply air utilizes condenser heat from the refrigeration plant, while the only form of heating for the non-heated air is extract air heat recovery.
- The extract air heat recovery unit is a cross-flow air-to-air plate heat exchanger instead of a rotary heat exchanger, as in AHU1.1
- The supply and exhaust fans are rated up to 5 m^3/s correspond to 2.5 L/sm^2.
- The exhaust/extract fan is located inside the recirculation loop; in full recirculation mode, both fans need to be operated.

In the hall space, the heated supply air is directed towards the stands and outside the rink, while the non-heated portion serves as ventilation for the ice pad area. The extract air terminal is located below the ceiling level approximately in the center of the space. A cross-section of the space, along with the air distribution arrangement, can be examined in Figure 4.

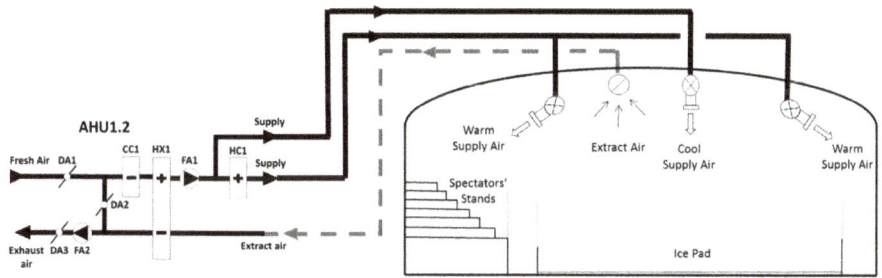

Figure 4. Schematic view of AHU1.2 and its corresponding air distribution system.

AHU2.2 differs from the other presented units in that it is not fully VAV-compatible. It is operated as a two-speed unit, namely half- and full-speed, but both speed options can be programmed to any percentage of the fan's maximum capacity. The supply fan is rated up to 4 m^3/s, and the exhaust fan up to 2 m^3/s. Like AHU1.2, the unit is equipped with regenerative exhaust air heat recovery outside the recirculation loop, and like AHU1.2, the supply air is split into heated and non-heated airflows. The non-heated flow is untreated after the condensing dehumidifier, making its temperature lower than in AHU2.1. The heating coil utilizes condenser heat from the refrigeration plant. A schematic view of AHU2.2 is available in Figure 5.

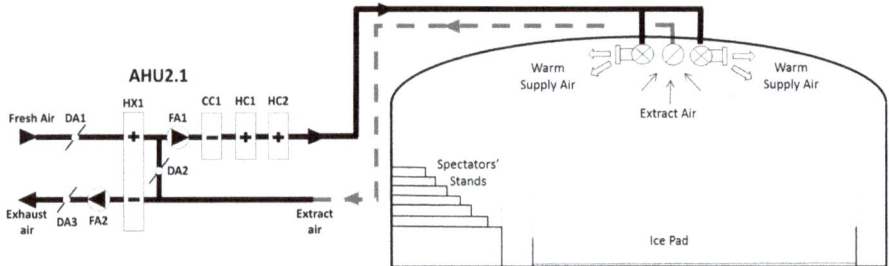

Figure 5. Schematic view of AHU2.1 and its corresponding air distribution system.

The air distribution system corresponding to AHU2.2 is unlike the other presented systems. Non-heated supply air enters the hall space from terminals connected to small holes drilled to the sideboards of the rink. The idea is to ventilate the occupational zone above the rink without compromising the quality of the ice pad with heated air. Heated supply air is distributed at an angle towards the stands, while the extract air terminal is located towards the end of the hall below the ceiling. The system is presented in Figure 6.

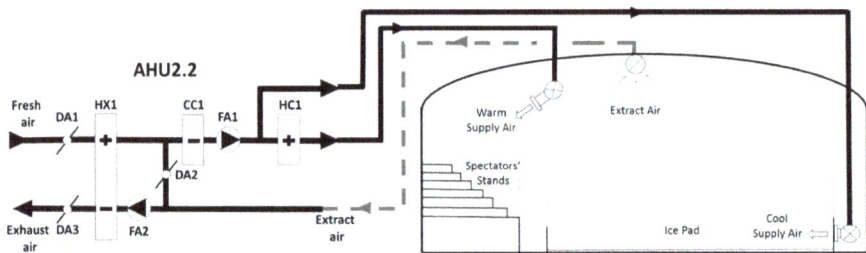

Figure 6. Schematic view of AHU2.2 and its corresponding air distribution system.

It is important to explain that the four models described above are used for 24 h measurements, but there are temperature gradients measured in only three of them. Therefore, those three, which are similar, as shown in Figure 2a–c, were used to validate the simulation results.

2.2. Measurements

For each AHU and its corresponding hall space, a series of measurements were carried out. The measurement periods lasted between six and eight days, and the measurement interval was five minutes. The measurement plan of each AHU could not be perfectly implemented, due to differences in the air handling units' space coverage, capacity, and accessibility to the measurement locations. The missing measurements were compensated with measurements performed from the building automation system, when possible. All of the measurements were carried out within May and June 2016.

The indoor air temperature and relative humidity were measured and logged with T/RH-loggers (THERMADATA MALLI) before and after each AHU component, i.e., before and after the heating coils, the cooling coils, and the heat exchangers. The indoor air CO_2 levels were measured with T/RH/CO_2-loggers at the supply and extract air or extracted air positions. Meanwhile, fresh air was assumed to have a constant CO_2 level of 400 ppm. For airflow rates, the pressure difference over the fan was measured and logged, and it could then be converted into an airflow rate by using a unit-specific k-factor. An overview of the conducted measurements for each AHU is presented in Table 1.

Table 1. Overview of the conducted measurements. M = measured, M* = measured short-term, AS = measured by the automation system, E = estimated, C = calculated, - = not valid for said AHU.

Measured Parameters	Airflows						Temperature and RH Change over:					CO_2 Level		
Measured at the Section	Fresh	Supply			Extract	Exhaust	CC	HC1	HC2	HX	Fresh	Supply	Extract
Machines		Total	Heated	Non-Heated									
AHU 1.1	C	M	-	-	E	M	M	M	M	M	E	M	M
AHU 2.1	C	M	E	E	M	C	M	M	-	M	E	M	M
AHU 1.2	C	E	-	-	E	C	M	M	M	M	E	M	M
AHU 2.2	C	M	C	M*	E	C	M	M/AS	-	M	E	M	M

As Table 1 states, a series of estimations had to be made, especially regarding the airflow rates. When the extract airflow could not be measured, due to the fan's location or the lack of available pressure differential measuring points, the extract airflow rate was estimated to match the supply airflow rate. For AHU 1.2, the airflow measurements could not be performed at all without interfering with the unit's operation. However, examining the temperature and RH changes in the supply air made it clear the unit was operated in an on–off fashion. For example, the temperature of the supply air after the cooling coil would periodically lower to a constant value for a while, and then rise to another value that was constant along the whole unit. Based on this behavior, it could be estimated that the air inside the unit was partially moving and partially standing still. For when the air was moving, it was estimated that the unit worked at full capacity, and when the air was evidently standing still, the airflow rate was set to 0 m^3/s. The resulting average airflow rate was in line with measured average rates from the other AHUs.

For AHU 2.1, as the ratio between the heated and non-heated supply airflows could not be experimentally determined, and the flow rates were estimated at 80% and 20% of the total supply airflow, respectively. The estimated ratio resulted in a total supply air average temperature that was in line with the produced thermal conditions inside the hall space. For AHU2.2, the flow rate of the non-heated supply air was determined in short-term measurements, and the calculated ratio between the heated and non-heated supply flows was estimated to stay constant throughout the measurement period.

3. Simulation Setup and Building Model

To highlight the core differences between the studied AHUs, and to exclude any external variables affecting their performance, a version of each AHU was modeled, and its performance was simulated by using IDA ICE v. 4.7.1 with the Ice Rinks and Pools 0.912 add-on, for a period of one year, with typical meteorological conditions for Helsinki, Finland. The simulated demand-control-strategy, based on temperature, relative humidity, and CO_2-measurements from the hall space, was unmodified across the modeled AHUs. Three of the built simulation models were validated by using experimental data. For comparison's sake, Seghouani presented, modeled and simulated a modified VAV-version of the AHU to study its performance [8].

3.1. Building Specifications

We used a rather simplified approach that was common to all simulation models. We used a one-zone airspace with a size of 65 × 35 × 7 m, and one door with a size of 3.5 × 5.0 m, which was opened seven times a day for 10 min each time. The external walls of the building were made of Aluminum 0.003 m, light insulation 0.2 m and aluminum 0.003 m. The Roof was made of Aluminum 0.003 m, light insulation 0.3 m, and renders 0.01 m. The external floor was made of floor coating 0.05 m and 0.2 m concrete. The main door was made of 0.003 m aluminum. There were no thermal bridges formed in the building. Infiltration through the building was constant, with 0.03 ACH.

The cooling pipes were submerged 2 cm into the concrete slab underneath the ice pad. The rest of the 0.2 m concrete slab and an insulation layer of 0.1 m formed the base layer underneath the ice pad. Heating pipes are located in the soil beneath the insulation layer. The cooling and heating powers were 200 W/m^2 and 40 W/m^2, respectively. The ice layer thickness was 3.5 cm, and the ice temperature set-point was $-5\ °C$.

3.2. Control Strategy

The zone was ventilated and heated by the AHU, which was controlled based on the measured indoor and extract air conditions. The AHU maximizes the recirculation air usage for energy conservation. The supply and the exhaust fans were controlled by responding to the measured temperature, relative humidity and CO_2 values of the zone, with boundaries of 4–6 °C (corresponding to output signals of either 1 or 0 respectively), 60–70% RH and 1000–1100 ppm CO_2 (both corresponding to outputs of 0 and 1, respectively). The supply air temperature was adjusted according to the zone average temperature with simplified set points of 30 °C when the indoor air temperature was below 3 °C, and 3 °C when the air temperature was above 7 °C (Figure 4). The heat recovery unit was always on. The cooling coil cooled and dehumidified by reducing the air temperature to +1 °C when the moisture content of the air exceeded 3.65 g/kg dry air. The fresh air intake was controlled by the CO_2 concentration of the extract air, according to a setpoint range of 1000–1100 ppm, corresponding to outputs of 0.037 and 1, respectively. The minimum fresh air intake was set to 3.7%, as reported by Toomla [14]. The extract air CO_2 concentration with set points of 1000–1100 ppm, corresponded to signals of 0 to 1, respectively. The CO_2 concentration of extract air controlled the exhaust fan as well. Both fans were rated up to 4 m^3/s (2.0 L/s/m^2) capacity, according to ASHRAE 90.1, with the Specific Fan Power (SFP) set to 1.23 (kW/m^3/s), and the efficiency to 0.6.

3.3. Assumptions and Parameters for the Simulation Models

The supply fan was operated based on the zone signal. Smooth functions (from 0 to 1) for high-temperature HI 6 and low-temperature LO 4, RH HI 0.7 LO 0.6, CO2 HI 1100 LO 1000, and MAX signals of these three controlled the supply fan speed. The exhaust fan was controlled by the CO^2 content in the extract air. A smooth function of 0 to 1 was set with HI 1100 and LO 1050, i.e., therefore, the exhaust fan only ran when the CO_2 level was high.

The recirculation of indoor air or outdoor air intake was controlled by the extract air CO_2 content, with a smooth function 0.037 (3.7%) to 1 (100%), with LO 1000 and HI 1100.

Heat exchangers always function with an effectiveness of 0.85 and an unknown capacity. The minimum allowed leaving temperature was +1 °C. Drying with the cooling coil was controlled so that the temperature set point was the minimum from either the incoming temperature or the incoming humidity control, so that the cooling coil temperature set points were 4 °C below 3.15 g/kg, and 1 °C above 3.65 g/kg. The cooling coil effectiveness causes a liquid-side temperature rise of 5 °C. The cooling was simulated as district cooling, to show the cooling demands of the dehumidification.

The heating coil effectiveness was 1, and the liquid-side temperature drop was 20 °C. The heating coil set-point temperature for the supply air was controlled by the zone average air temperature, according to the curve presented in the Figure 7.

The indoor air temperature gradient was set, based on the three variants' measured values, 1 °C/m, 1.5 °C/m, and 2 °C/m.

Lighting was carried out with 20 × 400 W (4.0 W/m^2) units with a luminous efficacy of 12 lm/W, and a convective fraction of 0.5. The lighting was used only when players were present. Inside the zone, there was an ice pad (60 × 30 m) with Freezium as the coolant and heating medium.

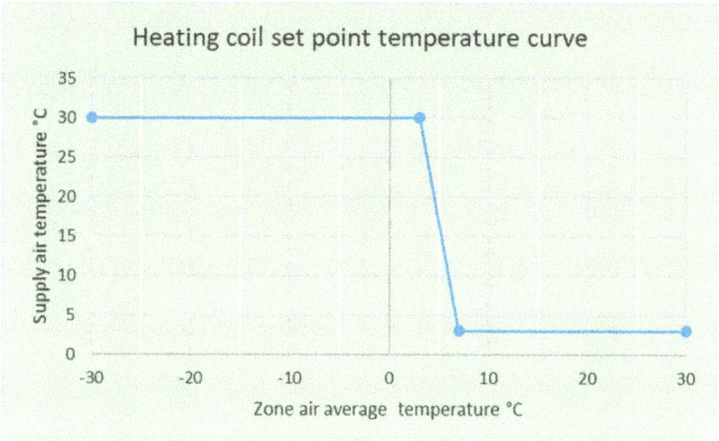

Figure 7. Heating coil set-point temperature curve to control the supply air.

Figures 8 and 9 are a few examples of how the system and control set-points were set in the simulation software. All of the properties of the system were simply set in the IDA-ICE simulation software, as shown in the Figure 8, where the operation set points of the refrigeration plant, the ice pad, subfloor heating, and further details were set. The indoor air control set-points (air flow, temperature, relative humidity) were also set in the IDA-ICE (4.7.1, EQUA, Stockholm, Sweden), as shown in Figure 9.

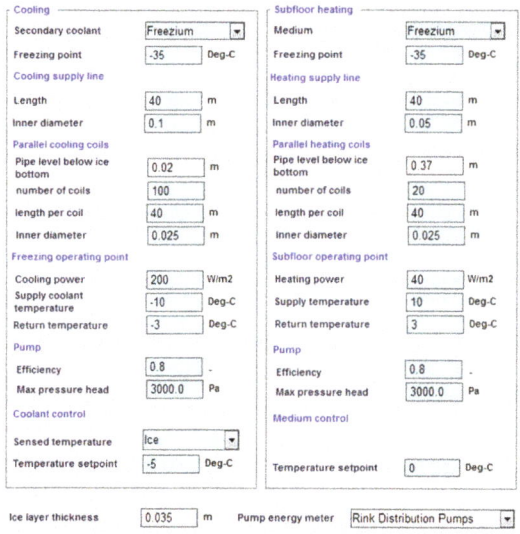

Figure 8. Refrigeration plant and the ice pad set points.

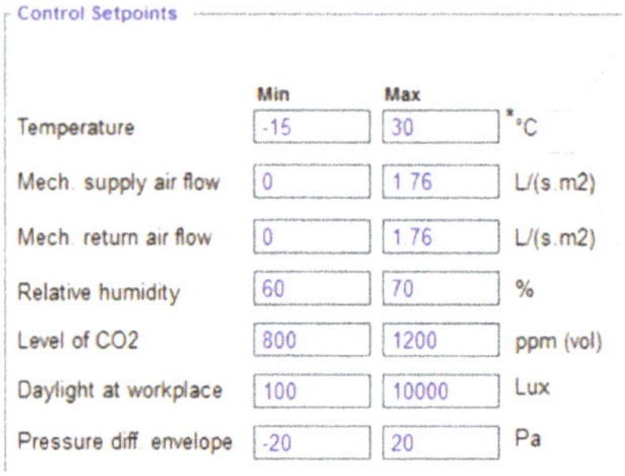

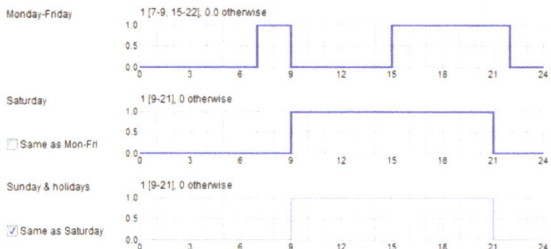

Figure 9. Indoor air ventilation quantity and quality set points.

The internal gains of the zone were the players, spectators, and lighting. The internal load of the players was set based on 20 players with an activity level of 5 Metabolic Equivalent of Task (MET), and scheduled as presented in Figure 10. On weekdays, the players were present from 7:00 to 9:00 a.m., and also from 3:00 to 10:00 p.m. On weekends, they were present from 9:00 a.m. to 9:00 p.m.

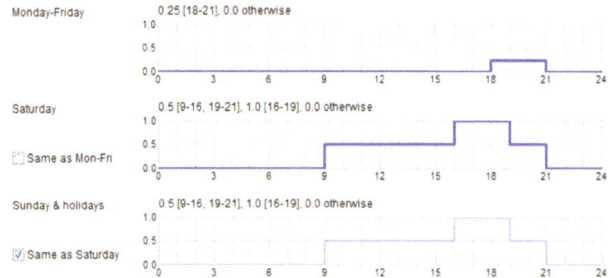

Figure 10. Scheduled internal loads of players.

Players were present according to the following schedule. The players' heat load was 20, with an activity level of 5 MET according to the schedule below, a maximum of 100 spectators, and 20 × 400 W lamps.

The spectator attendance was modeled as 25 persons from 6:00 to 9:00 p.m. on weekdays, and as 50 persons from 9:00 a.m. to 9:00 p.m. on weekends, with a peak of 100 persons between 4:00 and 7:00 p.m.

The heat load caused by the spectators was set based on a maximum of 100 spectators, with an activity rate of 1.5 MET. The X factor, which is the percentage of spectators' occupancy in different days/times was implemented in the simulation according to the schedule presented in the following Figure 11:

Figure 11. Scheduled internal loads of spectators.

The heat load of lighting with 20 × 400 W lamps with a convective fraction of 0.5, was similar as player's schedule.

Finally, the temperature gradient values for the simulation models were set as 2.0 °C/m and 1.5 °C/m, to represent the average ice arenas, similar to the measured air stratification in real cases. In addition, the stratification value of 1 °C/m was set to describe an arena with a lower indoor air temperature gradient as an ultimate condition, which would be a significant improvement in comparison to the currently measured arenas.

3.4. Theoretical Heat Exchange and Airflow Principles

3.4.1. Ice surface Modeling

In order to calculate the heat that is exchanged between the ice surface and the indoor air, we needed to concentrate on the transient model above the ice. To do so, it is initially required to determine the heat transfer coefficients of the air layer on the ice. Theoretical challenges on how accurate the model calculates the U_FILM, the H$_{Conv}$, and the condensation heat transfer through the ice surface to the hall space, are described as:

$$P_{in} = 10^5 \exp\left(17.391 - \frac{6142.83}{273.15 + T_{in}}\right) \quad (1)$$

$$P_{ice} = 10^5 \exp\left(17.391 - \frac{6142.83}{273.15 + T_s}\right) \quad (2)$$

The relative humidity at the height of h = 0.1 m above the ice surface are calculated as follows:

$$RH_h = \left(\frac{h}{1,5}\right) \times (90 - RH_{1,5}) \quad (3)$$

$$dp = \left(\frac{RH_h}{100}\right) \times (p_h - p_{ice}) \quad (4)$$

$$dp_{atm} = \left(\frac{dp_{pa}}{101325}\right) \quad (5)$$

The heat transfer coefficient for condensation is also calculated as:

$$hd = 1750 \times h_{conv} \times \frac{\Delta P}{\Delta T}(RH\_h/100) \quad (6)$$

$$q_{cond} = h_d \times (T_{in} - T_{ice}) \quad (7)$$

3.4.2. Airflow Balance Equations

The calculated and measured airflow rates, along with the measured temperature and the RH changes over the components, were used to calculate the component theoretical energy output over the measurement periods. The heating powers of the heating coil and the heat exchanger were calculated as:

$$P_{heat} = q_{air}\rho_{air}c_{air}\Delta T_{air} \quad (8)$$

and the cooling coil's cooling powers as:

$$P_{cool} = q_{air}\rho_{air}\Delta h_{air} \quad (9)$$

where the enthalpy of air can be expressed as:

$$h_{air} = c_{air}T_{air} + x_{air}(c_w T_{air} + h_{we}) \quad (10)$$

The fresh air intake of the AHU was calculated based on CO_2-level differences between the extract, supply, and fresh air. Any decrease in CO_2 level from the extract to supply air meant that a portion of the supply air was fresh air, since it is reasonable to assume no other CO_2 sources within the unit exist. Fresh air intake can be calculated as:

$$q_{fresh} = q_{sup}\left(\frac{C_{ext} - C_{sup}}{C_{ext} - C_{fresh}}\right) \qquad (11)$$

The resulting flow rate for fresh air intake serves more as an approximation rather than an exact value, but its accuracy is sufficient to determine when the unit is operating in full or partial recirculation mode.

4. Experimental Results

4.1. Temperature Gradient Measurements

The vertical temperature profiles in various ice rinks in Finland were measured in previous studies [10–14], and its outcomes as temperature gradient curves are used in the current paper, as energy consumption of the same ice rinks has been measured to describe case arenas. The set air stratification intensity for the simulation is based on experimental measurements conducted in three ice rink arenas in Finland. The procedure of the measurements is subsequently described, and the measurement results are presented in Figure 12. The actual, non-linear temperature stratification was linearized into a gradient factor describing the temperature increase as degrees Celsius per meter. The cause for this simplification was the limitation of the simulation software.

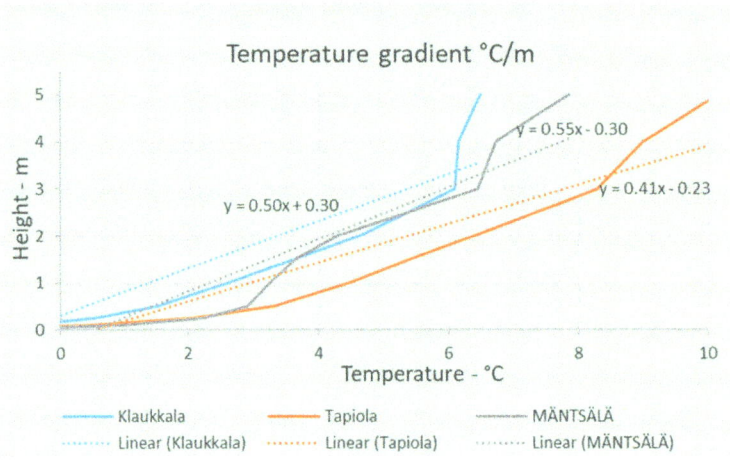

Figure 12. Air stratification measurement results of three ice arenas in Finland.

In order to understand the differences between the observed energy performances of each AHU, a perspective with regard to their respective outside air and produced indoor air conditions needed to be established. The 24-h periods of the AHUs were evaluated based on the maximum similarity of the outside air temperature and the hall space occupational load. It is noteworthy that neither the produced indoor air conditions nor the outside air humidity, which both affected the AHU's performance, were the same across the studied arenas. This limitation in the experimental setup will be taken into account when the results are discussed.

4.2. The 24-h Outdoor and Indoor Air Measurements

The outside air temperature and relative humidity measurements were implemented for a selected 24-h period in close proximity to each case study arena, inside the arena hall space in the skating zone. The measurement results are presented in Figure 13a. The lower graphs always represent the temperature (left vertical axis) and the higher graphs represent the relative humidity (right vertical axis). The average temperatures were between 14.4 °C and 16.2 °C, while the average relative humidity has a larger range, 43.5% to 83.2%. The average indoor air temperatures were 3.5 °C to 8.8 °C, and the corresponding average relative humidity was 64.5% to 82%. Both are presented in Figure 13b. It is noteworthy that AHU1.2 produced the warmest and most humid conditions, even though temperature and relative humidity are inversely correlated with each other.

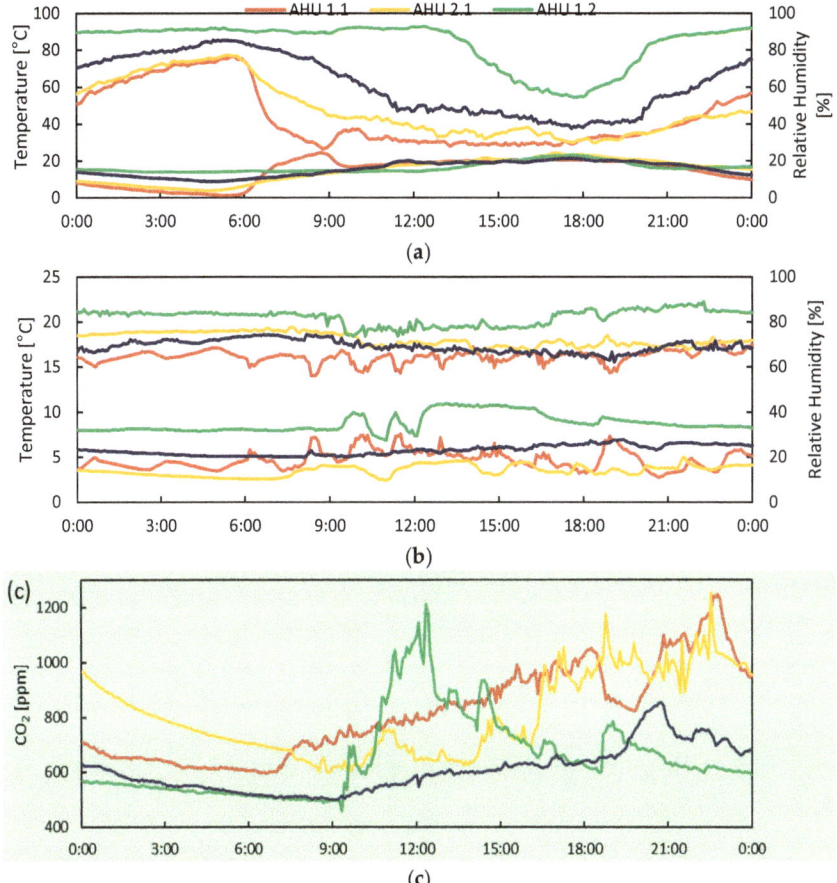

Figure 13. Indoor and outdoor air temperature, relative humidity and CO_2 measurements (**a**) Outside air temperature and relative humidity measurements; (**b**) Indoor air temperature and relative humidity measurements; (**c**) Indoor air CO_2 levels.

Hall space CO_2 levels are presented in Figure 13c. The CO_2 measurements are required to understand how indoor air CO_2 level changes against occupancy variations within a 24 h working period. It is particularly important to have a realistic perception about fresh air requirements, in order to keep indoor air CO_2 levels in an acceptable range, which is necessary for the control settings of the simulation models.

AHUs 1.1, 2.1, and 2.2 followed an approximately similar air distribution system, where the indoor air CO_2 level more or less steadily increased towards the end of the day, while for AHU1.2, the peak was reached at midday. The calculated fresh air fraction of the supply air ranged from effectively 0% for AHUs 1.2 and 2.2, to 10% for AHU2.1 and 19% for AHU1.1. The fraction was observed to stay relatively constant for each AHU, regardless of the indoor air CO_2 level, leading to the conclusion that each AHU operated in what was set as its maximum allowed extract air recirculation rate.

4.3. Energy Measurements at AHU Sections

The total external heating and cooling powers for each AHU are presented in Figure 14. The external power is defined as the power supplied to the supply air by the CCs and HCs. The heat exchanger was not considered, as it utilized internal heating power removed from the extract air. Heating power-wise, AHU 1.1 and 2.1 operated on a similar scale, with averages of 33 kW and 36 kW, respectively. AHU2.2 had a higher average of 45.3 kW, while the heating power of AHU1.2 was substantially larger, averaging at 81.3 kW. For the cooling power, the on–off type control of the CC of AHU2.2, as shown in 14b, led to the smallest average cooling power of 7.6 kW. The averages for AHU1.1, 2.1, and 1.2 were 13.6 kW, 25.6 kW, and 42.5 kW, respectively.

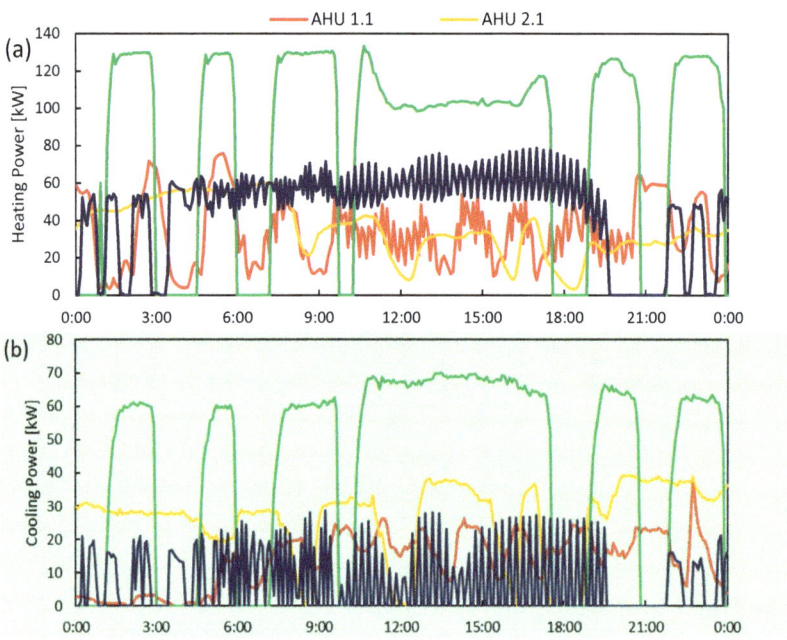

Figure 14. Total external (**a**) heating and (**b**) cooling power used by the AHUs for supply air treatment.

Figure 15 presents the total heating and cooling energy consumption by the supply air of the AHUs during the selected 24 h period, including heating energy supplied by the heat exchanger. Where the heat exchanger could be utilized despite the extract air recirculation, i.e., AHU 1.1 and 2.1, the heating energy supplied by the heat exchanger represented approximately 40% of the total heating energy demand. Total amounts of heating and cooling energy consumed by the supply air treatment ranged between 1088 kWh and 1951 kWh for heating, and between 182 kWh and 1021 kWh for cooling, as shown in Figure 15.

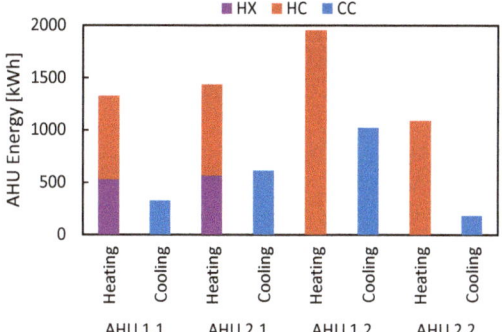

Figure 15. Total heating and cooling energy consumption of each AHU for the selected 24 h period.

5. Validating Simulation Models

The total heating and cooling energy demands in three different ice rink arenas with various AHUs are presented in Table 2. As shown, the energy demand results are provided from two different bases, one from the experimental measurements, and the other from the running simulations, and they are compared.

Table 2. The measurements and the simulation results of the three ice rink arenas.

Air Handling Units	AHU 2.2		AHU 1.1		AHU 2.1	
Temp. gradients Measurements date&time	1.5 °C/m 2016-05-20T14:55– 2016-05-27T10:00		1.6 °C/m 2016-05-03T10:50– 2016-05-12T9:00		2 °C/m 2016-06-15T14:50– 2016-06-21T10:20	
-	Heating	Cooling	Heating	Cooling	Heating	Cooling
Measurement result	C: Mäntsälä		B: Klaukkala		A: Tapiola	
	9808 MWh	3755 MWh	8592.8 MWh	2907.6 MWh	10,695.5 MWh	4939.78 MWh
Simulation result	9485 MWh	3843 MWh	8731 MWh	2864 MWh	10,214 MWh	5179 MWh
Deviation	3.3%	2.3%	1.6%	1.5%	4.5%	4.8%

The measurements performed during May and June and the simulations correspondingly ran for similar periods of time. The simulations were carried out while the AHU layouts, building specification as well as control strategies similar as the measurements, were used. The simulation results, compared to the measurement results showed that the simulation models nearly always corresponded with the real measurements with less than 5% fault, as presented in Table 2. Therefore, the simulation models were verified to represent the energy demand behaviors of the ice rink arenas with an acceptable range of accuracy. The reason for such models is because it is not easy to measure the yearly energy demands of ice rinks, particularly with the variety of AHU layouts or various temperature gradients, which were required for this study. Therefore, it is necessary to validate the simulation models according to the measurements, and then run the simulation models for the entire yearly period, to obtain the results for various combinations.

6. Simulation Results

The heat exchanger and the cooling coil energy demands were independently studied, in order to highlight the significance of the AHU configurations. Table 3 and Figure 16 present heating and cooling energy demands by using two different AHU layouts, to clarify the impact of the AHU layouts on energy consumption. The simulation results of the AHUs indicate that approximately a reduction of 60% for cooling energy demands and a reduction of 21% for heating energy demands can be achieved by precisely planning the AHU layout.

Table 3. The yearly energy consumption results of the simulation for AHU1.1 and AHU2.1.

kWh	AHU2.1 (Old Layout) kWh/(m²a)	AHU1.1 (Energy-Efficient Layout) kWh/(m²a)	Reduced Energy %
Zone heating	5.6	5.6	−1.2%
Zone cooling	212.6	214.0	0
AHU heating	207.6	164.0	−21%
AHU cooling	69.3	28.1	−59.5%
DHW heating	33.6	33.6	0
Total	523.1	445.2	-

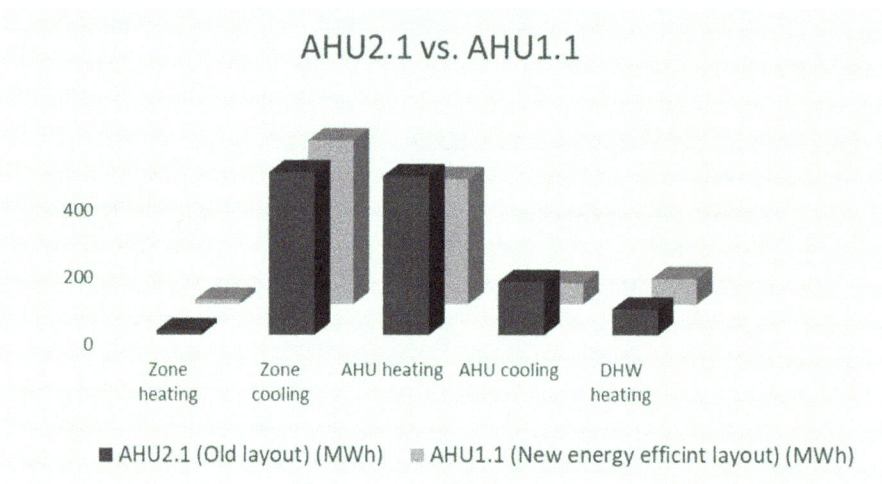

Figure 16. Comparison of cooling and heating energy demands between AHU2.1 and AHU1.1.

The simulation results of the cooling and heating energy demands are presented in Table 4 and Figure 17, while various temperature gradients have been applied, in order to study the impacts of temperature gradients on energy consumption.

Table 4. Simulation results of the yearly energy consumption of ice rinks with different temperature gradients.

Temperature Stratifications	Temperature Stratification 2 (°C/m)	Temperature Stratification 1.5 (°C/m)	Reduction	Temperature Stratification 1 (°C/m)	Reduction
Annual Energy consumption	kWh/(m²a)	kWh/(m²a)	%	kWh/(m²a)	%
Zone heating	10.2	7.4	27	7.9	−22
AHU heating	227.1	208.1	8	185.7	18
Zone cooling	274.6	179.8	35	170.0	38
AHU cooling	34.3	29.3	15	26.1	24
Electricity consumption of refrigeration plant	93.0	61.7	34	57.9	38
Condenser heat	367.7	241.5	34	227.9	38
In the case of using 50% of the condenser heat	183.8	120.7	34	113.9	38

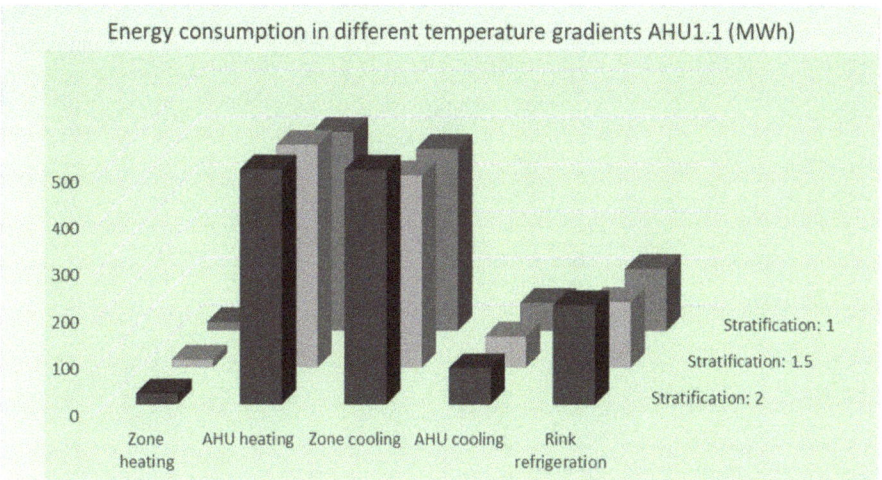

Figure 17. Energy consumption of different temperature gradients, AHU1.1.

There were three temperature gradient values, 2, 1.6, and 1.5 °C/m, measured on the three ice rinks, in which two of them were selected to be used in the simulation as high (2) and medium (1.5) temperature gradient values. The models were also simulated with an additional temperature gradient value equal to one, as an ultimate ideal condition.

Some of the measured cases included two supply air temperatures, warm and cool supply. However, this was instead simulated by using an average supply air temperature. The temperature gradient parameter in the building component takes in to account the effects of different air distribution solutions that create various temperature gradients in the simulation.

As presented in the Table 4, the energy demands for AHU cooling and AHU heating were decreased by 24% and 18%, respectively. The zone cooling of the ice-pad, as well as the electricity consumption requirements of the refrigeration process were both reduced by 38%. Finally, the overall results demonstrated clearly and concisely how energy can be significantly saved through re-planning the AHU layout and by reducing the indoor vertical temperature gradient.

7. Discussion

The most crucial challenge was how to implement the air distribution system in order to form a less stratified indoor air temperature. The ideal condition is to approach a temperature gradient of 1 °C/m. To do so, creating two thermally separated virtual zones should be considered. This means that two different temperatures are maintained in two warmer and cooler zones. The warmer zone is for the spectators, and the cooler zone is for the players. Therefore, it is reasonable to supply a more customized and localized air conditions to each zone, and then extract them from the same zone.

Figure 18 illustrates the air distribution strategies proposed by this study. As shown, the warmer air is supplied into the spectators' zone, and the cooler air into the players' zone. The air is extracted from the same zones similarly. The supply air terminals have to be as close to the occupants of the zones as possible. Two virtually separated zones are then created and subsequently, two different average temperatures are formed in each zone. The air distribution solutions reduce the risk of mixing the air within the zones. The virtual zones are indicated via dashed line boxes in the proposed air distribution models shown in Figure 18. As discussed earlier, such air distribution models more likely tend to approach the ideal temperature gradient of 1 °C/m on average. This study also verified that the lower temperature gradient results in lower cooling and heating energy demands, leading to more efficient planning of the AHU and air distribution systems. This is done by planning two separate supply and exhaust ducts, to avoid the mixing of warmer and cooler air in the main ducts. Therefore, the cooler air may not need to go through the heating coil. Moreover, it justifies the planning of two completely separate AHUs, one for the player's zone, and the other for the spectator's zone. A further advantage of this solution is that the spectators' AHU does not need to run continuously. It may run conditional to the spectators' presence, with the speed control being proportional to the number of spectators.

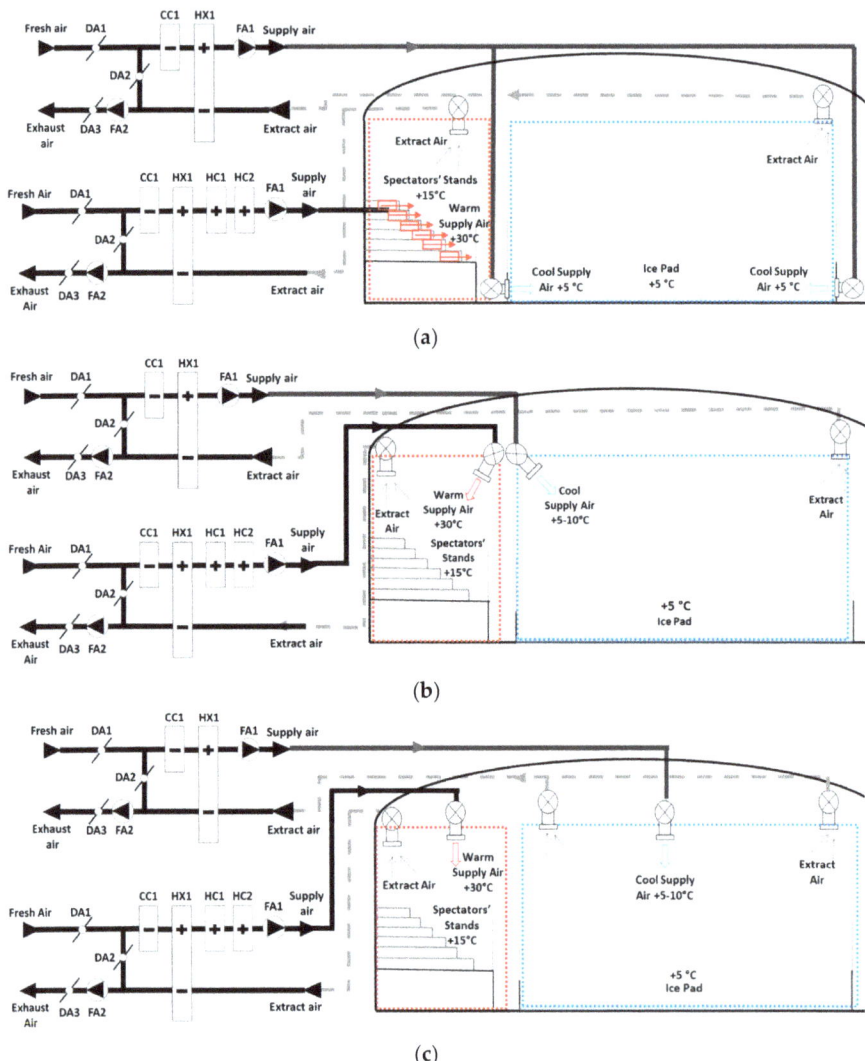

Figure 18. Proposed air distribution strategies to reduce the indoor temperature gradient. (**a**) Horizontal supply air; (**b**) inclined supply air; (**c**) vertical supply air.

8. Conclusions

This study points out the feasibility of reducing the heating energy required for space heating by approximately 21%, and reducing the cooling energy demand for dehumidification by about 60%. These results are achieved by carefully designing the AHU layouts. Furthermore, the more significant result of the study are the impacts of indoor air temperature gradients on energy demand. Both the simulation and measurement results verify that the smaller the temperature gradient, the lower heating and cooling energy demands. The results indicate that the cooling power required for refrigeration process can be reduced by up to 38% by reducing the indoor temperature stratification from 2 °C/m to nearly 1 °C/m.

Considering the aforementioned conclusion necessitates careful design for both AHU configurations and air distributions. There are no precise air distribution models for creating any specific indoor air temperature gradient. However, as in the earlier examples proposed in Figure 18, more customized air distribution models tend to be more likely to reduce indoor air temperature gradient and this consequently leads to a more energy efficient system of air distribution. To do so, the heights and the directions of the airflows have to be more carefully planned, so that the heated or non-heated air is delivered right to the occupied zone where it is needed.

Finally, for the sake of energy conservation, it is proposed that common AHUs should not be planned for the entire arena. Instead, it is more intelligent to plan various AHUs for the spectator's zone and the rink zone, so that each AHU circulates air within its own thermal zone. The supply and exhaust air terminals have to be vertically placed in such a position as to prevent mixing of the warmer and the cooler air within the zones. If mixing of cooler and warmer air is avoided, then supplying additional heating will subsequently be avoided. The additional advantages of such a system are the control of the utilization of the spectator's AHU or its running speed based on the occupancy percentage in the spectator's zone.

Author Contributions: Conceptualization, methodology, data curation, software, visualization and writing-original draft preparation S.T., L.L. and M.T.; Validation, formal analysis and writing-review & editing M.T.; Supervision, project administration and funding acquisition, J.K.

Funding: This study was financed and supported by the Finnish Ministry of Education and Culture, through the (OKM) project, and by the Estonian Centre of Excellence in Zero Energy and Resource Efficient Smart Buildings and Districts, ZEBE, grant 2014-2020.4.01.15-0016 funded by the European Regional Development Fund.

Conflicts of Interest: The authors declare no conflict of interest.

References

1. Pisello, A.L.; Bobker, M.; Cotana, F. A building energy efficiency optimization method by evaluating the effective thermal zones occupancy. *Energies* **2012**, *5*, 5257–5278. [CrossRef]
2. Domínguez, S.; Sendra, J.J.; León, A.L.; Esquivias, P.M. Towards energy demand reduction in social housing buildings: Envelope system optimization strategies. *Energies* **2012**, *5*, 2263–2287. [CrossRef]
3. Laurier Nichols, P. *Improving Efficiency in Ice Hockey Arenas*. ASHRAE Journal, USA. June 2009. Available online: https://www.stantec.com/content/dam/stantec/files/PDFAssets/2017/Improving%20Efficiency%20in%20Ice%20Hockey%20Arenas.pdf (accessed on 20 February 2019).
4. Rogstam, J.; Dahlberg, M.; Hjert, J. *Stoppsladd fas 3-Energianvändning i svenska ishallar; En studie av Svenska Ishallar i syfte att Främja Teknikutveckling och Hållbar Energianvändning*; Energy Kylanal. svenska kyltekniska föreningen: Älvsjö, Sweden, 2012.
5. Rogstam, J.; Dahlberg, M.; Hjert, J. *Stoppsladd fas 2 Energianvändning i Svenska ishallar; En studie av Svenska Ishallar i syfte att Främja Teknikutveckling och Hållbar Energianvändning*; svenska kyltekniska föreningen: Stockholm, Sweden, 2011.
6. Rojas, G.; Grove-Smith, J. Improving Ventilation Efficiency for a Highly Energy Efficient Indoor Swimming Pool Using CFD Simulations. *Fluids* **2018**, *3*, 92. [CrossRef]
7. Daoud, A.; Galanis, N.; Bellache, O. Calculation of refrigeration loads by convection, radiation and condensation in ice rinks using a transient 3D zonal model. *Appl. Therm. Eng.* **2008**, *28*, 1782–1790. [CrossRef]
8. Seghouani, L.; Daoud, A.; Galanis, N. Prediction of yearly energy requirements of indoor ice rinks. *Energy Build.* **2009**, *41*, 500–511. [CrossRef]
9. Seghouani, L.; Daoud, A.; Galanis, N. Yearly simulation of the interaction between an ice rink and its refrigeration system: A case study. *Int. J. Refrig.* **2011**, *34*, 383–389. [CrossRef]

10. Daoud, A.; Galanis, N. Prediction of airflow patterns in a ventilated enclosure with zonal methods. *Appl. Energy* **2008**, *85*, 439–448. [CrossRef]
11. Omri, M.; Galanis, N. Prediction of 3D Airflow and Temperature Field in an Indoor Ice Rink with Radiant Heat Sources. *Build. Simul.* **2010**, *3*, 153–163. [CrossRef]
12. Lestinen, S.; Koskela, H.; Jokisalo, J.; Kilpeläinen, S.; Kosonen, R. The use of displacement and zoning ventilation in a multipurpose arena. *Int. J. Vent.* **2016**, *15*, 151–166. [CrossRef]
13. Omri, M.; Barrau, J.; Moreau, S.; Galanis, N. Three-Dimensional Transient Heat Transfer and Airflow in an Indoor Ice Rink with Radiant Heat Sources. *Build. Simul.* **2016**, *9*, 175–182. [CrossRef]
14. Toomla, S.; Lestinen, S.; Kilpeläinen, S.; Leppä, L.; Kosonen, R.; Kurnitski, J. Experimental investigation of air distribution and ventilation efficiency in an ice rink arena. *Int. J. Vent.* **2018**. [CrossRef]
15. Palmowska, A.; Lipska, B. Experimental study and numerical prediction of thermal and humidity conditions in the ventilated ice rink arena. *Build. Environ.* **2016**, *108*, 171–182. [CrossRef]
16. Pennanen, A.S.; Salonen, R.O.; Aim, S.; Jantunen, M.J.; Pasanen, P. Characterization of air quality problems in five Finnish indoor ice arenas. *J. Air Waste Manag. Assoc.* **1997**, *47*, 1079–1086. [CrossRef] [PubMed]
17. Ouzzane, M.; Zmeureanu, R.; Scott, J.; Sunyé, R.; Giguere, D.; Bellache, O. Cooling Load and Environmental Measurements in a Canadian Indoor Ice Rink. *ASHRAE Trans.* **2006**, *112*, 538–546.
18. Piché, O.; Galanis, N. Thermal and economic evaluation of heat recovery measures for indoor ice rinks. *Appl. Therm. Eng.* **2010**, *30*, 2103–2108. [CrossRef]

 © 2019 by the authors. Licensee MDPI, Basel, Switzerland. This article is an open access article distributed under the terms and conditions of the Creative Commons Attribution (CC BY) license (http://creativecommons.org/licenses/by/4.0/).

Article

Towards the EU Emission Targets of 2050: Cost-Effective Emission Reduction in Finnish Detached Houses

Janne Hirvonen [1,*], Juha Jokisalo [1], Juhani Heljo [2] and Risto Kosonen [1]

1. Department of Mechanical Engineering, Aalto University, 00076 Aalto, Finland; juha.jokisalo@aalto.fi (J.J.); risto.kosonen@aalto.fi (R.K.)
2. Department of Civil Engineering, University of Tampere, 33014 Tampere, Finland; juhani.heljo@tuni.fi
* Correspondence: janne.p.hirvonen@aalto.fi; Tel.: +358-431-5780

Received: 22 October 2019; Accepted: 15 November 2019; Published: 19 November 2019

Abstract: To mitigate the effects of climate change, the European Union calls for major carbon emission reductions in the building sector through a deep renovation of the existing building stock. This study examines the cost-effective energy retrofit measures in Finnish detached houses. The Finnish detached house building stock was divided into four age classes according to the building code in effect at the time of their construction. Multi-objective optimization with a genetic algorithm was used to minimize the life cycle cost and CO_2 emissions in each building type for five different main heating systems (district heating, wood/oil boiler, direct electric heating, and ground-source heat pump) by improving the building envelope and systems. Cost-effective emission reductions were possible with all heating systems, but especially with ground-source heat pumps. Replacing oil boilers with ground-source heat pumps (GSHPs), emissions could be reduced by 79% to 92% across all the studied detached houses and investment levels. With all the other heating systems, emission reductions of 20% to 75% were possible. The most cost-effective individual renovation measures were the installation of air-to-air heat pumps for auxiliary heating and improving the thermal insulation of external walls.

Keywords: deep renovation; energy retrofit; detached house; multi-objective optimization; greenhouse gas emissions; heat pump; genetic algorithm

---

1. Introduction

European Union (EU) climate goals aim to reduce greenhouse gas emissions by 80% by 2050, compared to the level of 1990 [1]. Energy use in buildings accounts for 40% of energy consumption and a similar fraction of greenhouse gas emissions in the European Union (EU). This is why the Energy Performance of Buildings Directive (EPBD) requires all new houses to be nearly zero-energy buildings by the end of the year 2020. In addition, the latest update to the EPBD calls for all EU member states to create a roadmap for the energy renovation of existing buildings as well. [2]. The requirement to improve building energy efficiency alongside other renovations has also been outlined in Finnish environmental regulation [3].

In Finland, 79% of buildings (according to the heated net area) were constructed before the year 2000, which highlights the need to focus on the existing building stock [4]. Detached houses make up 34% of all Finnish built floor area, with row houses accounting for another 7%. Together they make up 190 million square meters of heated floor area. Energy retrofit of these buildings can have a great effect on Finnish carbon emissions. In this vein, [5] showed that final energy use in Swedish detached houses could be reduced by 65%–75%. The aim of this article is to examine the retrofit potential in a Finnish context.

To drive the renovation of old residential buildings, a tool for retrofitting design was presented in [6]. The tool utilized preferences given by the user and then ranked different solution packages according to the importance given for different categories (environmental, economic, social). Another such tool was the monthly-based energy auditing tool presented by [7]. While not requiring much expert knowledge, it was able to estimate heating and cooling demands in several climates within 8% to 15% of a more detailed dynamic simulation tool, TRNSYS [8]. However, another study compared the energy efficiency improvements of a simple single zone model to those realized in an actual building and found out the real heating demand was up to 50% higher than the simulated demand [9]. The use of a more detailed multi-zone modeling was suggested. Simplified building modeling was also used in [10], where the whole German energy system was simulated in detail, but the retrofit of buildings was estimated by simple interpolation on a line with an assumed cost to energy savings ratio. To achieve the decarbonization of the whole energy sector in Germany required a 60% reduction in building energy use [11]. This highlights the importance of more detailed calculations in the building sector to find the best ways to reach the targets.

While estimating current energy use and testing pre-selected renovation packages can be beneficial, perhaps the most effective tool for designing building retrofits is to use multi-objective optimization, where many design variables can be freely changed to iteratively achieve an optimal retrofit solution. This has been utilized for old Finnish apartment buildings to minimize life cycle costs and primary energy use [12] or carbon emissions [13]. In a Korean study, a genetic algorithm (GA) was used for the optimization of the energy system of an elementary school [14]. Only heating and cooling systems were retrofitted, without any changes to the building envelope. Three objectives, life cycle cost, renewable energy penetration, and greenhouse gas emissions, were used in the optimization. A pairwise comparison between two objectives was used to help with the challenge of having three objectives. A genetic algorithm was also used in a Canadian case, where further efficiency improvements were planned to improve an old house that had already been improved before [15]. Three objectives were split into two separate optimizations, life cycle cost vs. peak load and life cycle cost vs. energy savings with the aim of improving building energy performance above minimum requirements of the national code. The two different objectives resulted in different focus points for the retrofit, even though their principal goal of environmental benefits was the same. It is also possible to combine multiple objectives into one by calculating the weighted sum of all the objectives. This was used in [16] to optimize the retrofit of building envelope and solar panels.

A Portuguese study examined the optimal retrofit of houses set in four different regions in Portugal [17]. Using a GA, the envelope and mechanical systems were optimized to maximize energy savings and minimize initial investment cost. The rebound effect was considered by reducing the realized savings from upgrades. The importance of simulation zones and energy use profiles was studied in [18]. Optimal retrofit configurations remained the same regardless of user profiles, but the achieved energy savings obtained by each configuration did change.

When deep renovation of old detached houses was studied in Estonia, the installation of mechanical ventilation with heat recovery was found to be the most effective energy retrofit option [19]. Adding thermal insulation to external walls proved to be too costly. A study made on many types of detached houses in Chicago revealed that most homes could have their energy consumption reduced by 50% over a 25-year payback period [20]. Optimization was made by targeting the most cost-effective configurations, i.e., by finding the highest savings per cost ratio. The optimization was made in two steps, first to optimize the envelope, then to optimize the energy system. In this case, wall insulation was deemed economical. Ekström et al. [21] reported that the cost-effectiveness of detached house energy retrofitting in Sweden was dependent on the heating system of the house. Generally, installing exhaust air heat pumps was cost-effective, while window retrofitting was not. Renovation up to passive house standards was cost-effective in houses with direct electric heating. Heat pumps have generally been found to reduce energy consumption and emissions. In Canadian studies, air-to-water heat pumps reduced the greenhouse gas emissions of the housing stock by 23% [22], while solar-assisted heat pumps reduced emissions by 19% [23]. Here, the key question is how the electricity to run the heat pumps is produced, as larger reductions are achievable with low emission electricity.

Bjørneboe et al. [24] confirmed through a year-long monitoring campaign that simulated energy savings of a renovated single-family house matched those of a real building. Heating energy consumption was reduced by over 50% while improving thermal comfort. In addition, 77% of the renovation expense was covered by an increase in house value. This means that the real cost of energy retrofitting can be lower than typically estimated.

Deep renovation that reduces heating demand can cause a risk of overheating. However, overheating may be reduced by solar shading systems and increased ventilation rates [25]. Another issue with deep renovation is whether it can be practically done. Significant emission reductions in Sweden could be achieved by deep renovation in principle, but reaching passive house levels may not be possible in most cases, due to the design of the building envelope or foundation [5]. Another issue with achieving emission reductions is the embodied energy of building materials. Low operational energy use requires more materials with their own emission footprint. The inclusion of all phases of the building life cycle in the emission assessment is thus suggested [26].

The perceptions of the benefits of energy renovations in Swedish single-family houses were found to vary according to motivation to perform the renovations [27]. The indoor environment was often found to be a more important reason for house renovations than reducing energy consumption. Lack of information and access to low-interest loans were found to impede energy renovation projects. Similar observations were made in a Danish study, which suggested that to increase the prevalence of energy renovation, the focus should be shifted from investment to non-energy benefits [28]. In addition, subsidies should be enhanced, renovation plans should be included in the energy performance certificates, and maximum allowed energy consumption in houses should be regulated. These findings were repeated in another study, which tried to offer energy efficiency packages to building owners [29]. Despite the rational basis of the packages, building owners were better motivated by indoor comfort and easy solutions that might improve the aesthetics or property value. Thus, more acceptable renovation packages were formulated, with less emphasis on energy savings. Similar results were found in a survey of 883 Danish single-family house owners, which emphasized that owners need more information on non-economical and non-energy related renovation possibilities [30]. House owners often lack knowledge of their own energy consumption, which further reduces interest in doing energy retrofits.

Typically, studies on building energy renovation are focused on energy or cost. Often, energy savings are reported without taking into account the energy source. However, there is a strong national aspect to the solutions, as the climate and energy generation infrastructure varies between countries. The novelty in this study is that we minimize greenhouse gas emissions in detached houses by taking into account the emissions of different heating systems and the seasonally variable emissions of electricity in Finland. The building envelopes and technical systems also vary by age and region; thus, there is no single universal solution to be used for every country. In this study, multi-objective optimization is utilized to find a balance between emissions and life cycle cost for deep renovation solutions in Finnish buildings built in different decades. An earlier study used similar methodology to search for cost-optimal solutions to minimize emissions in Finnish apartment buildings [13]. This study will cover the rest of the Finnish residential building stock by finding the emission reduction potential in single-family homes of four different age categories, which has not been done before in Finland. The novelty is in the optimization of a large combination of buildings and heating systems, covering existing buildings of various ages. All the retrofit parameters are adjusted freely without pre-selected packages. The study is part of a larger plan to optimize the deep renovation of all major building types in Finland and estimate the effect on the national energy system.

2. Materials and Methods

2.1. Simulation Setup

The Finnish detached houses were modeled on an hourly timescale using the dynamic simulation tool IDA-ICE [31], which has been validated, for example, in [32,33]. The weather file used in the simulations was the Test Reference Year for Southern Finland (TRY2012-Vantaa). The annual average outdoor air temperature is 5.6 °C, and the annual solar insolation on a horizontal surface is 970 kWh/m^2 [34]. The heating degree day value for the studied climate zone (at indoor temperature of 17 °C) is 3952 Kd [35].

The dynamic building simulation by IDA-ICE was combined with MATLAB for additional pre- and post-processing, and the MOBO tool [36] for multi-objective optimization. The optimization process is shown in Figure 1. The MOBO software runs the optimization loop, which provides trial configurations for building retrofit and feeds them to MATLAB. MATLAB then generates a simulation input file compatible with the building simulation tool IDA-ICE. After IDA-ICE has simulated the building performance, MATLAB calculates the cost and emissions of running the system and returns the answers to MOBO to prepare the next iteration of the optimization.

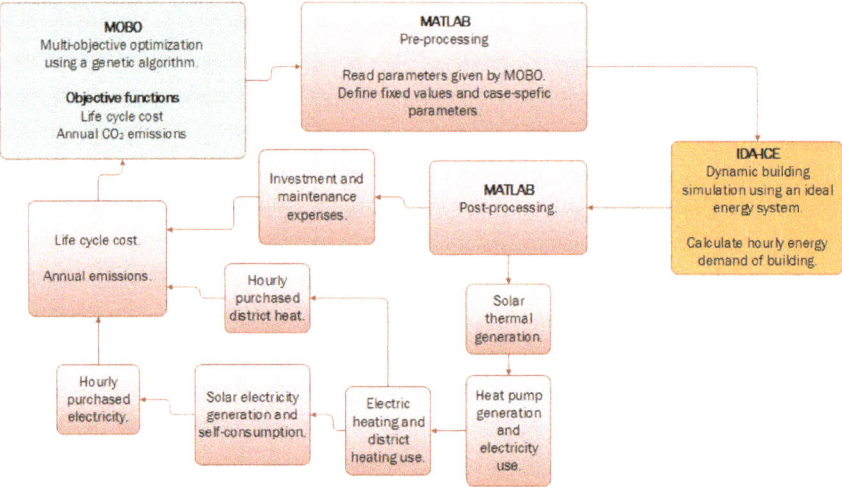

Figure 1. The optimization process that combines the three tools.

2.2. Building Descriptions

Because residential detached houses account for such a large fraction of Finnish building stock, their effect on national emissions is also great. Figure 2 shows the building stock age distribution for single-family houses (SH) [4]. The buildings were divided into four age categories, according to the Finnish building code of the time. The oldest group SH1 was built before any energy efficiency requirements were issued and is thus poorly insulated and uses natural ventilation. SH2 is a large group of old buildings with slightly improved insulation and mechanical exhaust ventilation (E. vent.) without heat recovery. SH3 is a group of well-insulated buildings equipped with mechanical supply and exhaust ventilation and heat recovery (S. & E. vent.). SH4 is very well insulated and has improved ventilation heat recovery efficiency. The details of the buildings are shown in Table 1.

Figure 2. Distribution of built floor area in single-family houses over different construction periods.

Table 1. Properties of the reference single-family houses. [37].

Building Age Class	SH1	SH2	SH3	SH4
Construction years	–1975	1976–2002	2003–2009	2010–
U-values of envelope (W/(m^{2}K))				
External wall	0.584	0.28	0.25	0.17
Floor	0.48	0.36	0.25	0.16
Ceiling	0.343	0.22	0.16	0.09
Doors	1.4	1.4	1.4	1.0
Windows	1.8	1.6	1.4	1.0
Total solar heat transmittance (g)	0.71	0.59	0.46	0.46
Direct solar transmittance (ST)	0.64	0.52	0.39	0.39
Air tightness				
n_{50}, (1/h)	6	4	3.5	2
q_{50} m^3/(h m^2)	15.6	10.4	9.1	5.2
Ventilation				
Type	Natural ventilation	Mech. E. vent.	Mech. S.&E. vent.	Mech. S.&E. vent.
Heat recovery temp eff	0	0	0.55	0.65
Ventilation rate (L/s/m^2)	0.30	0.33	0.36	0.36
Total air exchange rate (1/h)	0.41	0.46	0.5	0.5
SFP (kW/m^3/s)	0	1.5	2.5	2
Heating setpoint (°C)	22	22	21.5	21

2.3. Building Service Systems

The main heating systems used in single-family houses in Finland are shown in Figure 3. The information is based on the registry information from Statistics Finland but does not contain possible changes that have been made after the construction of the building [38]. In old buildings, wood and oil-based heating are the most common, but oil boilers are mostly phased out in new buildings. The fraction of wood-based heating also goes down for newer buildings. Direct electric heating is a common system in all age categories. District heating (DH) in Finnish detached houses is much less common than in apartment buildings, but it still covers 10%–15% of heating and has grown more common in new buildings. Ground-source heat pumps (GSHP) are especially common in new buildings. To follow this distribution, five main heating system types were modeled:

1. Wood boiler
2. Oil boiler
3. Direct electric heating
4. District heating
5. Ground-source heat pump

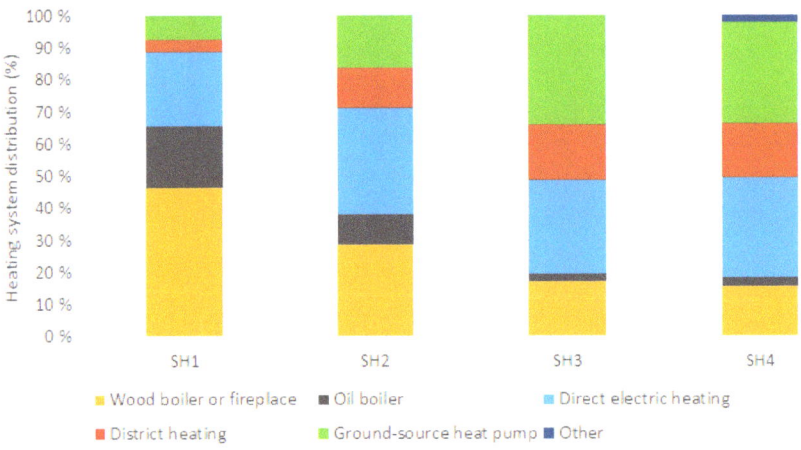

Figure 3. Heating systems in single-family houses. [38].

In addition, all systems (except the GSHP system) can be supplemented by an air-to-air heat pump (AAHP) as part of the building retrofit.

There is some uncertainty in the heating systems of detached houses, and, especially, the amount of wood fuel consumption is not accurately known, as some people use fireplaces as luxury items or support systems, while others use them as the main heating system. There are also different views concerning the emissions of wood-based energy production. EU policy dictates that biomass has no emissions, under the assumption that all emitted carbon is absorbed into new biomass growth. However, in practice, the immediate emissions are on the level of coal, and the assumed reabsorption time has an effect on the global warming potential. If biomass heating was assumed to have no emissions, then no emission benefits could be gained from energy efficiency retrofit. In this paper, wood-combustion is assumed to have non-zero emissions. Wood-based heating systems also include traditional fireplaces. For simplicity, they are modeled as wood pellet boilers with water-based heat distribution systems. Electric heating can be used with electric radiators or through water-based heat distribution systems and hot water storage tanks. However, the water-based or accumulating electric heating systems are so rare that only direct electric heating systems were modeled in this study. For the retrofitted buildings, solar thermal (flat-plate) and solar electric (PV) systems were also optionally included. The solar systems were always installed at a 40° angle, which provides the maximum annual generation in most of Finland. The roof area available for solar systems was 70 m^2, with each kW of PV panels taking up 6.5 m^2 of space. The roof was south-facing. In reality, some houses have roof designs where the available space is lower, or the direction is suboptimal for solar energy. However, detached houses typically have garages and storage sheds or even extra yard space, which could be used to install solar energy systems.

Heat distribution efficiency of water-based radiators was 80% for 70/40 temperatures, 85% for 40/30 temperatures, and 95% for direct electric heaters [39]. The coefficient of performance (COP) of the heat pumps was calculated as a function of heat source and heat distribution temperatures (Table 2). The performance of the AAHP was limited by air-distribution in the house, such that it could only cover a maximum of 40% of space heating demand [40].

Table 2. COP of the ground-source heat pump [41] at the standardized test conditions [42].

Temperature (°C/°C) (Source/Output)	COP
0/35	4.3
0/45	3.5
10/35	5.2
10/45	4.2

2.4. User Profiles and Internal Loads

The domestic hot water (DHW) use profile was based on measured data from Finnish buildings [43] and was normalized to 35 kWh/m^2 per year [44]. While the buildings of different ages had different space heating demands, the DHW demand was the same for all buildings. Lighting and electric appliance profiles were based on measured profiles from 1630 Finnish households [45], and the consumption was normalized to 5.3 kWh/m^2 and 15.9 kWh/m^2 per year, respectively [46].

2.5. Emissions of Different Energy Sources

Emissions of electricity generation in Finland vary seasonally, according to the balance between electricity demand and the availability of low emission energy sources. Electricity consumption is higher in winter, reducing the fraction of electricity generated with nuclear, hydro, and wind energy and thus increasing emissions. Historical emissions data from Finland between 2011 and 2015 was used to generate a monthly emission profile, where the emissions are higher in winter (173 g-CO_2/kWh) and lower in summer (81 g-CO_2/kWh) [47]. District heating was assumed to have constant emissions all year long, 164 g-CO_2/kWh [48]. This value includes the benefit distribution between heat and electricity generation when using combined heat and power (CHP). Practically all district heating in Finland is produced using combustion technologies, which is why there is little variation in emissions compared to electricity generation, where the production mix changes according to weather conditions and available power plants. District heating fuels do vary between regions, but here, a national average has been used.

Detached houses may also use on-site boilers for heat generation. In this case, the emissions for oil boilers were 263 g-CO2/kWh [49]. With wood fuels, the specific emissions during combustion were 403 g-CO2/kWh. These on-site systems are not part of the EU Emission Trading System (ETS), unlike district heating and electrical heating systems. In principle, this makes emission reductions of these systems more effective, because they do not release emission allowances for others to use. The emission factors for all heating systems are shown in Table 3.

Table 3. The emissions [49,50] and costs of different energy sources [51–55]. For electricity, the average cost is reported.

Energy Source	Specific Emissions (kg-CO_2/MWh)	Cost (€/MWh)
District heat	164	62.9
Wood fuel	403	56.1
Heating oil	263	104.4
Electricity import	81–173	120.2
Electricity export	0	48.8

2.6. Economic Assumptions and Cost of Retrofitting

The costs of all renovation measures were presented as life cycle cost (LCC) per heated floor area, calculated over 25 years. The LCC consisted of the initial investment and the lifetime energy, maintenance, and renewal costs, subtracted by the residual value of the upgraded components according to Equation (1):

$$LCC = C_{investment} + \sum_{t=1}^{25} \frac{C_{energy,t}}{(1+r_e)^t} + \sum_{t=1}^{25} \frac{C_{maintenance,t}}{(1+r)^t} + \sum_{t=1}^{25} \frac{C_{renewal,t}}{(1+r)^t} - \frac{R}{(1+r)^{25}} \quad (1)$$

where $C_{investment}$ is the initial investment cost of the building and system retrofits, $C_{energy,t}$ is annual electricity and heat cost, $C_{maintenance,t}$ is the annual maintenance cost, $C_{renewal,t}$ is the cost of system renewals, and R is the residual value of the retrofitted systems at the end of the 25 year period. The real interest rate (r) was 3%, and the annual rise in electricity and district heating cost was 2%. The escalated real interest rate (r_e) is used to take into account the rising energy cost alongside the real interest rate. The cost of energy generation with different fuels and devices is presented in Table 3, which shows the emission factors and the after-tax cost of heat and electricity. District heat and wood fuels are significantly cheaper than oil or imported electricity but have higher emissions. The cost of electricity varies hourly, but only the annual average cost is listed. The electricity price includes the Nord Pool spot price, distribution price, and the Finnish electricity tax.

The costs of the various retrofit options are shown in the following tables. All costs are total costs with equipment and installation and also include the 24% value added tax (VAT). Table 4 shows the cost of additional thermal insulation for external walls and roofs, while Table 5 shows the cost of window upgrades. The cost of installing new heat pumps and boilers are shown in Table 6. The GSHP needs to be partly renewed after 15 years, with the renewal cost shown in Table 6. The costs of solar energy systems are presented in Table 7. The solar thermal system was completely renewed after 20 years.

Table 4. Cost of improving the thermal insulation level of the building envelope. [56].

Insulation (mm)	Wall Cost (€/wall-m²)		Insulation (mm)	Roof Cost (€/roof-m²)	
	SH1	SH2, SH3, SH4		SH1	SH2, SH3, SH4
0	83.1	83.1	0	0.0	0.0
25	108.5	96.2	25	19.0	7.5
50	115.1	102.8	50	20.2	8.7
100	128.5	116.2	100	22.5	11.1
150	141.8	129.5	150	24.9	13.4
200	155.2	142.9	200	27.2	15.8
250	168.5	156.3	250	29.6	18.2
300	181.9	169.6	300	32.0	20.5

Table 5. Cost of a basic refurbishment of current windows or replacing them with new ones. [56].

Windows U-value (W/(m²K))	Cost (€/window-m²)			
	SH1	SH2	SH3	SH4
1.8	14.2	-	-	-
1.6	-	14.2	-	-
1.4	-	-	14.2	-
1	342.8	342.8	342.8	14.2
0.8	393.4	393.4	393.4	393.4
0.6	453.9	453.9	453.9	453.9

Table 6. Cost of heating devices (with installation). [57].

Boilers (20 kW$_{th}$)		GSHP		AAHP	
Fuel	Price (€)	Capacity (kW$_{th}$)	Price (€/kW$_{th}$)	Capacity (kW$_{th}$)	Price (€/kW$_{th}$)
Wood	5100	6	1925	1	1240
Oil	4300	14	1114	2	750
-	-	20	1000	3	570
-	-	GSHP renewal	231	4	468
-	-	-	-	5	406
-	-	-	-	6	367

Table 7. Solar energy system costs. The flat-plate solar thermal cost also includes the required thermal storage tank. [12,58].

Solar Electricity	
PV Capacity (kW)	Price (€/kW)
1	2400
3.25	1750
5.5	1400
10	1200
	Price (€/m²)
Solar thermal	675

The ventilation refurbishment costs are shown in Table 8. Buildings SH3 and SH4 have mechanical supply and exhaust ventilation with heat recovery (HR) by default. Demand-based ventilation using variable air volume (VAV) is only possible if a mechanical supply and exhaust ventilation system is installed. VAV reduces ventilation airflow according to occupation to a minimum of 40% airflow when the apartment is empty, reducing energy consumption.

The residual values of retrofitted components at the end of the 25-year calculation period were a fraction of their investment cost, discounted by 25 years. The fractional residual values are shown in Table 9.

Table 8. Cost of ventilation retrofit. [56] Variable air volume (VAV) can only be used with mechanical supply and exhaust ventilation.

Ventilation Measure	Cost (€/floor-m²)			
	SH1	SH2	SH3	SH4
Installation of a new mechanical supply and exhaust ventilation system	60.7	60.7	-	-
Improving the HR efficiency of existing ventilation system	-	-	16.9	16.9
VAV for mechanical supply and exhaust ventilation	10	10	10	10

Table 9. Residual values of components at the end of the 25-year period as percentage of investment cost [12].

Component	Residual Fraction (%)
Solar thermal	75
GSHP	50
Wall	37.5
Roof	37.5
Windows	37.5
Ventilation	32.5

2.7. Optimization

Multi-objective optimization with the genetic algorithm NSGA-II [59] was used to find the most cost-effective retrofit solutions for each building type and heating system. The objective was to minimize both the life cycle cost and carbon dioxide emissions by retrofitting existing buildings with better-insulated envelopes and improved energy systems. The objective values were reported relative to the heated floor area of the buildings. The optimization algorithm runs in the MOBO optimization software, which calls MATLAB and IDA-ICE to perform the actual building and energy simulation.

The genetic algorithm is a heuristic method that iteratively progresses toward better solutions by combining the features of previous solutions over many generations. First, it generates a random initial set of solution candidates, performs the building and energy simulation, and calculates their objective values (LCC and emissions). Second, the variable values of the best solutions of each generation are mixed to make new solution candidates (crossover). There is also a chance of randomly changing some variables (mutation). Because the two objectives are conflicting (lower emissions typically result in higher LCC), instead of a single optimal solution, a set of many Pareto optimal solutions (the Pareto front) is formed over many iterations. As a heuristic algorithm, NSGA-II is not guaranteed to find the true global optimum, and there are always some random elements in the results. Individual variables in otherwise very good solutions may be less than optimal because there is no mechanism to target specific variables for separate optimization. However, with enough iterations, NSGA-II does provide near optimal results, which are good enough for practical purposes.

The system retrofit paths are shown in Figure 4. Buildings with district heating or direct electric heating were assumed to keep their current heating system due to big investments already made to obtain the system or difficulty in switching to a different kind of heat distribution system. Oil heating systems were abandoned during renovation, to be replaced by wood-based heating or ground-source heat pumps. A reference GSHP case was calculated using a pre-installed heat pump sized to 70% of the peak heating demand, but buildings with an existing heat pump were not retrofitted further. The GSHP optimization cases assumed the installation of a new GSHP system to replace oil or wood boilers.

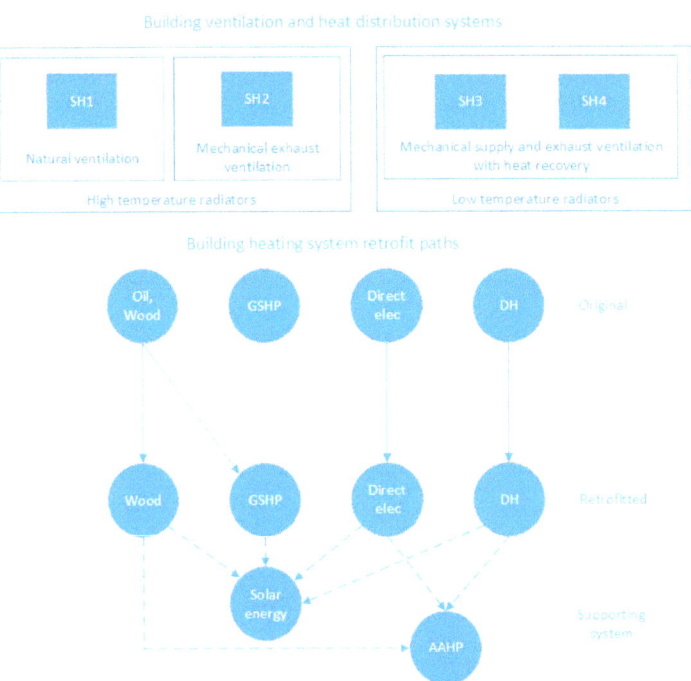

Figure 4. Retrofit paths for buildings with different heating systems.

A list of all optimization parameters is shown in Table 10. The value ranges depend on the building age class, as the starting levels are different. The air-source heat pump is not used with the ground-source heat pump. In old buildings, the ventilation system has to be retrofitted as the mechanical supply and exhaust system before demand-based variable air volume ventilation can be used. The retrofit will always include a high-efficiency heat recovery system. For new buildings with pre-existing supply and exhaust ventilation systems, the improved HR efficiency and VAV ventilation can be implemented separately.

Table 10. Optimization parameters for all building age classes.

Variable	Unit	Min	Max	Description
SH1				
Wall U-value	W/m²K	0.1	0.58	Only basic refurbishment or basic refurbishment with added thermal insulation.
Window U-value	W/m²K	0.6	1.8	Only basic refurbishment or installation of new windows.
Roof U-value	W/m²K	0.09	0.34	Only basic refurbishment or basic refurbishment with added thermal insulation.
Door U-value	W/m²K	0.8	1.4	No change or installation of new doors.
ST area	m²	0	20	Solar thermal collectors with daily storage.
PV capacity	kW$_e$	0	10	Solar electric panels.
AAHP capacity	kW$_{th}$	0	6	Air-to-air heat pump not used with ground-source heat pump.
GSHP capacity	kW$_{th}$	0	20	Ground-source heat pump only used for the specific optimization case.
Radiator temperature	°C	40	70	Retrofitting the heat distribution to allow lower radiator inlet water temperature. Not used with electric heating.
Ventilation	-	0	2	0: Natural ventilation, 1: Mech. S. and E. vent. with 75% HR, 2: Mech. S. and E. vent. with 75% HR and VAV
SH2				
Wall U-value	W/m²K	0.08	0.28	Only basic refurbishment or basic refurbishment with added thermal insulation.
Window U-value	W/m²K	0.6	1.6	Only basic refurbishment or installation of new windows.
Roof U-value	W/m²K	0.08	0.22	Only basic refurbishment or basic refurbishment with added thermal insulation.
Door U-value	W/m²K	0.8	1.4	No change or installation of new doors.
ST area	m²	0	20	Solar thermal collectors with daily storage.
PV capacity	kW$_e$	0	10	Solar electric panels.
AAHP capacity	kW$_{th}$	0	6	Air-to-air heat pump not used with ground-source heat pump.
GSHP capacity	kW$_{th}$	0	20	Ground-source heat pump only used for the specific optimization case.
Radiator temperature	°C	40	70	Retrofitting the heat distribution to allow lower radiator inlet water temperature. Not used with electric heating.
Ventilation	-	0	2	0: Mechanical exhaust ventilation, 1: Mech. S. and E. vent. with 75% HR, 2: Mech. S. and E. vent. with 75% HR and VAV
SH3				
Wall U-value	W/m²K	0.08	0.25	Only basic refurbishment or basic refurbishment with added thermal insulation.
Window U-value	W/m²K	0.6	1.4	Only basic refurbishment or installation of new windows.
Roof U-value	W/m²K	0.07	0.16	Only basic refurbishment or basic refurbishment with added thermal insulation.
Door U-value	W/m²K	0.8	1.4	No change or installation of new doors.
ST area	m²	0	20	Solar thermal collectors with daily storage.
PV capacity	kW$_e$	0	10	Solar electric panels.
AAHP capacity	kW$_{th}$	0	6	Air-to-air heat pump not used with ground-source heat pump.
GSHP capacity	kW$_{th}$	0	20	Ground-source heat pump only used for the specific optimization case.
Radiator temperature	°C	40	40	Low temperature radiators used by default.
Ventilation	-	1	4	1: Mech. S. and E. vent with 60% HR, 2: Mech. S. and E. vent. with 60% HR and VAV, 3: Mech. S. and E. vent. with 75% HR, 4: Mech. S. and E. vent. with 75% HR and VAV

Table 10. *Cont.*

Variable	Unit	Min	Max	Description
SH4				
Wall U-value	W/m²K	0.07	0.17	Only basic refurbishment or basic refurbishment with added thermal insulation.
Window U-value	W/m²K	0.6	1	Only basic refurbishment or installation of new windows.
Roof U-value	W/m²K	0.06	0.09	Only basic refurbishment or basic refurbishment with added thermal insulation.
Door U-value	W/m²K	0.8	1	No change or installation of new doors.
ST area	m²	0	20	Solar thermal collectors with daily storage.
PV capacity	kW$_e$	0	10	Solar electric panels.
AAHP capacity	kW$_{th}$	0	6	Air-to-air heat pump not used with ground-source heat pump.
GSHP capacity	kW$_{th}$	0	20	Ground-source heat pump only used for the specific optimization case.
Radiator temperature	°C	40	40	Low temperature radiators used by default.
Ventilation	-	1	4	1: Mech. S. and E. vent. with 65% HR, 2: Mech. S. and E. vent. with 65% HR and VAV, 3: Mech. S. and E. vent. with 75% HR, 4: Mech. S. and E. vent. with 75% HR and VAV

3. Results

Figure 5 shows the specific heating demand of the reference buildings of the four age classes. It shows the reduction of space heating demand between the age classes and how domestic hot water becomes a larger part of the total demand for newer buildings. The heat demand for ventilation is missing from SH1 and SH2 because there is no mechanical balanced ventilation in either building type. Therefore, any make-up air is heated through the space heating system in those buildings. The share of ventilation heating is small in SH3 and SH4 due to heat recovery systems. The following subsections will show the results of the optimization of deep renovation in all the building age classes, focusing on the life cycle cost and the achieved emissions. The energy consumption of the buildings is presented in Appendix A.

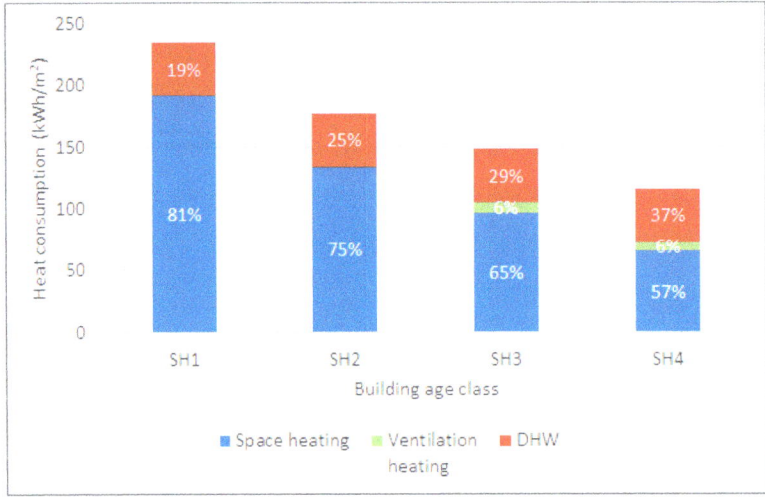

Figure 5. Specific heating demand in the reference buildings.

3.1. SH1–Buildings Built before 1976

The emissions and life cycle costs were calculated for the reference cases with all different heating systems of the oldest single-family house SH1. Optimization was then performed for all except oil heated systems. Figure 6 shows all the results from the optimization of SH1 with a district heating system. All solutions reduce emissions, but many of them also increase costs significantly. Two separate clusters of solutions can be identified. The cluster with higher LCC is formed of solutions where the natural ventilation system was replaced by mechanical supply and exhaust ventilation with heat recovery. In the scope of this optimization study, the upgrade is not economical because the monetary value of the better indoor climate achieved with a mechanical ventilation system was not taken into account in the optimization. However, the literature shows that people often give more value to indoor comfort than energy savings, so these solutions should be applied to enhance the indoor climate.

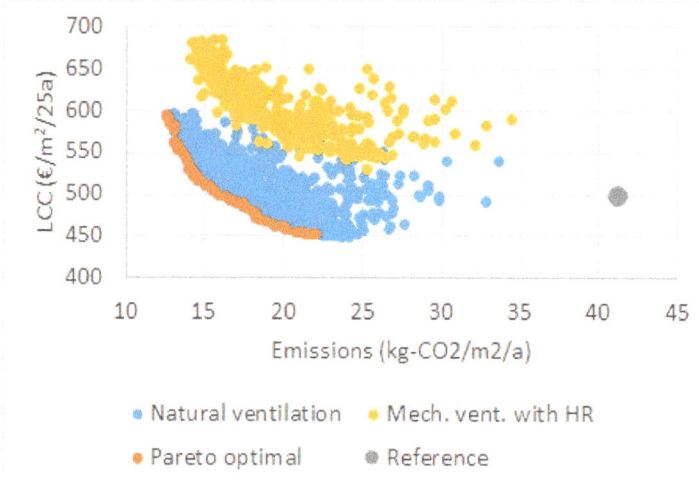

Figure 6. Results from the optimization of SH1 with district heating. Mechanical ventilation cases are shown in a separate cluster due to their higher investment costs. Mechanical supply and exhaust ventilation improve indoor air quality, which is not considered in the optimization.

The total air exchange rate in the building increased after the installation of the new ventilation system. The air exchange rate was 0.14 1/h with natural ventilation and 0.5 1/h with mechanical ventilation. While the heat recovery system reduced the ventilation heating demand, the high investment cost, higher total airflow, and increased electricity expenses made the refurbishment a non-economical choice. However, this refurbishment improves the indoor air conditions. With natural ventilation, there is no guarantee that enough fresh air is available in all seasons, unlike with mechanical ventilation. In addition, there is no filtering to protect the residents from outdoor pollutants such as particulate matter or pollen. Mechanical supply and exhaust ventilation also allow the heating and cooling of the supply air, which can improve thermal comfort. However, these issues were not considered in this study.

Figure 7 shows the breakdown of the life cycle costs for all the Pareto optimal renovation configurations of SH1 with the district heating system. Each bar represents one optimal solution from Figure 6. Marked on the chart is the first time each measure is used (read from right to left). Cost of energy dominates the life cycle cost in the low impact solutions (on the right), while investments take an even share in the high impact solutions (on the left).

Electricity demand went down when PV panels were installed, and the district heating demand was affected by all the other system refurbishments. Thermal insulation was added to the roof and external walls in all cases. New windows were installed for 2/3 of the solutions, and door insulation was improved in 1/3 of the cases. Solar thermal investments increased from minimum to maximum levels as emissions lowered. An air-to-air heat pump was always utilized.

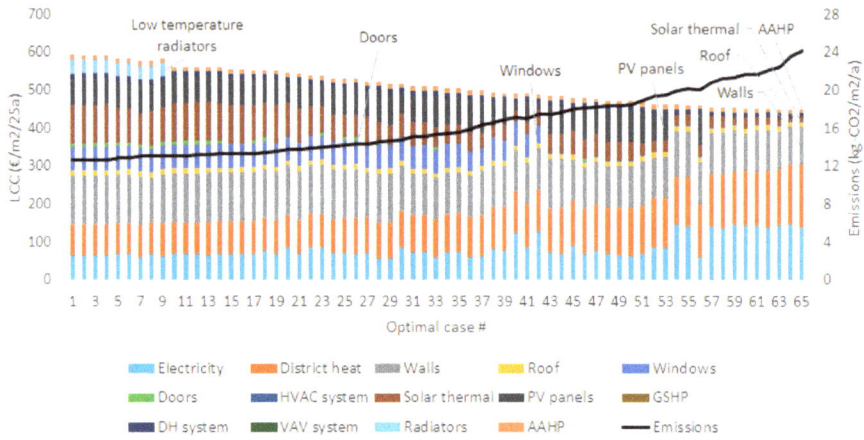

Figure 7. Breakdown of life cycle cost for SH1 with district heating.

Figure 8 shows the emissions and LCC for the reference cases alongside the Pareto optimal retrofit solutions of the optimization. Comparing oil and wood boiler systems, the reference case for oil had lower emissions and higher LCC than the wood boiler. In both cases, optimal energy renovation could reduce both emissions and costs. The same was true for district heating and the electric heating system. With oil and electricity, all optimal configurations were both more economical and less emission-intensive compared to the reference case. The ground-source heat pump was a very effective solution even without any other improvements, as shown by the case with a pre-installed GSHP. This case had reduced investment costs compared to the optimized cases, which had a new heat pump installed as they converted away from oil or wood boilers. Also highlighted in the figure are specific optimal cases A to D, which are shown in more detail in Table 11.

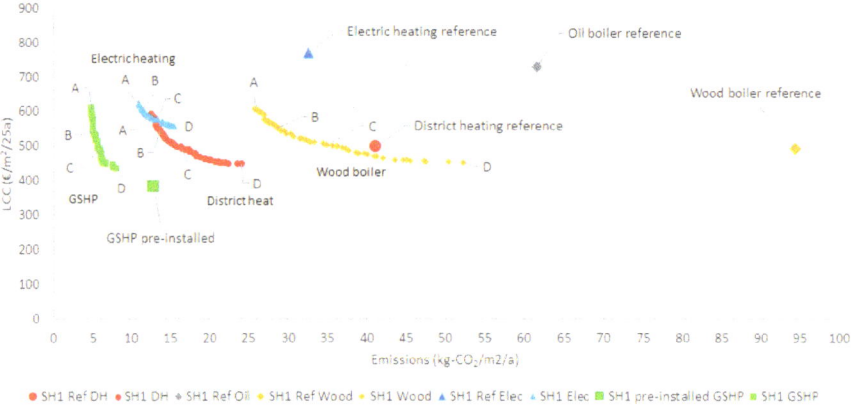

Figure 8. Reference cases and optimal results for the oldest single-family house SH1 with different heating systems. The cost and emissions of the reference cases are shown with large individual symbols. The corresponding optimal results are shown as Pareto fronts with the same color. Four individual solutions were chosen from each Pareto front, identified with the letters A to D. A is the most expensive solution, and D the least cost solution, with B and C equally distributed between them.

Table 11. Selected optimal cases for SH1 with all heating systems. Cost-effective solutions are marked in green. Cases D to A are optimized solutions from the lowest to the highest life cycle cost (LCC). For the ventilation system: 0 is natural ventilation, 1 is mechanical supply and exhaust ventilation with 75% heat recovery (HR) efficiency, and 2 is mechanical supply and exhaust ventilation with 75% HR and VAV.

Solution	Emissions kg-CO_2/m²/a	Emission Reduction kg-CO_2/m²/a	Relative Reduction %	Reduction Cost €-LCC/kg-CO_2/m²	LCC €/m²/25a	Investment €/m²	U-values (W/m²K) Walls	U-values (W/m²K) Roof	U-values (W/m²K) Doors	U-values (W/m²K) Windows	ST m²	PV kW$_p$	Ventilation System	Dist. Temp. °C	GSHP kW$_{th}$	AAHP kW$_{th}$
DH Ref	41.3	-	-	-	497.7	110.5	0.58	0.34	1.4	1.8	0	0	0	70	0	0
DH D	24.2	17.1	41.4	−2.8	449.6	166.0	0.2	0.1	1.4	1.8	2	0	0	70	0	2
DH C	16.4	24.9	60.3	0.0	498.1	335.9	0.12	0.1	1.4	1	8	9	0	70	0	6
DH B	13.8	27.5	66.7	1.7	544.9	391.8	0.1	0.1	1.4	0.6	18	7	0	70	0	3
DH A	12.6	28.7	69.5	3.3	593.5	449.6	0.1	0.09	0.8	0.6	20	7	0	40	0	6
Wood Ref	94.4	-	-	-	491.9	106.7	0.58	0.34	1.4	1.8	0	0	0	70	0	0
Wood D	52.3	42.1	44.6	−0.9	452.6	179.3	0.2	0.1	1.4	1.8	2	0	0	70	0	2
Wood C	35.0	59.4	62.9	0.2	503.6	331.0	0.15	0.09	1.4	0.6	6	6	0	70	0	5
Wood B	28.7	65.7	69.6	0.9	552.4	389.5	0.1	0.09	1.4	0.6	16	5	0	70	0	5
Wood A	25.9	68.5	72.6	1.7	608.7	462.9	0.1	0.09	0.8	0.6	20	7	0	40	0	6
Elec Ref	32.6	-	-	-	768.5	78.4	0.58	0.34	1.4	1.8	0	0	0	-	0	0
Elec D	12.1	20.5	62.9	−10.2	559.5	312.9	0.15	0.09	1.4	1.8	6	9	2	-	0	2
Elec C	9.4	23.1	71.1	−7.8	587.5	396.8	0.12	0.09	1.4	0.6	8	7	2	-	0	2
Elec B	8.4	24.2	74.3	−6.2	617.7	447.4	0.1	0.09	1	0.6	14	8	2	-	0	3
Elec A	8.0	24.6	75.4	−5.1	643.7	468.3	0.1	0.09	0.8	0.6	20	7	2	-	0	6
Oil Ref	61.6	-	-	-	729.8	102.3	0.58	0.34	1.4	1.8	0	0	0	70	0	0
GSHP D	8.1	53.5	86.8	−5.5	437.3	271.5	0.2	0.12	1.4	1.8	0	10	0	70	7	0
GSHP C	5.7	55.9	90.7	−4.2	496.5	391.8	0.12	0.09	1.4	1	0	10	0	40	7	0
GSHP B	5.1	56.5	91.7	−3.2	550.2	453.7	0.1	0.1	1.4	0.6	8	9	0	40	8	0
GSHP A	4.8	56.8	92.1	−2.1	611.5	491.9	0.1	0.09	1	0.6	20	6	0	40	8	0

Table 11 shows the properties of four cases (A–D) from each Pareto front, selected according to the achieved emission and cost levels. This way, the specific means to obtain different emission reductions can be seen. The cases were chosen based on the LCC: Case A is the highest cost-optimal solution, Case D is the lowest cost-optimal solution, and Cases B and C are evenly distributed between these points, cost-wise.

Observations from Figure 7 can be confirmed in Table 11, which shows the measures used to reach different emission levels. In all renovated cases, regardless of the heating system, the building envelope was improved by adding more thermal insulation to the roof and external walls. Improvement of thermal insulation level of windows and doors was mainly made on B and A levels. The natural ventilation system was not replaced and, thus, variable flow ventilation was never used either. Solar electricity was not used in the cheapest DH and boiler cases but was used in all electrically heated cases. In the GSHP cases, as more efficiency measures were implemented, the amount of PV panels went down. The installation of the mechanical supply and exhaust ventilation system was only used in the electrically heated case, due to its high energy cost. In all other cases, the investment cost of the measure was too high to be beneficial.

3.2. SH2–Buildings Built between 1976 and 2002

Figure 9 shows the optimization results of SH2 as Pareto fronts alongside the reference cases. While each case has a range of values, the solution sets can be ranked according to their emissions levels. Lowest emissions come from the ground-source heat pump systems, followed by direct electric heating, then district heating, and, finally, the wood boiler. Absolute reduction potential was greatest in the wood boiler case. Similar to SH1, the reference case for GSHP included a pre-installed heat pump, but the optimal solutions added a new heat pump to convert from oil and wood boilers.

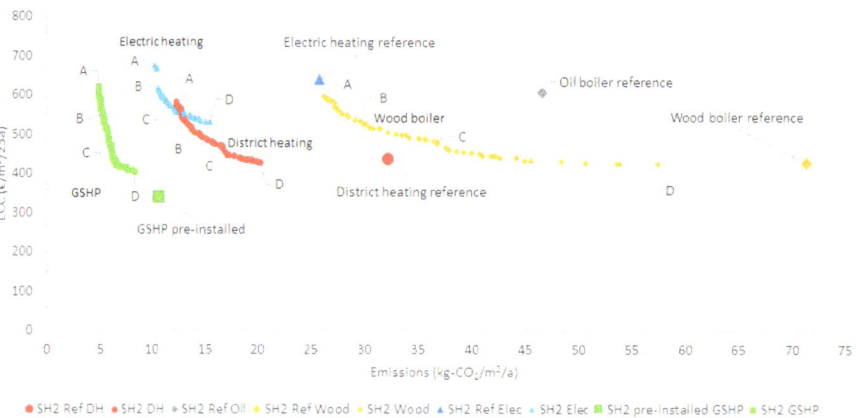

Figure 9. Reference cases and optimal results of retrofitting for the single-family house SH2 with different heating systems.

Table 12 shows the details of the selected solutions of SH2. The selection was made based on the life cycle cost, with four evenly distributed solutions chosen from all the optimal solutions of each heating system. The air-to-air heat pump was included in all optimal solutions (except the GSHP), which increased electricity consumption. However, both emissions and primary energy demand were reduced in each case. Emissions went down 20% to 63% between cases D and A without a GSHP and 82% to 90% with a GSHP. In most cases, thermal insulation was added to the roof and external walls. New windows and doors were not installed in the low-cost D cases. When new windows were installed, they were often better than the minimum requirements, i.e., below the U-value of

1 W/m^{2}K [3]. Solar electricity had a larger role in the electric heating and heat pump cases, while it was not included in the D cases with boiler or DH. Low-temperature radiators were installed in the A cases of all water-based heating systems and in B and C cases as well when GSHP was utilized. In only one case, Elec A, the mechanical exhaust ventilation system was replaced by mechanical supply and exhaust ventilation with heat recovery. The A cases represent the highest possible emission reductions, which includes the use of very costly measures. This is also seen in the high heat pump capacity of the GSHP A case, where the high thermal power is only utilized briefly in peak demand situations. The high capacity value can also be an artifact of the genetic algorithm since it is much higher than the capacity in the nearest other solution, and there is always some randomness in optimization with a genetic algorithm.

3.3. SH3–Buildings Built between 2003 and 2009

The optimal results for SH3 are shown in Figure 10. This building included a mechanical supply and exhaust ventilation by default. The Pareto fronts are similar to the older buildings, with GSHP having the lowest emissions and very steeply rising costs for minimal additional gains. The direct electric cases are somewhat similar to the DH case in emissions but have higher costs.

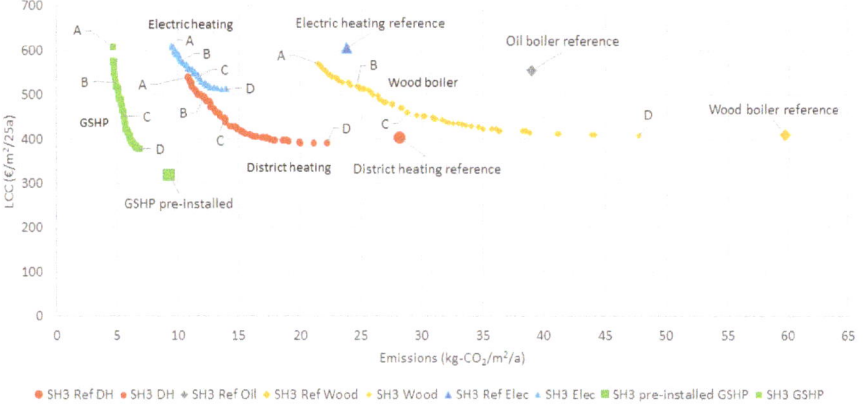

Figure 10. Reference cases and optimal results for the single-family house SH3 with different heating systems.

Additional details on four specific optimal solutions of each heating system are shown in Table 13. Additional thermal insulation was added to the roof in all cases. For external walls, additional thermal insulation was not added in the D cases, except in the electric heating case, which had the highest unit cost of heating energy. Thanks to the existing supply and exhaust ventilation system, adding demand-based VAV ventilation control was not expensive, and the upgrade was made in all cases. Cases C to A also included the installation of a new air-handling unit (AHU) with higher heat recovery efficiency. Solar thermal capacity was maximized in the A cases, being much lower in the other situations.

Table 12. Selected optimal cases for SH2 with all heating systems. Cost-effective solutions are marked in green. Cases D to A are optimized solutions from the lowest to the highest LCC. For the ventilation system: 0 is mechanical exhaust ventilation, 1 is mechanical supply and exhaust ventilation with 75% HR efficiency, and 2 is mechanical supply and exhaust ventilation with 75% HR and VAV.

Solution	Emissions kg-CO$_2$/m^2/a	Emission Reduction kg-CO$_2$/m^2/a	Relative Reduction %	Reduction Cost €-LCC/kg-CO^2/m^2	Investment €/m^2	LCC €/m^2/25a	U-values (W/m^{2}K) Walls	U-values (W/m^{2}K) Roof	U-values (W/m^{2}K) Doors	U-values (W/m^{2}K) Windows	ST m^2	PV kW$_p$	Ventilation System -	Dist. Temp. °C	GSHP kW$_{th}$	AAHP kW$_{th}$
DH Ref	32.2	-	-	-	93.4	435.0	0.28	0.22	1.4	1.6	0	0	0	70	0	0
DH D	24.0	8.3	25.7	-1.7	131.0	421.4	0.19	0.08	1.4	1.6	0	0	0	70	0	3
DH C	15.9	16.3	50.6	2.4	305.7	475.0	0.12	0.08	1.4	0.6	6	7	0	70	0	6
DH B	13.3	19.0	58.9	4.9	371.3	528.4	0.1	0.08	1	0.6	18	7	0	70	0	5
DH A	12.3	19.9	61.8	7.4	424.8	581.4	0.08	0.08	1	0.6	20	5	0	40	0	5
Wood Ref	71.3	-	-	-	106.7	428.4	0.28	0.22	1.4	1.6	0	0	0	70	0	0
Wood D	57.4	13.9	19.5	-0.3	125.3	424.8	0.28	0.08	1.4	1.6	0	0	0	70	0	3
Wood C	36.4	34.9	48.9	1.6	304.2	482.8	0.12	0.09	1.4	1	8	7	0	70	0	3
Wood B	29.1	42.2	59.2	2.6	359.8	536.8	0.12	0.08	1.4	0.6	20	5	0	70	0	5
Wood A	26.2	45.2	63.3	3.7	438.2	597.3	0.08	0.08	1	0.6	20	5	0	40	0	5
Elec Ref	25.8	-	-	-	78.4	639.9	0.28	0.22	1.4	1.6	0	0	0	-	0	0
Elec D	15.5	10.3	39.9	-10.7	192.1	530.2	0.19	0.08	1.4	1.6	6	7	0	-	0	5
Elec C	11.5	14.2	55.2	-4.3	347.0	578.5	0.08	0.08	1.4	0.6	10	9	0	-	0	4
Elec B	10.5	15.2	59.1	-1.5	383.9	616.6	0.08	0.08	0.8	0.6	20	7	2	-	0	4
Elec A	10.3	15.5	60.1	2.3	454.6	675.1	0.08	0.08	0.8	0.6	20	7	2	-	0	4
Oil Ref	46.6	-	-	-	102.3	605.1	0.28	0.22	1.4	1.6	0	0	0	70	0	0
GSHP D	8.3	38.2	82.1	-5.3	225.0	404.3	0.28	0.08	1.4	1.6	0	10	0	70	7	0
GSHP C	5.9	40.7	87.3	-3.2	339.6	473.3	0.12	0.09	1.4	1.6	8	9	0	40	9	0
GSHP B	5.3	41.3	88.7	-1.3	433.5	551.7	0.08	0.08	1	0.8	12	7	0	40	6	0
GSHP A	4.9	41.7	89.5	0.4	503.9	620.0	0.08	0.08	0.8	0.6	18	8	0	40	17	0

Cost-effective (green) LCC values: DH D 421.4; Wood D 424.8; Elec D 530.2, Elec C 578.5, Elec B 616.6; GSHP C 473.3, GSHP B 551.7.

Table 13. Selected optimal cases for SH3 with all heating systems. Cost-effective solutions are marked in green. Cases D to A are optimized solutions from the lowest to the highest LCC. For the ventilation system: 1 is mechanical supply and exhaust ventilation with 60% HR efficiency, 2 is mech. S & E ventilation with 60% HR and VAV, 3 is mech. S & E ventilation with 75% HR, and 4 is mech. S & E ventilation with 75% HR and VAV.

Solution	Emissions kg-CO$_2$/m²/a	Emission Reduction kg-CO$_2$/m²/a	Relative Reduction %	Reduction Cost €-LCC/kg-CO$_2$/m²	LCC €/m²/25a	Investment €/m²	U-values (W/m²K)				ST m²	PV kW$_p$	Ventilation System -	Dist. Temp. °C	GSHP kW$_{th}$	AAHP kW$_{th}$
							Walls	Roof	Doors	Windows						
DH Ref	28.2	-	-	-	401.8	78.4	0.25	0.16	1.4	1.4	0	0	1	40	0	0
DH D	22.2	6.0	21.2	-1.8	391.1	110.6	0.25	0.08	1.4	1.4	2	0	2	40	0	1
DH C	13.9	14.3	51.7	3.1	446.3	263.5	0.14	0.08	1.4	1.4	16	7	4	40	0	2
DH B	12.2	16.0	56.8	5.7	492.4	344.2	0.1	0.09	1.4	0.8	14	7	4	40	0	5
DH A	10.8	17.4	61.7	8.0	541.4	399.9	0.08	0.08	1	0.6	20	6	4	40	0	5
Wood Ref	59.7	-	-	-	409.1	106.7	0.25	0.16	1.4	1.4	0	0	1	40	0	0
Wood D	47.7	12.0	20.1	-0.1	407.4	138.9	0.25	0.08	1.4	1.4	2	0	2	40	0	1
Wood C	28.8	30.9	51.8	1.7	460.5	303.4	0.1	0.08	1.4	1.4	10	9	4	40	0	3
Wood B	24.8	34.9	58.4	3.1	516.6	350.1	0.1	0.07	1	0.6	10	2	4	40	0	5
Wood A	21.5	38.3	64.0	4.2	569.1	428.3	0.08	0.08	1	0.6	20	6	4	40	0	5
Elec Ref	23.8	-	-	-	605.9	78.4	0.25	0.16	1.4	1.4	0	0	1	-	0	0
Elec D	14.0	9.9	41.5	-9.5	512.4	212.2	0.17	0.07	1.4	1.4	6	9	2	-	0	3
Elec C	11.6	12.2	51.3	-4.9	545.5	282.3	0.12	0.08	1.4	1.4	16	8	4	-	0	6
Elec B	10.2	13.7	57.4	-2.0	578.5	361.1	0.1	0.07	1.4	0.6	14	8	4	-	0	4
Elec A	9.5	14.4	60.3	0.3	610.8	405.4	0.08	0.07	0.8	0.6	18	8	4	-	0	4
Oil Ref	39.0	-	-	-	555.1	102.3	0.25	0.16	1.4	1.4	0	0	1	40	0	0
GSHP D	6.9	32.1	82.4	-5.5	378.3	225.1	0.25	0.1	1.4	1.4	0	10	2	40	6	0
GSHP C	5.5	33.5	85.8	-3.1	451.0	339.3	0.08	0.07	1.4	1.4	4	10	4	40	7	0
GSHP B	4.9	34.1	87.6	-0.8	527.3	431.1	0.08	0.07	1.4	0.6	10	9	4	40	7	0
GSHP A	4.6	34.4	88.1	1.5	608.3	490.0	0.08	0.07	0.8	0.6	20	7	4	40	16	0

3.4. SH4–Buildings Built from 2010 onwards

The reference emission levels in SH4 were lower than in the other building age classes, but the shapes of the Pareto fronts (Figure 11) did not differ from the other buildings. The steepest slope and lowest emissions were found for the GSHP case, such that the installation of the heat pump has the greatest effect, and other means mainly add to the cost with minimal emission benefit. With district heating and boiler systems, the improvements to the envelope and the installation of solar energy systems are very beneficial. Replacing an oil boiler by a wood boiler will usually reduce costs, but if other changes are not made, it can increase emissions.

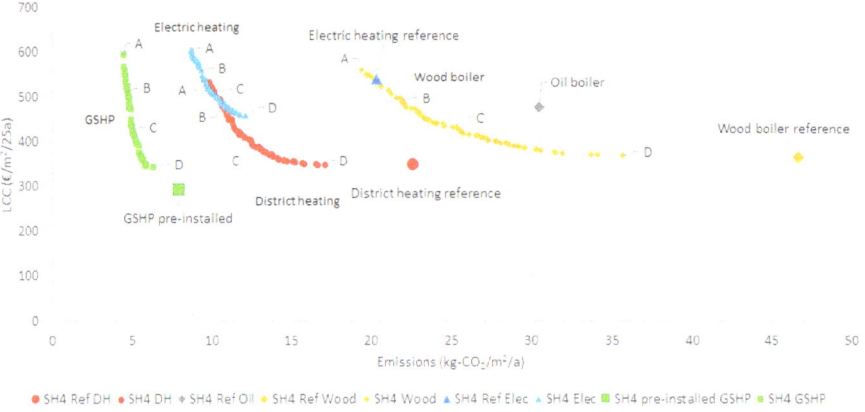

Figure 11. Reference cases and optimal results for the single-family house SH4 with different heating systems.

The detailed results for chosen optimal cases of SH4 are shown in Table 14. Switching from oil boiler to wood needs level C measures to reduce both cost and emissions. Despite the initially good U-values for the building envelope, solutions of level C and above still contained improvements to the thermal insulation level of external walls.

3.5. Emission Reduction Potential in the Detached House Building Stock

The previous sections outlined the possible emission levels obtainable in buildings of different ages with various heating systems. This section shows a rough estimate of how much the emissions could be reduced on the national level if all detached houses were renovated in Finland. Assuming the building stock size per age group to be equal to Figure 2, the scenarios were separated according to the renovation levels A to D. All buildings in all groups will be renovated to the same level. The renovation paths are chosen according to Figure 4 so that DH and direct electric heating systems do not change. Half of the oil and wood boilers are changed to GSHP systems, while the other half changes to (or keeps using) wood boilers. Buildings with pre-existing ground-source heat pumps will keep using them without any other renovation measures. The initial breakdown of different systems was taken as in Figure 3. "Other" heating systems are taken as a mixture of all the rest. The new system distribution is shown in Figure 12, and the effect of the different scenarios on absolute emissions is shown in Figure 13.

Table 14. Selected optimal cases for SH4 with all heating systems. Cost-effective solutions are marked in green. Cases D to A are optimized solutions from the lowest to the highest LCC. For the ventilation system: 1 is mechanical supply and exhaust ventilation with 65% HR efficiency, 2 is mech. S & E ventilation with 65% HR and VAV, 3 is mech. S & E ventilation with 75% HR, and 4 is mech. S & E ventilation with 75% HR and VAV.

Solution	Emissions kg-CO$_2$/m^2/a	Emission Reduction kg-CO$_2$/m^2/a	Relative Reduction %	Reduction Cost €-LCC/kg-CO$_2$/m^2	LCC €/m^2/25a	Investment €/m^2	U-values (W/m^{2}K) Walls	Roof	Doors	Windows	ST m^2	PV kW$_p$	Ventilation System	Dist. Temp. °C	GSHP kW$_{th}$	AAHP kW$_{th}$
DH Ref	22.7	-	-	-	348.0	78.4	0.17	0.09	1	1	0	0	1	40	0	0
DH D	17.1	5.5	24.4	−0.2	347.1	110.3	0.17	0.09	1	1	4	0	2	40	0	1
DH C	12.1	10.6	46.6	5.7	407.8	248.6	0.09	0.07	1	1	12	7	2	40	0	5
DH B	10.8	11.8	52.2	10.3	469.4	321.6	0.07	0.07	0.8	1	18	7	4	40	0	5
DH A	9.9	12.8	56.5	14.3	531.3	404.6	0.07	0.06	0.8	0.6	20	7	4	40	0	4
Wood Ref	46.7	-	-	-	364.5	106.7	0.17	0.09	1	1	0	0	1	40	0	0
Wood D	35.7	11.0	23.5	0.4	368.5	138.6	0.17	0.09	1	1	4	0	2	40	0	1
Wood C	25.1	21.6	46.2	3.1	431.1	248.8	0.13	0.09	0.8	1	16	5	2	40	0	5
Wood B	21.8	24.8	53.2	5.3	495.5	343.2	0.07	0.06	0.8	1	20	7	2	40	0	5
Wood A	19.5	27.2	58.3	7.2	559.9	433.0	0.07	0.06	0.8	0.6	20	7	4	40	0	4
Elec Ref	20.4	-	-	-	538.8	78.4	0.17	0.09	1	1	0	0	1	-	0	0
Elec D	12.2	8.2	40.3	−10.0	457.1	187.9	0.17	0.07	1	1	6	9	2	-	0	3
Elec C	10.2	10.2	50.1	−3.2	505.9	268.3	0.11	0.08	1	1	16	8	4	-	0	5
Elec B	9.4	11.0	54.0	2.0	561.3	354.1	0.08	0.06	1	0.6	14	7	4	-	0	5
Elec A	8.7	11.6	57.0	5.6	604.1	405.0	0.07	0.06	0.8	0.6	20	7	4	-	0	4
Oil Ref	30.4	-	-	-	475.8	102.3	0.17	0.09	1	1	0	0	1	40	0	0
GSHP D	6.3	24.1	79.3	−5.5	344.0	198.5	0.17	0.09	1	1	0	10	1	40	5	0
GSHP C	5.0	25.4	83.5	−2.1	423.6	300.2	0.11	0.07	1	1	10	9	2	40	8	0
GSHP B	4.7	25.7	84.6	1.5	513.6	377.4	0.08	0.07	1	1	20	7	4	40	14	0
GSHP A	4.4	26.0	85.4	4.6	595.0	482.4	0.07	0.06	0.8	0.6	20	7	4	40	12	0

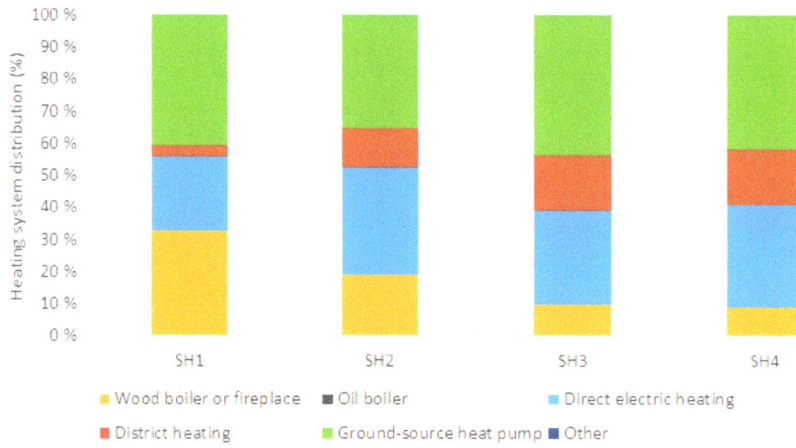

Figure 12. Assumed distribution of heating systems in the optimized scenarios.

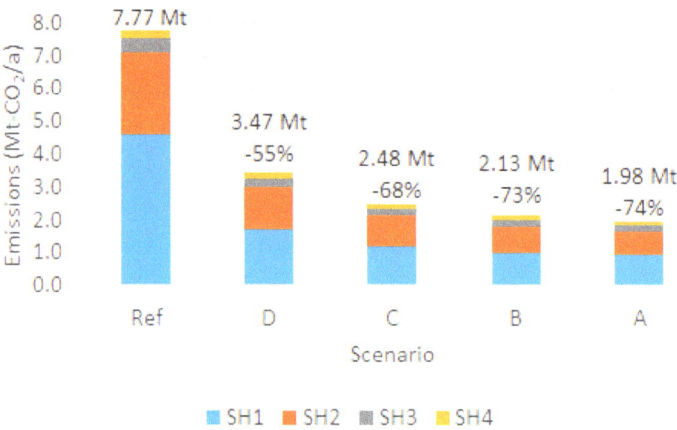

Figure 13. Absolute carbon emissions of the whole detached house building stock in Finland in the reference and optimized scenarios. The numbers show the remaining total CO_2 emissions and the relative reduction compared to the reference case.

The most important improvements were those to SH1 and SH2, as the newer buildings are responsible for only a small fraction of the total emissions. The total annual emission reductions for the whole detached house building stock were 4.3, 5.3, 5.6, and 5.8 Mt-CO_2 when all Finnish detached houses were renovated to levels D, C, B, and A, respectively. Even scenario D caused a 55% reduction in total emissions. This is mainly due to the significant fraction of wood and oil boilers in the original heating systems, as shown in Figure 3. The emissions of wood combustion are so high that when wood boilers are even partly replaced by heat pumps, the drop in total emissions is significant. Scenario C resulted in a 68% drop in emissions, which is still a significant improvement over D. Diminishing returns set in after that, as scenario B reduced emissions by 73% and scenario A by 74%.

The total investment cost of renovating all Finnish detached houses was 35.9, 52.6, 60.3, and 66.4 billion euros for scenarios D, C, B, and A, respectively. The unit cost of reducing emissions in scenarios D to A was 8300 to 11,500 €/t-CO_2 when the investment cost was simply divided by the reduction in annual emissions. However, all the renovation measures have a long-term impact, and thus the reduction costs can be calculated over a 25-year period of cumulative emissions. This way, the reduction cost was 334 to 459 €/t-CO_2 in scenarios D to A. The cost for only SH1 buildings was lower, 225 to 339 €/t-CO_2 because emission reductions are cheaper to do when the starting level is high. For comparison, the cost of an EU emission allowance in 2019 is about 25 €/t-CO_2 [60], which implies that even though many building renovation measures were cost-effective, it would still be more effective to make the reductions in the sectors covered by the EU emission trading system.

4. Discussion

Finding general solutions for the optimal energy renovation of existing single-family homes is a challenging task. Many assumptions have to be made concerning the mix of energy systems, the performance levels of the reference buildings as well as the cost of improvements. For example, one cause for uncertainty is the use of wood fuels. Many houses have fireplaces, but there is no reliable information available whether they are used as the main heating system, as support for peak demand, or only used for mood setting or not at all. For simplicity, all wood-based systems were assumed to be wood boilers working as the main heating system.

All houses were also assumed to be constructed of wood, which is the most common construction material for detached houses. The cost of wall-related renovations would be different for houses with brick facades. No attempts were made to calibrate the energy demand of the building stock to match the national estimates. Instead, the energy demand of each building was simulated based on the minimum requirements defined in the building code of their time. A limited amount of building types had to be chosen to make optimization feasible and, thus, only one building size and usage profile was utilized. Only the weather file for southern Finland was used, which covers the climate zones of southern (Zone I) and western (Zone II) Finland, but these areas contain 75% of the buildings and are therefore very representative of national emission reduction potential as well [34]. The optimization only takes into account the cost and annual CO_2 emissions. Other benefits of renovation, such as improved indoor air quality or rising property values, were not taken into account. As over 70% of deep renovation expenses may be translated into rising house values [24], the effective cost of the renovations goes down.

Building orientation was not considered in the study, as it cannot be changed after construction. The roof of the simulated building was assumed to be south-facing. The roofs of all buildings in the building stock are not south-facing, so extrapolating the results to the national scale would somewhat overestimate the potential of solar energy. However, in many cases, solar panels can be installed on other surfaces, such as garages. If the available roof space for solar installations was lower, the maximum PV capacity of the solar electric system would go down. However, for systems above 5 kW nominal power, most of the solar electricity will be exported to the grid anyway, so the on-site emission reduction potential of solar energy would likely not be greatly influenced.

This study found the ground-source heat pump to be the most effective way to reduce emissions in Finnish detached houses. This is similar to the results obtained in the previous study concerning Finnish apartment buildings [13]. The result relies strongly on the low emission factor of the Finnish energy system. It was assumed that any building could utilize the average emission factor of each month, but the case could also be made that any increase in electricity consumption would be met by high emission marginal generation. The energy consumption values of the examined cases are shown in Appendix A. Regardless, the importance of low emissions electricity becomes clear when electric heating is compared with heating based on combustion, such as using district heating or building-specific boilers. Similar results should be achievable in other countries with large amounts of low emissions electricity generation, such as Sweden and Norway. Of course, the emissions of district heating could also be brought down by using centralized heat pumps and solar energy [61] or small modular nuclear reactors [62,63].

On-site wood or oil boilers are not part of the European emission trading system (ETS). Thus, reducing the amount of on-site wood or oil burning anywhere translates totally into global emission reduction as well. Reducing the use of district heating or electricity also releases emission allowances in the EU emission trading system (ETS), which allows someone else to use the emission allowances instead. On the other hand, switching from outside the ETS (wood boiler) into the ETS (heat pump) can have an amplified effect as it will increase the demand for emission allowances and, thus, deny someone else the opportunity to emit CO_2. On the European level, on-site fossil fuel use still accounts for 66% of heat consumption, while district heating and electricity only account for 12% each [64]. If these on-site systems were included in the EU ETS, the demand for emission allowances would surge, increasing the price of allowances. This would make more efficiency improvements economical and incentivize the construction of low emission energy generation.

Unlike in the previous study focusing on apartment buildings [13], additional thermal insulation of external walls was a common renovation option. The unit costs in the case of detached houses were estimated lower than with apartment buildings. The prices here were for houses with a wooden envelope. The costs might be different for brick houses. There was no consideration of how thick a layer of insulation can practically be added to an existing building. This might limit some of the suggested insulation options.

The effects on the whole detached house building stock were examined, and this preliminary analysis shows that the buildings built after the strict efficiency regulations of 2002 do not need a significant focus due to their small number and resulting low impact on the national level. The problem is the large stock of older buildings. Because significant emission reductions are possible in an economical way, the government should consider enforcing and incentivizing such improvements. One option would be to provide access to low-interest loans such that little initial capital is needed by the residents. This ensures that even low-income people will be able to benefit from energy savings. Even direct monetary support or tax cuts could be used, since these may provide enough employment and new tax income to cancel out any apparent costs to the government. Campaigns to share information and showcase building retrofits done by other people might also boost interest in making energy renovations [65]. Shifting the emission reduction burden away from individual building owners is another possibility. Seasonal thermal energy storage has shown significant potential in reducing emissions in Finnish communities of single-family houses when it is combined with solar thermal collectors [66] or distributed PV panels and centralized heat pumps [67]. The economics of solutions at different scales (individual house, district, country) need to be compared to find the best solutions.

5. Conclusions

Optimal energy renovation plans were found for Finnish detached houses of four different ages. Cost-effective renovations were possible for every type, reducing CO_2 emissions significantly. The obtained emission reductions were 41% to 92% for the oldest building SH1 and 24% to 85% for the newest building SH4, depending on the selected heating system and investment level. For the whole building stock of detached houses, a 55% emission reduction was possible cost-effectively and a 74% reduction with the highest cost non-economical investments.

The heating energy source had a big influence on the emissions of the buildings. Electricity has a low emission factor, which made even direct electric heating an option for low emission heating. The main heating systems can be ranked in order from the lowest to highest emissions: ground-source heat pump, direct electric heating, district heating, wood boiler. The final emission ranges for the optimally renovated cases for all building age classes together were 10 to 24 kg-CO_2/m^2/a for district heating, 19 to 52 kg-CO_2/m^2/a for wood boiler, 9 to 15 kg-CO_2/m^2/a for electric heating, and 4 to 8 kg-CO_2/m^2/a for the ground-source heat pump.

Improving the building envelope was also an effective way to reduce emissions. While, in some cases, additional external wall insulation was not always included in the cheapest solutions, it was still included in a great majority of the optimally retrofitted cases. The newer the building was, the less likely the installation of lower U-value windows. Solar thermal capacity was low or zero for the lowest cost quarter of solutions, but always maximized for the cases with the most ambitious emission reductions. Solar electricity was used in the majority of optimal cases: for all electrically heated cases but not for the lowest cost cases with other heating systems. An air-to-air heat pump was added as auxiliary heating in all optimal solutions.

In this optimization study, replacing natural or mechanical exhaust ventilation by mechanical supply and an exhaust ventilation system was not a cost-effective solution in most of the cases because the monetary value of better indoor air quality achieved with a balanced ventilation system was not taken into account. However, in the buildings that already had such ventilation systems (SH3 and SH4) it was beneficial to both install a new AHU with a better heat recovery efficiency and install demand-based ventilation to reduce the heating of unoccupied houses. Installation of a mechanical supply and exhaust ventilation system was cost-effective in the oldest building (SH1) with direct electric heating, due to the very high energy expense.

Cost-effective energy retrofitting was found to be possible in detached houses of every age. For district heating, local boiler and electric heating retrofit solutions with lower LCC compared to the reference cases were found. The GSHP was by default so effective that additional improvements may not be necessary. Replacing boilers by heat pumps had the biggest emission impact. However, these solutions rely on the availability of low emissions electricity. A more detailed study must be done on the building stock to understand how fast the renovations can be done and to estimate how much clean electricity is actually available for use in newly electrified buildings.

Author Contributions: Conceptualization, J.H. (Janne Hirvonen), J.J., J.H. (Juhani Heljo), R.K.; methodology, J.H. (Janne Hirvonen), J.J., J.H. (Juhani Heljo), R.K.; software, J.H. (Janne Hirvonen); validation, J.H. (Janne Hirvonen).; formal analysis, J.H. (Janne Hirvonen); investigation, J.H. (Janne Hirvonen); resources, J.H. (Janne Hirvonen), J.J., J.H. (Juhani Heljo), R.K.; data curation, J.H. (Janne Hirvonen); writing—original draft preparation, J.H. (Janne Hirvonen); writing—review and editing, J.H. (Janne Hirvonen), J.J., R.K.; visualization, J.H. (Janne Hirvonen); supervision, J.J., R.K.; project administration, R.K.; funding acquisition, R.K.

Funding: The study was funded by the Academy of Finland, grant number 309064. It was made as part of the project: Optimal transformation pathway towards the 2050 low-carbon target: integrated buildings, grids and national energy system for the case of Finland.

Conflicts of Interest: The authors declare no conflict of interest. The funders had no role in the design of the study; in the collection, analyses, or interpretation of data; in the writing of the manuscript, or in the decision to publish the results.

Abbreviations

AAHP	Air-to-air heat pump
AHU	Air handling unit
DH	District heating
DHW	Domestic hot water
GSHP	Ground-source heat pump
HR	Heat recovery
LCC	Life cycle cost
Mech. E. vent	Mechanical exhaust ventilation
Mech. S. & E. vent.	Mechanical supply and exhaust ventilation
NSGA	Non-dominated sorting genetic algorithm
SH	Single-family house
VAV	Variable air volume ventilation

Appendix A

Table A1. Energy demand in the optimized and reference cases. The primary energy factor is 1.0 for oil boiler, 0.5 for wood boiler and DH, and 1.2 for electricity.

Solution	SH1 Electricity Demand kWh/m²	SH1 DH/boiler Demand kWh/m²	SH1 Primary Energy kWh/m²	SH2 Electricity Demand kWh/m²	SH2 DH/boiler Demand kWh/m²	SH2 Primary Energy kWh/m²	SH3 Electricity Demand kWh/m²	SH3 DH/boiler Demand kWh/m²	SH3 Primary Energy kWh/m²	SH4 Electricity Demand kWh/m²	SH4 DH/boiler Demand kWh/m²	SH4 Primary Energy kWh/m²
DH Ref	20.6	234.3	141.8	23.2	177.0	116.4	28.4	148.3	108.2	25.8	115.8	90.1
DH D	34.4	117.5	100.0	38.6	112.6	102.6	34.0	106.6	94.1	31.6	77.6	76.6
DH C	26.2	76.3	69.6	27.0	73.0	68.9	25.6	62.2	61.8	24.3	52.3	55.3
DH B	23.7	62.7	59.8	25.5	58.1	59.6	24.9	52.3	56.0	23.6	45.2	50.9
DH A	23.4	55.8	55.9	25.0	52.8	56.4	23.9	44.8	51.0	22.6	40.2	47.2
Wood Ref	20.6	234.3	141.8	23.2	177.0	116.4	28.4	148.3	108.2	26.8	115.8	90.1
Wood D	34.4	117.5	100.0	39.9	121.1	108.5	34.0	106.6	94.1	31.6	77.6	76.6
Wood C	26.3	77.3	70.2	27.2	76.1	70.7	25.8	62.2	62.0	25.4	53.1	57.1
Wood B	25.0	62.2	61.0	26.5	59.1	61.3	26.7	52.1	58.1	23.5	45.8	51.0
Wood A	23.4	55.8	55.9	25.0	52.8	56.4	23.9	44.8	51.0	22.6	40.2	47.2
Elec Ref	224.8	0	269.8	179.2	0	215.0	166.6	0	199.9	142.6	0	171.1
Elec D	81.6	0	97.9	104.6	0	125.5	94.2	0	113.0	82.0	0	98.4
Elec C	63.6	0	76.3	77.4	0	92.8	77.7	0	93.2	68.0	0	81.6
Elec B	56.4	0	67.7	70.4	0	84.5	68.2	0	81.8	62.8	0	75.4
Elec A	53.8	0	64.6	68.8	0	82.6	63.5	0	76.2	58.7	0	70.4
Oil Ref	20.6	234.3	259.0	23.2	177.0	204.9	28.4	148.3	182.3	26.8	115.8	147.9
GSHP D	55.1	0	66.1	56.7	0	68.0	47.0	0	56.5	43.4	0	52.1
GSHP C	39.3	0	47.2	40.1	0	48.1	38.0	0	45.6	34.2	0	41.1
GSHP B	34.8	0	41.7	35.7	0	42.8	33.2	0	39.8	32.1	0	38.5
GSHP A	32.8	0	39.3	33.2	0	39.8	31.7	0	38.0	30.4	0	36.5

References

1. European Commission. EU Climate Action—2050 Long-Term Strategy. 2016. Available online: https://ec.europa.eu/clima/policies/strategies/2050_en (accessed on 3 February 2019).
2. European Parliament. Directive (EU) 2018/844 of the European Parliament and of the Council amending Directive 2010/31/EU on the energy performance of buildings and Directive 2012/27/EU on the energy efficiency. *Off. J. Eur. Union* **2018**, *L 156*, 75–91.
3. Ministry of the Environment. *4/13 Asetus rakennuksen energiatehokkuuden parantamisesta korjaus—ja muutostöissä (4/13 decree on improving energy performance in renovations)*. Available online: http://www.ym.fi/download/noname/%7B924394EF-BED0-42F2-9AD2-5BE3036A6EAD%7D/31396 (accessed on 7 January 2019).
4. Statistics Finland. *Number of Buildings by Intended Use and Year of Construction on 31 December 2015*; Statistics Finland: Helsinki, Finland, 2016.
5. Ekström, T.; Blomsterberg, Å. Renovation of Swedish Single-family Houses to Passive House Standard – Analyses of Energy Savings Potential. *Energy Procedia* **2016**, *96*, 134–145. [CrossRef]
6. Taillandier, F.; Mora, L.; Breysse, D. Decision support to choose renovation actions in order to reduce house energy consumption—An applied approach. *Build. Environ.* **2016**, *109*, 121–134. [CrossRef]
7. Dermentzis, G.; Fabian, O.; Marcus, G.; Toni, C.; Dietmar, S.; Wolfgang, F.; Chiara, D.; Roberto, F.; Chris, B. A comprehensive evaluation of a monthly-based energy auditing tool through dynamic simulations, and monitoring in a renovation case study. *Energy Build.* **2019**, *183*, 713–726. [CrossRef]
8. Thermal Energy System Specialists, LLC TRNSYS: Transient System Simulation Tool. 2019. Available online: http://www.trnsys.com/ (accessed on 12 September 2019).
9. Hamburg, A.; Kalamees, T. How well are energy performance objectives being achieved in renovated apartment buildings in Estonia? *Energy Build.* **2019**, *199*, 332–341. [CrossRef]
10. Henning, H.-M.; Palzer, A. A comprehensive model for the German electricity and heat sector in a future energy system with a dominant contribution from renewable energy technologies—Part I: Methodology. *Renew. Sustain. Energy Rev.* **2014**, *30*, 1003–1018. [CrossRef]
11. Palzer, A.; Henning, H.-M. A comprehensive model for the German electricity and heat sector in a future energy system with a dominant contribution from renewable energy technologies—Part II: Results. *Renew. Sustain. Energy Rev.* **2014**, *30*, 1019–1034. [CrossRef]
12. Niemelä, T.; Kosonen, R.; Jokisalo, J. Cost-effectiveness of energy performance renovation measures in Finnish brick apartment buildings. *Energy Build.* **2017**, *137*, 60–75. [CrossRef]
13. Hirvonen, J.; Jokisalo, J.; Heljo, J.; Kosonen, R. Towards the EU emissions targets of 2050: Optimal energy renovation measures of Finnish apartment buildings. *Int. J. Sustain Energy* **2018**, *38*, 649–672. [CrossRef]
14. Ko, M.J.; Kim, Y.S.; Chung, M.H.; Jeon, H.C. Multi-objective optimization design for a hybrid energy system using the genetic algorithm. *Energies* **2015**, *8*, 2924–2949. [CrossRef]
15. Tokarik, M.S.; Richman, R.C. Life cycle cost optimization of passive energy efficiency improvements in a Toronto house. *Energy Build.* **2016**, *118*, 160–169. [CrossRef]
16. Fan, Y.; Xia, X. A multi-objective optimization model for energy-efficiency building envelope retrofitting plan with rooftop PV system installation and maintenance. *Appl. Energy* **2017**, *189*, 327–335. [CrossRef]
17. Eskander, M.M.; Sandoval-Reyes, M.; Silva, C.A.; Vieira, S.M.; Sousa, J.M.C. Assessment of energy efficiency measures using multi-objective optimization in Portuguese households. *Sustain. Cities Soc.* **2017**, *35*, 764–773. [CrossRef]
18. Ren, G.; Heo, Y. Sunikka-Blank Investigating an adequate level of modelling for retrofit decision-making: A case study of a British semi-detached house. *J. Build. Eng.* **2019**, *26*, 100837. [CrossRef]
19. Alev, Ü.; Allikmaa, A.; Kalamees, T. Potential for Finance and Energy Savings of Detached Houses in Estonia. *Energy Procedia* **2015**, *78*, 907–912. [CrossRef]
20. Leinartas, H.A.; Stephens, B. Optimizing Whole House Deep Energy Retrofit Packages: A Case Study of Existing Chicago-Area Homes. *Buildings* **2015**, *5*, 323–353. [CrossRef]
21. Ekström, T.; Bernardo, R.; Blomsterberg, Å. Cost-effective passive house renovation packages for Swedish single-family houses from the 1960s and 1970s. *Energy Build.* **2018**, *161*, 89–102. [CrossRef]
22. Asaee, S.R.; Ugursal, V.I.; Beausoleil-Morrison, I. Techno-economic feasibility evaluation of air to water heat pump retrofit in the Canadian housing stock. *Appl. Therm. Eng.* **2017**, *111*, 936–949. [CrossRef]

23. Asaee, S.R.; Ugursal, V.I.; Beausoleil-Morrison, I. Techno-economic assessment of solar assisted heat pump system retrofit in the Canadian housing stock. *Appl. Energy* **2017**, *190*, 439–452. [CrossRef]
24. Bjørneboe, M.G.; Svendsen, S.; Heller, A. Evaluation of the renovation of a Danish single-family house based on measurements. *Energy Build.* **2017**, *150*, 189–199. [CrossRef]
25. Psomas, T.; Heiselberg, P.; Duer, K.; Bjørn, E. Overheating risk barriers to energy renovations of single family houses: Multicriteria analysis and assessment. *Energy Build.* **2016**, *117*, 138–148. [CrossRef]
26. Beccali, M.; Cellura, M.; Fontana, M.; Longo, S.; Mistretta, M. Energy retrofit of a single-family house: Life cycle net energy saving and environmental benefits. *Renew. Sustain. Energy Rev.* **2013**, *27*, 283–293. [CrossRef]
27. Azizi, S.; Nair, G.; Olofsson, T. Analysing the house-owners' perceptions on benefits and barriers of energy renovation in Swedish single-family houses. *Energy Build.* **2019**, *198*, 187–196. [CrossRef]
28. Bjørneboe, M.G.; Svendsen, S.; Heller, A. Initiatives for the energy renovation of single-family houses in Denmark evaluated on the basis of barriers and motivators. *Energy Build.* **2018**, *167*, 347–358. [CrossRef]
29. Pornianowski, M.; Antonov, Y.I.; Heiselberg, P. Development of energy renovation packages for the Danish residential sector. *Energy Procedia* **2019**, *158*, 2847–2852. [CrossRef]
30. Mortensen, A.; Heiselberg, P.; Knudstrup, M. Economy controls energy retrofits of Danish single-family houses. Comfort, indoor environment and architecture increase the budget. *Energy Build* **2014**, *72*, 465–475. [CrossRef]
31. EQUA Simulation AB, IDA ICE—Simulation Software. 2019. Available online: https://www.equa.se/en/ida-ice (accessed on 8 August 2019).
32. EQUA Simulation AB. *Validation of IDA indoor climate and energy 4.0 with respect to CEN standards EN 15255-2007 and EN 15265-2007*; EQUA Simulation AB: Solna, Sweden, May 2010.
33. Loutzenhiser, P.; Manz, H.; Maxwell, G. Empirical validations of shading/daylighting/load interactions in building energy simulation tools. *IEA Int. Energy Agency* **2007**, *7*, 1426–1431.
34. Kalamees, T.; Jylhä, K.; Tietäväinen, H.; Jokisalo, J.; Ilomets, S.; Hyvönen, R.; Saku, S. Development of weighting factors for climate variables for selecting the energy reference year according to the EN ISO 15927-4 standard. *Energy Build.* **2012**, *47*, 53–60. [CrossRef]
35. Finnish Meteorological Institute. *Test Reference Years for Energy Calculations*; Finnish Meteorological Institute: Helsinki, Finland, 19 November 2018. (In Finnish)
36. Palonen, M.; Hamdy, M.; Hasan, A. MOBO a new software for multi-objective building performance optimization. In Proceedings of the 13th Conference of International Building Performance Simulation Association, Chambéry, France, 2013.
37. Ministry of the Environment. Directive on building energy certificates, Attachment 1. Available online: https://www.finlex.fi/data/sdliite/liite/6822.pdf (accessed on 12 December 2019). (In Finnish).
38. Statistics Finland. *Email Query about Building Heating System Statistics*; Statistics Finland: Helsinki, Finland, 2017.
39. Haakana, M. *Ympäristöministeriön Asetus Rakennuksen Energiatodistuksesta*; Finlex: Helsinki, Finland, 2017.
40. Laitinen, A. *Air-to-Air Heat Pump Energy Consumption Effect in Detached Houses*; VTT: Finland, Espoo, 2016. (In Finnish)
41. NIBE AB. *NIBE F1345 Heat Pump Installer Manual*; NIBE AB: Sweden, Markaryd, 2017.
42. CEN European Committee for Standardization. Standard EN 14511-3:2013. *Air Conditioners, Liquid Chilling Packages and Heat Pumps with Electrically Driven Compressors for Space Heating and Cooling. Part 3: Test methods*. 2013. Available online: https://www.en-standard.eu/din-en-14511-1-air-conditioners-liquid-chilling-packages-and-heat-pumps-for-space-heating-and-cooling-and-process-chillers-with-electrically-driven-compressors-part-1-terms-and-definitions/ (accessed on 8 January 2019).
43. Koivuniemi, J. Lämpimän käyttöveden mitoitusvirtaama ja lämpötilakriteerit veden mikrobiologisen laadun kannalta kaukolämmitetyissä taloissa [Domestic hot water design flow and temperature criteria as pertains to water microbiological quality in district heated houses, In Finnish]. Mster's Thesis, Helsinki University of Technology, Espoo, Finland, 2005.
44. Ministry of the Environment. Ympäristöministeriön asetus uuden rakennuksen energiatehokkuudesta (1010/2017) (Decree of the Ministry of the Environment on the energy performance of the new building (1010/2017); 2018. Available online: https://www.finlex.fi/fi/laki/alkup/2017/20171010 (accessed on 14 December 2018). (In Finnish).

45. Degefa, M.Z. *Project Report of SGEM Task 6.11: Spatial load analysis*; Aalto University School of Electrical Engineering: Espoo, Finland, 2012.
46. Ministry of the Environment. *D3—Rakennusten energiatehokkuus*; Ministry of the Environment: Helsinki, Finland, 2012. (In Finnish)
47. Finnish Energy. *Emission factors of Finnish Electricity, Email Contact*; Finnish Energy: Helsinki, Finland, 2017.
48. Motiva Oy, CO2 emission factors in Finland. 2019. Available online: https://www.motiva.fi/ratkaisut/energiankaytto_suomessa/co2-laskentaohje_energiankulutuksen_hiilidioksidipaastojen_laskentaan/co2-paastokertoimet (accessed on 15 January 2019). (In Finnish).
49. Statistics Finland. *Fuel Classification 2019*; Statistics Finland: Helsinki, Finland, 2019. (In Finnish)
50. Finnish Energy, 2018. Available online: https://energia.fi/en (accessed on 8 Augest 2019).
51. Fortum, District Heating Price [In Finnish]. 2019. Available online: https://www.fortum.fi/kotiasiakkaille/lammitys/kaukolampo/kaukolammon-hinnat-pientaloille (accessed on 1 January 2019).
52. Vapo Oy, Wood Fuel Price. 2019. Available online: https://kauppa.vapo.fi/tuotteet/500-kg-pellettisakki/ (accessed on 15 April 2019).
53. Lämpöpuisto Oy, Heating Oil Price [In Finnish]. 2019. Available online: https://www.lampopuisto.fi/fi/yksityisille/lammitysoljyn-hinta (accessed on 16 April 2019).
54. Nord Pool, Historical Electricity Market Data. 2018. Available online: https://www.nordpoolgroup.com/historical-market-data/ (accessed on 14 January 2019).
55. Caruna Oy, Electricity Distribution Pricing. 2019. Available online: https://www.caruna.fi/en/our-services/products-and-rates/electricity-distribution-rates (accessed on 13 February 2019).
56. Lindberg, R.; Kivimäki, C.; Sahlstedt, S. *Korjausrakentamisen kustannuksia 2019 [Cost of building renovation, in Finnish]*; Rakennustieto Oy: Helsinki, Finland, 2019.
57. Taloon.com, Taloon.com web store. 2019. Available online: https://www.taloon.com (accessed on 3 December 2019).
58. Motiva Oy, Solar electric system prices. Aurinkosähköä kotiin, 2019. Available online: https://aurinkosahkoakotiin.fi/tarjoukset/ (accessed on 12 August 2019). (In Finnish).
59. Deb, K.; Pratap, A.; Agarwal, S.; Meayarivan, T. A fast and elitist multiobjective genetic algorithm: NSGA-II. *IEEE Trans. Evol. Comput.* **2002**, *6*, 182–197. [CrossRef]
60. Business Insider, European Emission Allowance prices. *markets.businessinsider.com*. 2019. Available online: https://markets.businessinsider.com/commodities/co2-european-emission-allowances (accessed on 12 September 2019).
61. Paiho, S.; Reda, F. Towards next generation district heating in Finland. *Renew. Sustain. Energy Rev.* **2016**, *65*, 915–924. [CrossRef]
62. Värri, K.; Syri, S. The possible role of modular nuclear reactors in district heating: Case Helsinki region. *Energies* **2019**, *12*, 2195. [CrossRef]
63. Partanen, R. Nuclear District Heating in Finland—The Demand, Supply and Emissions Reduction Potential of Heating Finland with Small Nuclear Reactors. Think Atom, February. 2019. Available online: https://thinkatomnet.files.wordpress.com/2019/04/nuclear-district-heating-in-finland_1-2_web.pdf (accessed on 7 June 2019).
64. Persson, U.; Werner, S. *Quantifying the Heating and Cooling Demand in Europe*; D 2.2; Halmstad University: Halmstad, Sweden, 2015.
65. Hrovatin, N.; Zorić, J. Determinants of energy-efficient home retrofits in Slovenia: The role of information sources. *Energy Build.* **2018**, *180*, 42–50. [CrossRef]
66. Hirvonen, J.; Rehman, H.; Sirén, K. Techno-economic optimization and analysis of a high latitude solar district heating system with seasonal storage, considering different community sizes. *Sol. Energy* **2018**, *162*, 472–488. [CrossRef]
67. Hirvonen, J.; Sirén, K. A novel fully electrified solar heating system with a high renewable fraction—Optimal designs for a high latitude community. *Renew. Energy* **2018**, *127*, 298–309. [CrossRef]

© 2019 by the authors. Licensee MDPI, Basel, Switzerland. This article is an open access article distributed under the terms and conditions of the Creative Commons Attribution (CC BY) license (http://creativecommons.org/licenses/by/4.0/).

Article

Evaluation Method for the Hourly Average CO$_{2eq.}$ Intensity of the Electricity Mix and Its Application to the Demand Response of Residential Heating

John Clauß [1,*], Sebastian Stinner [1], Christian Solli [2], Karen Byskov Lindberg [3,4], Henrik Madsen [5,6] and Laurent Georges [1,6]

1. Norwegian University of Science and Technology NTNU, Department of Energy and Process Engineering, Kolbjørn Hejes vei 1a, 7491 Trondheim, Norway; sebastian.stinner@rwth-aachen.de (S.S.); laurent.georges@ntnu.no (L.G.)
2. NTNU, Property Division, Høgskoleringen 8, 7034 Trondheim, Norway; christian.solli@ntnu.no
3. NTNU, Department of Electric Power Engineering, 7491 Trondheim, Norway; karen.lindberg@sintef.no
4. SINTEF Building and Infrastructure, Pb 124 Blindern, 0314 Oslo, Norway
5. DTU Technical University of Denmark, DTU Compute, Asmussens Allé, Building 303B, 2800 Kgs. Lyngby, Denmark; hmad.dtu@gmail.com
6. Research Center on Zero Emission Neighbourhoods in Smart Cities, Trondheim 7491, Norway
* Correspondence: john.clauss@ntnu.no

Received: 1 March 2019; Accepted: 2 April 2019; Published: 8 April 2019

Abstract: This work introduces a generic methodology to determine the hourly average CO$_{2eq.}$ intensity of the electricity mix of a bidding zone. The proposed method is based on the logic of input–output models and avails the balance between electricity generation and demand. The methodology also takes into account electricity trading between bidding zones and time-varying CO$_{2eq.}$ intensities of the electricity traded. The paper shows that it is essential to take into account electricity imports and their varying CO$_{2eq.}$ intensities for the evaluation of the CO$_{2eq.}$ intensity in Scandinavian bidding zones. Generally, the average CO$_{2eq.}$ intensity of the Norwegian electricity mix increases during times of electricity imports since the average CO$_{2eq.}$ intensity is normally low because electricity is mainly generated from hydropower. Among other applications, the CO$_{2eq.}$ intensity can be used as a penalty signal in predictive controls of building energy systems since ENTSO-E provides 72 h forecasts of electricity generation. Therefore, as a second contribution, the demand response potential for heating a single-family residential building based on the hourly average CO$_{2eq.}$ intensity of six Scandinavian bidding zones is investigated. Predictive rule-based controls are implemented into a building performance simulation tool (here IDA ICE) to study the influence that the daily fluctuations of the CO$_{2eq.}$ intensity signal have on the potential overall emission savings. The results show that control strategies based on the CO$_{2eq.}$ intensity can achieve emission reductions, if daily fluctuations of the CO$_{2eq.}$ intensity are large enough to compensate for the increased electricity use due to load shifting. Furthermore, the results reveal that price-based control strategies usually lead to increased overall emissions for the Scandinavian bidding zones as the operation is shifted to nighttime, when cheap carbon-intensive electricity is imported from the continental European power grid.

Keywords: predictive rule-based control; hourly CO$_{2eq.}$ intensity; demand response; energy flexibility

1. Introduction

A transition to a low-carbon energy system is necessary to reduce its environmental impact in the future. This implies a reduction of CO$_{2eq.}$ emissions on the electricity supply side by making use of intermittent renewable energy sources. To fully exploit the electricity generated from these intermittent energy sources, deploying demand side flexibility is crucial [1]. Regarding building heating systems, demand side flexibility is the margin by which a building can be operated while still fulfilling its functional requirements [2]. From a global perspective, potential emission savings from the building sector are large since the building sector is responsible for 30% of the total energy use [3]. Several studies point out the importance of demand side management to improve the interaction between buildings and the electricity grid [1,4–8]. Progressively decreasing prices for sensing, communication, and computing devices will open up possibilities for improved controls for demand response (DR) as future management systems will be more affordable. A number of studies have investigated the building energy flexibility with a special focus on building heating systems [9–16]. In those studies, DR measures have been applied to electric heating systems, such as heat pump systems or direct electric heating systems.

DR measures can be applied to control the electricity use for the heating of buildings depending on power grid signals [12]. The most common signal for DR is the electricity spot price [15–20], but the $CO_{2eq.}$ intensity of the electricity mix [15,21–25], the share of renewables in the electricity mix [9,26] or voltage fluctuations [27] are applied as well.

DR measures are implemented into control strategies, such as predictive rule-based controls (PRBC) or more advanced controls, e.g., optimal control or model-predictive controls (MPC). PRBCs rely on pre-defined rules to control the energy system, where temperature set-points (TSP) for space heating (SH) or domestic hot water (DHW) heating are usually varied to start or delay the operation of the heating system depending on the control signal. These rules are rather straightforward to implement, but a careful design of the control rules is necessary. MPC solves an optimization problem but is more expensive to develop, for instance the identification of a model used for control is acknowledged as the most critical part in the design of an MPC [28,29]. PRBCs can be a good compromise to advanced controls because PRBC is simpler, but can still be effective in reducing operational costs or saving carbon emissions [29].

By applying the carbon intensity as a penalty signal for indirect control, the operation of a heating system can be shifted to times of low $CO_{2eq.}$ intensity in the grid mix using thermal storage. In general, the $CO_{2eq.}$ intensity can be used as an indicator of the share of renewable energies in the electricity mix. Applying the $CO_{2eq.}$ intensity as a penalty signal to operate building energy systems can help to reach the emission targets of the European Union. In Norway, electricity is mostly generated from hydropower. However, an increased interaction between the continental European and the Norwegian power grids is expected in the future [30]. As Norway has a very limited number of fossil fuel power plants for electricity generation, the hourly average $CO_{2eq.}$ intensity of the electricity mix already strongly depends on the electricity exchanges with neighboring bidding zones (BZ). Generally, the CO_2 price is seen as an essential driver for the transition to a low-carbon society [30]. This CO_2 price is expected to increase in the future so that the application of a $CO_{2eq.}$ intensity signal for control purposes is likely to gain importance. Compared to the electricity spot price, the use of the $CO_{2eq.}$ intensity of the electricity mix as a control signal is not as common because this control signal is not readily available. To the author's knowledge, in Europe, only the transmission system operator (TSO) for Denmark, Energinet, provides information on the average $CO_{2eq.}$ intensity of the electricity mix with an hourly resolution and at no charge. This may also be the reason why most studies that use $CO_{2eq.}$ intensity as a penalty signal for heating were performed for Denmark and conducted only rather recently: Vogler-Finck et al. [21], Pedersen et al. [22], Hedegaard et al. [23], and Knudsen et al. [24] use the data from Energinet in their DR studies for the heating of residential buildings.

They European power grid is highly interconnected. The leading power market in Europe is NordPool, which provides a day-ahead as well as an intra-day market. NordPool mainly operates in Scandinavia, UK, Germany, and the Baltics [31]. It is owned by the Scandinavian TSOs (Norway, Sweden, Denmark, and Finland) and the Baltic TSOs (Estonia, Lithuania, and Latvia) [32]. In order to avoid bottlenecks in the transmission system, BZs are created with different electricity prices. One country can have several BZs [33]. Norway consists of five BZs, each of them having physical connections to neighboring BZs that enables electricity imports and exports (Figure 1).

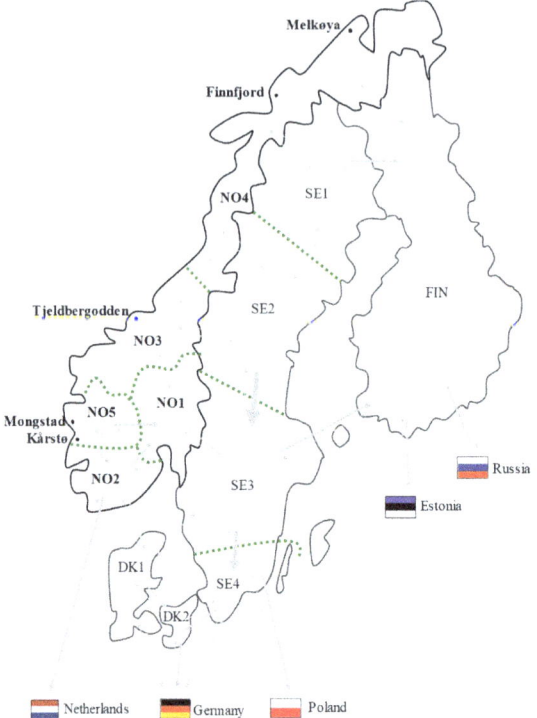

Figure 1. Overview of the Scandinavian power market bidding zones, also including gas-fired power plants in Norway (adjusted from [30]).

The main contributions of this paper are twofold. Firstly, a methodology for calculating the hourly average $CO_{2eq.}$ intensity of the electricity mix in an interconnected power grid is developed. The methodology is generic and takes into account the hourly average $CO_{2eq.}$ intensities of the electricity traded between neighboring BZs (imports and exports). The proposed method resorts to the logic of multi-regional input–output models (MRIO) [34]. Input–output models are usually used to perform energy system modeling in combination with an economic analysis considering different industry sectors [35–37]. They are based on the assumption that there always is a balance between consumption and generation for the whole system. In the present work, this logic can be applied for electricity where BZs are used instead of industry sectors. As an example, this paper evaluates the hourly average $CO_{2eq.}$ intensity for several Scandinavian BZs using the electricity production data for the year 2015. Emissions related to an electricity generation technology are considered on a life-cycle perspective. Electricity losses in the transmission and distribution grid are neglected. The input data required to calculate the average $CO_{2eq.}$ intensity is not readily available. Therefore, this work does not only provide the methodology to determine the $CO_{2eq.}$ intensities but also guidelines on where to retrieve and how to structure the input data to apply the proposed methodology.

Secondly, studies that focus on the evaluation of hourly $CO_{2eq.}$ intensities usually do not consider the detailed control of the HVAC system, whereas most studies that specifically focus on the heating system control do not comment on the methodology for determining the $CO_{2eq.}$ intensity. Therefore, the paper investigates how the characteristics of the $CO_{2eq.}$ intensity used as control signal influence the overall emission savings. This is done using the case study of residential heating where DR is performed using the $CO_{2eq.}$ intensity of several Scandinavian BZs. A detailed description of the HVAC system and its control is provided which has been spotted as a major short-coming of the other existing approaches that primarily focus on the $CO_{2eq.}$ intensity evaluation.

2. Review of Existing Evaluation Methods for $CO_{2eq.}$ Intensity

Generally, evaluation methods for the hourly $CO_{2eq.}$ intensities of the electricity mix can be categorized as presented in Table 1. In a de-coupled approach, the electricity demand and supply sides do not influence each other. On the contrary, in a coupled approach, the interaction between the demand and supply sides is taken into account. For example, if a large number of buildings would apply the average $CO_{2eq.}$ intensity as a penalty signal, the resulting electric load could be affected and thus the predicted generation would not be optimized for this load anymore. Ideally, a coupled approach should be used to take into account DR in the prediction of the electricity generation and the respective $CO_{2eq.}$ emissions [26,38,39]. Furthermore, average and marginal $CO_{2eq.}$ intensities are two distinct concepts. Marginal emissions are the emissions from one additional kWh generated/consumed and, consequently, it results from a single power plant. On the contrary, the average $CO_{2eq.}$ intensity is the $CO_{2eq.}$/kWh emitted on average from the entire electricity generation of the BZ. It thus results from a mix of power plants. On the one hand, it could be argued that the marginal $CO_{2eq.}$ intensity is most coherent for the control of a limited number of buildings because they will rather affect a single plant than the overall production. On the other hand, average $CO_{2eq.}$ intensities (or factors) have been used extensively in the past for buildings exporting electricity to the grid, e.g., Zero Emission Buildings or Nearly Zero Energy Buildings. Studies mostly focusing on the life-cycle assessment (LCA) of buildings often use average $CO_{2eq.}$ intensities rather than marginal intensities.

Table 1. Categorization of methodologies to evaluate $CO_{2eq.}$ intensities of the electricity mix (marked as 'CO_2') or determine the optimal dispatch and unit commitment in electricity grids (marked as 'EL').

De-Coupled Approach		Coupled Approach	
Average	Marginal	Average	Marginal
Energinet [40] (CO_2) Vandermeulen et al. [9] (CO_2) Milovanoff et al. [41] (CO_2) Roux et al. [42] (CO_2) Tomorrow [43] (CO_2)	Bettle et al. [44] (CO_2) Hawkes [45] (CO_2) Peán et al. (based on Hawkes) [25] (CO_2) Corradi [46] (CO_2)	Graabak [47] (CO_2)	Patteeuw et al. [26] (EL) Arteconi et al. (based on Patteeuw) [38] (EL) Graabak et al. [47] (CO_2, EL) Askeland et al. [48] (EL) Quoilin et al. [49] (EL)

2.1. De-Coupled Approach

The evaluation methodology of the TSO Energinet for determining the $CO_{2eq.}$ intensity of the Danish electricity mix has two distinct simplifications: (1) the methodology considers only the operational phase (meaning without the life-cycle perspective) and (2) the $CO_{2eq.}$ intensity of the imports from neighboring countries are assumed to be constant. For example, electricity imports from Norway are assumed to be 9 g/kWh, from Germany 415 g/kWh and from Sweden 28 g/kWh [40].

Vandermeulen et al. [9] aim at maximizing the electricity use of residential heat pumps at times of high shares of renewable energy generation in the Belgian power grid. Data from the Belgian TSO, Elia, is used. It is not stated whether electricity imports are considered.

Milovanoff et al. [41] determine the environmental impacts of the electricity consumption in France for the years 2012 to 2014. They calculate 'impact factors' for electricity generation and consumption, where the impact factor for consumption agrees with the environmental impacts per kWh consumed including country-specific electricity production and trades. Regarding the electricity trades with neighboring BZs, it is assumed that the impact of the electricity imports to France equals the impact of the electricity generated in the BZ France is importing from, without taking exports and imports between France' neighboring BZs into account. Even though the methodology does not take into account the electricity trades between neighboring BZs, it is pointed out that dynamic data of electricity imports and exports should be considered for calculating $CO_{2eq.}$ intensities of the electricity consumption. Milovanoff et al. as well as Roux et al. [42] make use of the Ecoinvent data base version 3.1 which provides CO_2 factors for electricity generation technologies considering the whole life-cycle of a technology. Roux et al. calculate hourly average $CO_{2eq.}$ intensities of the electricity generation and consumption in France for the year 2013. Their study focuses on the magnitude of errors when a yearly-average factor for the electricity mix is used instead of an hourly-average factor varying throughout the year. However, it is not stated how the $CO_{2eq.}$ intensities of the imports are considered in the study.

The company Tomorrow launched a website which shows the hourly average $CO_{2eq.}$ intensity for most European countries in real time [50]. Data on the electricity generation per production technology in a BZ is taken from ENTSO-E. Furthermore, trading between BZs as well as time-varying $CO_{2eq.}$ intensities are considered in their models (Tomorrow, 2018), but a license has to be purchased to get access to the data. Their method is based on the electricity balance between the supply and demand side. The hourly $CO_{2eq.}$ intensities of each BZ are considered as vectors in a linear equation system, which is solved for the $CO_{2eq.}$ intensity for each hour of year.

Bettle et al. [44] calculate marginal emission factors for Wales and England for the year 2000, based on 30 min data of a full year for all generating power plants considering all plants individually. Imports are not considered in their study. They found that the marginal emission factor is usually higher than the average emission factor (up to 50% higher). Thus, the average factor is likely to underestimate the carbon-savings potential.

Hawkes [45] estimates marginal CO_2 emission rates for Great Britain, based on data from 2002 to 2009. The emission rate is termed marginal emission factor and corresponds to the CO_2 intensity of the electricity not used as a result of a DR measure. The approach follows a merit-order approach and thus applies only to countries where the electricity generation is primarily based on fossil-fuel technologies. The methodology does not consider electricity imports from neighboring BZs. They point out a clear correlation between the total system load and the marginal emission factor, showing high emission factors during times of high system loads.

Peán et al. [25] determine marginal emission factors for Spain for the year 2016 based on the methodology proposed by Hawkes [45]. Similar to Hawkes, electricity imports are not considered.

Corradi [46] aims at calculating the marginal $CO_{2eq.}$ intensity of the electricity using machine learning. Flow tracing [51] is used to trace the flow of the electricity back to the area where the marginal electricity is generated. Following Corradi's approach, in the end, the total marginal emissions of a BZ are the weighted average of the emissions from the marginal electricity generation plant of that BZ and the marginal electricity generation plant of the imports (in case of imports), using the percentage of origin as a weight.

2.2. Coupled Approach

Graabak et al. [47] determine yearly average marginal emission factors for the European power grid and marginal emission factors for the Norwegian power grid based on a European Multi-area Power-market Simulator (EMPS). EMPS is a stochastic optimization model for hydro-thermal electricity markets where the electricity market is arranged so that electricity prices balance supply and demand in each area and time step. It is used by all main actors in the Nordic power market, such as the TSOs, and for energy system planning, production scheduling, and price forecasting. Average emission factors are estimated for several scenarios of production portfolio for the European power grid. They point out that an average emission factor should be used for the control of a large number of buildings while a marginal emission factor should be applied for the control of a limited number of buildings.

Patteeuw et al. [26], Arteconi et al. [38], Askeland et al. [48] and Quoilin et al. [49] determine optimal dispatch and unit commitment in electricity grids. None of the models calculates $CO_{2eq.}$ intensities. Nevertheless, their models could determine the carbon intensity by considering CO_2 factors for each electricity generation technology. Patteeuw et al. [26] developed an approach to model active demand response with electric systems while considering both the supply and demand sides. The model takes thermal comfort and techno-economic constraints of both sides of the power system into account. Formulated as an optimization problem, it minimizes the overall operational cost of the electricity generation. The case study is based on the Belgian power system using the year 2010. This model enables to investigate the influence of different levels of market penetration of DR on the decision of the marginal generation technology. The approach does not consider imports. Arteconi et al. [38] apply the same model for the choice of the generation technology and use a low-order resistance–capacitance building model as a case study to investigate the DR potential of a building.

Askeland et al. [48] developed an equilibrium model for the power market that couples the demand and supply sides. The model provides time series for the electricity demand as well as for the renewable energy generation. A case study for the Northern European power system is performed where the effect of DR on the potential shift in the generation mix is studied. Quoilin et al. [49] developed a model called Dispa-SET, which is an open-source model that solves the optimal dispatch and unit commitment problem at the EU level, applying one node per country, instead of per BZs.

3. Evaluating the Hourly Average $CO_{2eq.}$ Intensity

In the following, the sources for required input data are first presented. The assumptions for processing the data are then stated. Finally, a step-by-step guidance is given to determine the hourly average $CO_{2eq.}$ intensity of the electricity mix per BZ. In the case study, hourly average $CO_{2eq.}$ intensities are calculated for Scandinavia, meaning Norwegian, Swedish, and Danish bidding zones, but also for Finland, Germany, and the Netherlands.

3.1. Data Retrieval and Pre-Processing

3.1.1. Electricity Use per Bidding Zone

Data regarding the electricity generation per production type within a BZ can be retrieved from ENTSO-E, which is the European Network of TSOs for Electricity. The European TSOs are supposed to provide data to ENTSO-E to promote market transparency and closer cooperation across the TSOs. A free user account at ENTSO-E has to be set up to download the data. Furthermore, data may also be retrieved directly from national TSOs, if the dataset from ENTSO-E is incomplete for specific BZs. For instance, hourly electricity generation data for the Swedish BZs was here obtained from the Swedish TSO (as this data was not available from ENTSO-E).

The ENTSO-E dataset contains hourly values for the total electricity generated per production type within a BZ. Thermal power plants are included as a production type. In practice, these power plants can consume most of the generated electricity onsite. Therefore, it is necessary to have knowledge about the sales licenses to the grid of these power plants. When onsite-generated electricity is not sold to the grid, it should ideally not be considered for the average $CO_{2eq.}$ intensity of the electricity mix. Information regarding sales licenses can usually be obtained from either the TSO, state directorates or directly from the company owning the power plant. For our case study of Scandinavia, these sales licenses were only determined for Norway. The Norwegian thermal power plants with the highest installed capacities are shown in Figure 1. Only the power plant of Mongstad sells electricity to the grid. Comprehensive information on data treatment and assumptions is provided in [39].

The input data can be arranged in the following way:

$$BZ = \begin{bmatrix} P_{BZ_j, EGT_1}(t_1) & \cdots & & P_{BZ_j, EGT_m}(t_1) \\ & \ddots & \vdots & \\ \vdots & & P_{BZ_j, EGT_i}(t_g) & \vdots \\ & & \vdots & \ddots \\ P_{BZ_j, EGT_1}(t_{8760}) & \cdots & & P_{BZ_j, EGT_m}(t_{8760}) \end{bmatrix} \quad (1)$$

where

- i is the "index of EGTs" ranging from 1 to m
- j is the "index of a specific BZ"
- g is the "hour of the year" ranging from 1 to 8760 (or 8784)

The matrix *BZ* (where 'BZ' stands for bidding zone) includes the electricity generation from each electricity generation technology (EGT_i) at every hour of the year (t_g). EGT_i represents electricity generation from each technology in the BZ, but, by extension, it also includes the imports from first tier BZs. In other words, imports from other BZs are considered as an EGT in the matrix.

3.1.2. Emission Factors per Electricity Generation Technology

The average $CO_{2eq.}$ intensity of a BZ depends directly on the CO_2 factor that is associated to each EGT. This choice of the CO_2 factor strongly affects the final result of the evaluation. Two possible ways of acquiring CO_2 factors of a given EGT are the IPCC report [52,53] or the Ecoinvent database [54]. Ecoinvent provides a vast variety of CO_2 factors for electricity generation from a given fuel type depending on the type of power plant and the specific country. A license is necessary to use Ecoinvent, whereas the IPCC report is available free of charge. CO_2 factors can differ between references mainly due to different allocations of emissions, especially for combined heat-and-power plants. More detailed information regarding emission allocations are given in [52–54]. Regarding annual average $CO_{2eq.}$ intensities of a country (see Table A1 in Appendix A), they can be calculated from Ecoinvent or taken from the website of the European Environment Agency [55].

For the sake of the simplicity, our case study considers a same CO_2 factor per EGT for every country. Therefore, it assumes that the CO_2 factor is independent of the country and the specific power plant as long as it uses the same EGT (disregarding the thermal efficiency or the age of the power plant). The proposed methodology is however more general. If a different $CO_{2eq.}$ factor should be considered for a same EGT but for a different plant efficiency, plant age, or country, a new EGT should be defined for each different $CO_{2eq.}$ factor considered.

In this study, data from the Ecoinvent database has been used. The phases that were considered in the CO_2 factor per EGT are the extraction of fuels, the construction of the power plant (including infrastructure and transport), its operation and maintenance as well as the end of use of the power plant. The CO_2 factor for hydro pumped storage is here assumed constant in time, 62 $gCO_{2eq.}/kWh$. In fact, it is dependent on the $CO_{2eq.}$ intensity of the electricity mix used when pumping water into the storage reservoir. Unlike Norway, this assumption can be critical for BZs that usually have a relatively high $CO_{2eq.}$ intensity. Strictly speaking, if the constant CO_2 factor considered for hydro pumped storage is much lower than the $CO_{2eq.}$ intensity of the electricity mix when water is pumped into the storage, it would correspond to a 'greenwashing' of the electricity mix. Furthermore, ENTSO-E defines an EGT category called 'other' with no further specifications. It is thus difficult to allocate a CO_2 factor for this production type and it could also differ among different countries. In this work, the factor is assumed to be the average of the fossil fuel technologies. Table A1 gives an overview of typical CO_2 factors for different fuel types.

3.2. Calculation Methodology

The electricity mix is assumed homogeneous in each BZ, meaning that a same $CO_{2eq.}$ intensity is used for the entire BZ at each hour of the year. For neighboring countries where the $CO_{2eq.}$ intensities are not evaluated (here called 'boundary BZ'), a yearly-averaged $CO_{2eq.}$ intensity is considered. In our case study, these countries are Great Britain, Belgium, Poland, Estonia, and Russia. It is nonetheless reasonable to assume that they have a limited impact on the Norwegian electricity mix. They are either 2nd tier countries (i.e., which do not have a direct grid connection to Norwegian BZs) or have limited electricity export to Norway (which is typically the case for Russia). By extension, BZs for which the average $CO_{2eq.}$ intensity is calculated for every hour of the year are hereafter called computed BZs.

Matrix T(t) (where 'T' stands for technology) includes the electricity generation from all EGTs in all BZs and is calculated for each hour of the year (t):

$$T(t) = \begin{bmatrix} P_{BZ_1,EGT_1}(t) & \cdots & & & P_{BZ_n,EGT_1}(t) \\ & \ddots & & & \\ \vdots & & P_{BZ_j,EGT_i}(t) & & \vdots \\ & & & \ddots & \\ P_{BZ_1,EGT_m}(t) & & \cdots & & P_{BZ_n,EGT_m}(t) \end{bmatrix} \quad (2)$$

where

- i is the "index of EGTs" ranging from 1 to m
- j is the "index of BZs" ranging from 1 to n

The size of matrix $T(t)$ depends on the number of EGTs and number of BZs that are considered in a respective study.

The next step is a normalization to 1 MWh by dividing by the total hourly electricity generation (i.e., summing on all the EGT_i, also considering imports) in a bidding zone (BZ_j) during each hour (t) of the year. This step will be necessary to determine the $CO_{2eq.}$ intensity for 1 MWh of generated electricity in a respective BZ. Matrix $N(t)$ (where 'N' stands for normalization) is set up as follows:

$$N(t) = \begin{bmatrix} \frac{1}{\sum_i P_{BZ_1,EGT_i}}(t) & 0 & 0 & 0 & 0 \\ 0 & \ddots & 0 & 0 & 0 \\ 0 & 0 & \frac{1}{\sum_i P_{BZ_j,EGT_i}}(t) & 0 & 0 \\ 0 & 0 & 0 & \ddots & 0 \\ 0 & 0 & 0 & 0 & \frac{1}{\sum_i P_{BZ_n,EGT_i}}(t) \end{bmatrix} \quad (3)$$

where

- i is the "index of EGTs" ranging from 1 to m
- j is the "index of a specific BZ" ranging from 1 to n

The matrix $P(t)$ (where 'P' stands for production) is the share of each EGT on the total hourly electricity generation in a respective BZ, still considering electricity imports as an EGT. Regarding the electricity imports to a BZ, it is distinguished between imports from boundary BZs with a fixed $CO_{2eq.}$ intensity of the electricity mix ($P_{Import,fix,t}$) and imports from computed BZs with a variable electricity mix ($P_{Import,var,t}$). At this stage, $P(t)$ only considers the share of each EGT in the electricity mix of a specific BZ, but does not consider the share of each EGT in the imports. $P(t)$ is calculated by multiplying $T(t)$ and $N(t)$. The share of EGTs (located inside the BZ j) in the generation mix is called $P_{EGT,t}$.

$$P(t) = T(t) \cdot N(t) = \begin{bmatrix} P_{EGT,t} \\ P_{Import,fix,t} \\ P_{Import,var,t} \end{bmatrix} \quad (4)$$

$P_{EGT,t}$ and $P_{Import,fix,t}$ can be combined to define the matrix $P_{EGT,fix}$:

$$P_{EGT,fix} = \begin{bmatrix} P_{EGT,t} \\ P_{Import,fix,t} \end{bmatrix} \quad (5)$$

The next steps show how to include the share of each EGT in the electricity imports; in other words, how the electricity mix of a neighboring BZ influences the electricity mix of a BZ through imports.

In general, the balance between electricity consumption and electricity generation has to be satisfied at all times for each BZ. This idea is further generalized in MRIO models, where interdependencies within the whole system can be captured, while preserving regional differences [56]. For each BZ, the sum of electricity import and generation by a specific EGT should be consumed in the BZ, or exported. This complies with the logic of MRIO models which can be used to calculate consumption-based emissions for an entire country or region [34]. The electricity balance can then be expressed as

$$M = P_{EGT,fix} + M \cdot P_{Import,var,t}, \qquad (6)$$

with M (where 'M' stands for mix) representing the share of each EGT on the electricity use of BZ_j and the exports from BZ_j. Solving Equation (6) for M is done by

$$M(i,j) = P_{EGT,fix} \cdot (I - P_{Import,var,t})^{-1}. \qquad (7)$$

where I is an identity matrix. Matrix $M(i,j)$ contains the share of each EGT_i (and boundary BZ) on the electricity use and exports from BZ_j. A new matrix is computed for each hour of the year.

The average $CO_{2eq.}$ intensity of a BZ for every hour of the year is calculated by

$$e_j(t) = \sum_{i=1}^{m} ef_{EGT_i} \cdot M(i,j) \qquad (8)$$

where t is the hour of the year, i is the index of the EGT ranging from 1 to m, j is the index of a specific BZ ranging from 1 to n and ef_{EGT_i} is the emission factor of the EGT of index i.

3.3. Applicability of the Methodology

Comparing the proposed methodology with the other existing methodologies presented in Table 1, its major advantage is its simplicity. Nevertheless, it also has limitations. As a decoupled approach, the method can be used as long as the number of buildings participating in DR schemes is low; in other words, as long as the level of the market penetration of DR in buildings is still limited. This limitation counts for all decoupled approaches, not only for the proposed method. Furthermore, this approach is representative for most simulation-based studies in building energy flexibility. Yearly average $CO_{2eq.}$ intensities of the electricity mix have been used extensively in the past to calculate the emission balance of buildings that export electricity to the grid [57]. Studies mostly focusing on the LCA of buildings often use average $CO_{2eq.}$ intensities rather than marginal intensities. The use of hourly average $CO_{2eq.}$ intensities as a control signal for DR is thus remaining coherent with that approach. Also, in the EPBD method, average primary energy factors are often used to evaluate the performance of buildings.

3.4. $CO_{2eq.}$ Intensities in Scandinavian Bidding Zones

As a case study, the hourly average $CO_{2eq.}$ intensities for all Norwegian BZs are evaluated for the year 2015. Results are plotted in Figure 2a. NO5 has the highest annual average $CO_{2eq.}$ intensity which is due to the electricity generation from the thermal power plant in Mongstad (see Figure 1 and Table 2). Furthermore, it is obvious that the highest $CO_{2eq.}$ intensity peaks occur in NO2. Figure 2b presents the average $CO_{2eq.}$ intensities for NO2, NO3, SE1, SE4, DK1, and FIN. It is clear that the Norwegian electricity mix has low average $CO_{2eq.}$ intensities compared to non-Norwegian BZs. Resulting from the large differences in $CO_{2eq.}$ intensity, the average $CO_{2eq.}$ intensity in NO2 usually increases when electricity is imported from DK1, which also explains the $CO_{2eq.}$ peaks in the BZ. The impact of fossil fuel-based electricity imports on the average $CO_{2eq.}$ intensity is particularly strong for the case of Norway because the electricity generation in Norway is almost entirely from hydropower. The impact of carbon-intensive electricity imports on the average $CO_{2eq.}$ intensity of other countries, like Finland or

Denmark, is usually lower. A more detailed analysis of the correlation of the average $CO_{2eq.}$ intensity and electricity imports to Norway is provided in [39].

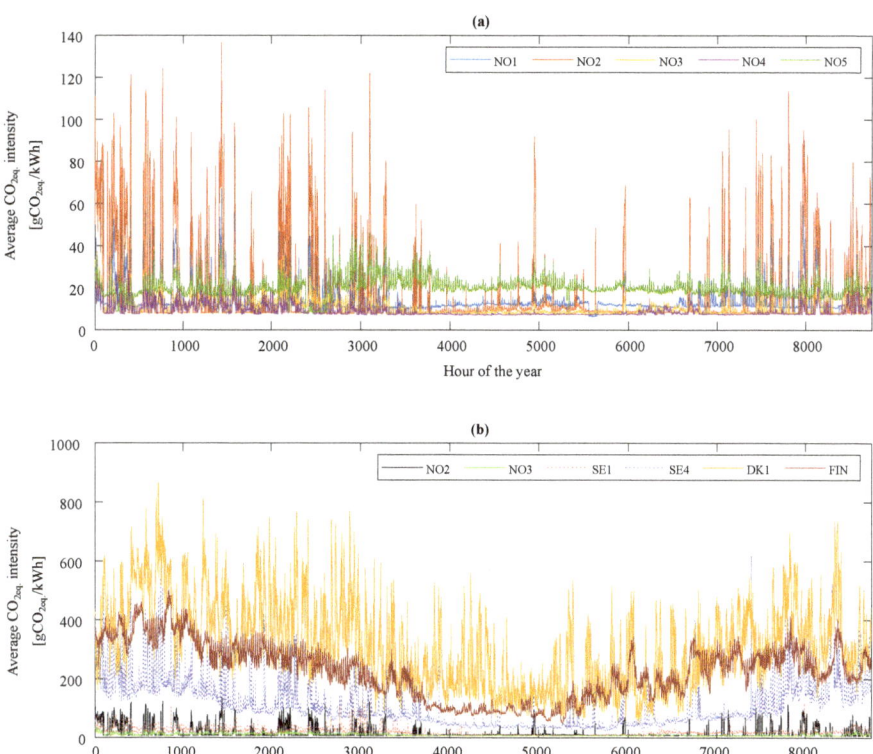

Figure 2. Hourly average $CO_{2eq.}$ intensity of the electricity mix for (**a**) the Norwegian bidding zones and (**b**) for several Scandinavian bidding zones.

Table 2. Comparison of the annual average $CO_{2eq.}$ intensities of the electricity mix for several Scandinavian bidding zones.

BZ	NO1	NO2	NO3	NO4	NO5	SE1	SE4	DK1	FIN
Average $CO_{2eq.}$ intensity ($gCO_{2eq.}$/kWh)	15	17	11	9	20	21	114	316	227
Average $CO_{2eq.}$ intensity without imports ($gCO_{2eq.}$/kWh)	7	8	8	7	20	21	259	461	241

An overview of the annual average $CO_{2eq.}$ intensities is provided in Table 2. DK1 has the highest annual average $CO_{2eq.}$ intensities, followed by FIN and SE4.

Figure 3 illustrates the relation between the average $CO_{2eq.}$ intensity and the electricity spot price for six Scandinavian BZs for an exemplary five-day period. For the Norwegian BZs NO2 and NO3 (see Figure 3a,b) it is shown that the average $CO_{2eq.}$ intensity is low, when the electricity spot price is high. In Norway, electricity is produced from hydropower in times of high demands. Electricity spot prices are high during high electricity demands, thus typically leading to low $CO_{2eq.}$ intensities. In general, the Norwegian hydropower reservoirs are operated in a cost-optimal way, so that electricity is imported during the night when electricity is cheap and exported to continental Europe during the day, when electricity is expensive [39]. This operation strategy can also be used to explain the $CO_{2eq.}$ intensities in the BZs shown in Figure 3. The correlation between the carbon intensity and the spot price for the Swedish BZs is similar to the Norwegian BZs.

SE1 (Figure 3c) is the northernmost Swedish BZ and relies primarily on electricity generation from hydro reservoirs and on-shore wind as well as on electricity imports from SE2. Thus, average $CO_{2eq.}$ intensities are generally rather low. Regarding the correlation with electricity spot prices, $CO_{2eq.}$ intensities are low when spot prices are high. This relation is also similar for SE4 (see Figure 3d) but to a lower extent. The average $CO_{2eq.}$ intensities in SE4 are several times higher compared to SE1 because SE4 trades a lot of electricity with Denmark, Poland, and Germany.

A rather weak correlation between the carbon intensity and the spot price can be seen for DK, which is in accordance with the findings from Knudsen et al. [24] who found that a low $CO_{2eq.}$ intensity does not necessarily occur concurrent with low costs. A difference between day and night is also visible for the $CO_{2eq.}$ intensity in DK1. A correlation between the carbon intensity and the spot price on an hour-by-hour basis is not obvious because the $CO_{2eq.}$ intensity varies much faster than the spot price. A clear relationship between the $CO_{2eq.}$ intensity and the spot price is not visible for FIN.

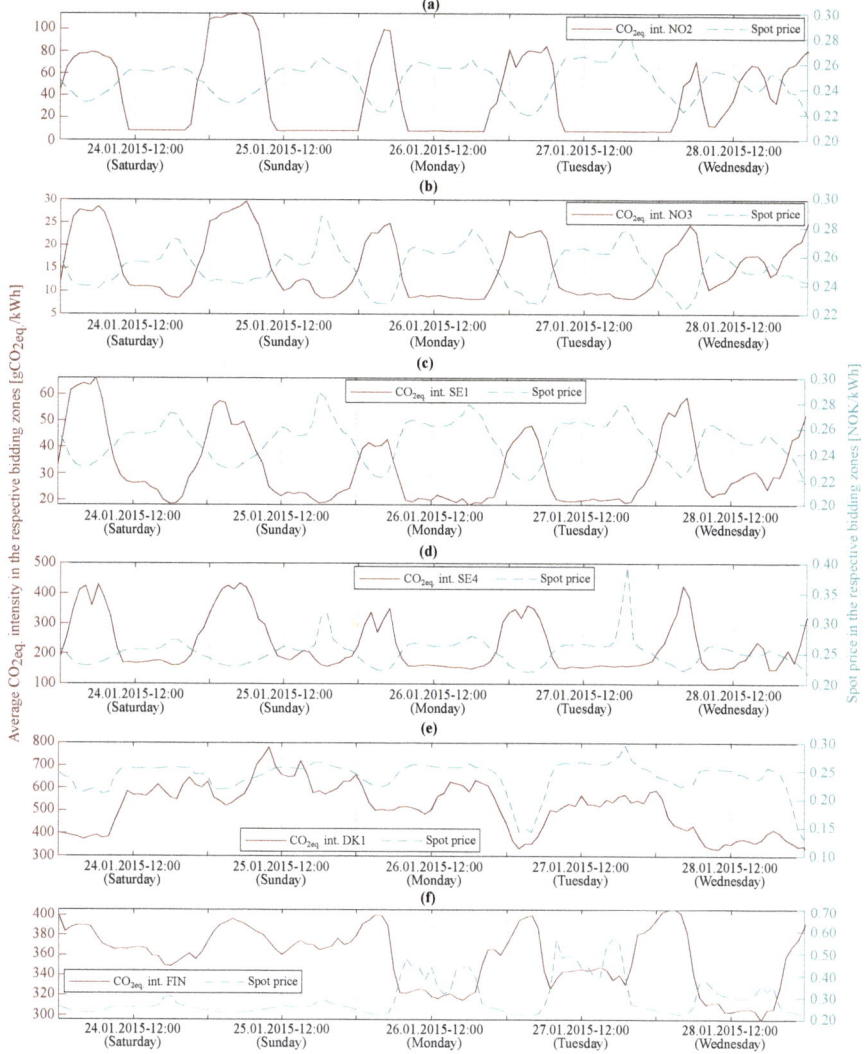

Figure 3. Average $CO_{2eq.}$ intensity and spot prices for five exemplary days in 2015 for six Scandinavian BZs: (**a**) NO2, (**b**) NO3, (**c**) SE1, (**d**) SE4, (**e**) DK1, and (**f**) FIN.

4. Case Study Using Demand Response for Heating

4.1. Case Building

To represent a large share of the Norwegian residential building stock, a single-family detached house built according to the building standard from the 1980s, TEK87, is chosen as a case [58]. The geometry of the building is based on the ZEB Living Lab which is a zero emission residential building located in Trondheim [59]. The envelope model of the real Living Lab has been calibrated with the help of dedicated experiments for the building performance simulation tool IDA ICE. However, the thermal properties of the building envelope for this case study comply with TEK87. The Living Lab has a heated floor area of 105 m^2, the floor plan being shown in Figure 4. An overview of the building properties is provided in Table 3. For the sake of the simplicity, natural ventilation, which was usually applied in TEK87 buildings, is modeled as balanced mechanical ventilation with a heat recovery effectiveness η_{HR} of 0%. This study considers the U-values of the building walls, but it should be noted that the building insulation level [15] as well as the thermal capacity of a building [60–62] influences the flexibility potential. The climate of Trondheim is also relevant for Norway in general.

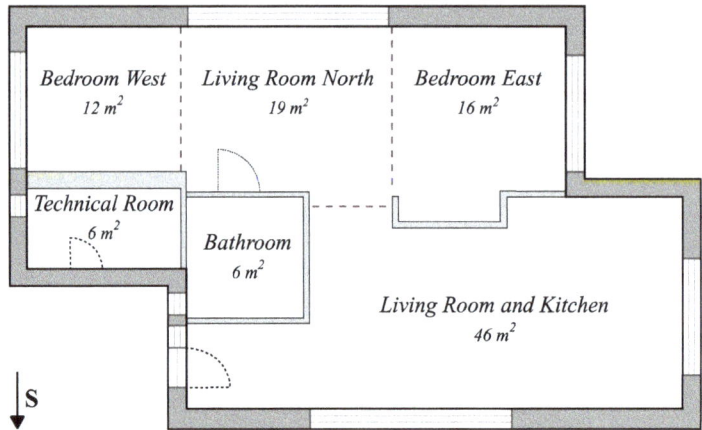

Figure 4. Floor plan of the studied building [15].

Table 3. Building envelope properties and energy system characteristics of the case study building (EW—external wall; IW—internal wall; n_{50}—air changes per hour at 50 Pa pressure difference; U_{Total}—the total U-value of the windows including the glazing and the frame; HR—heat recovery; ER—electric radiator; AHU—air handling unit; HDS—heat distribution system).

	Building Envelope				Thermal Bridges	Infiltration	Windows	AHU	HDS	SH Needs
Symbol	U_{EW}	U_{IW}	U_{Roof}	U_{Floor}			U_{Total}	η_{HR}	ER	
Unit		W/(m²·K)			W/(m²·K)	n_{50}	W/(m²·K)	%	W/m²	kWh/m²
	0.35	0.34	0.23	0.30	0.05	3.0	2.1	0	93	172

A detailed multi-zone model of the building is created using the software IDA ICE version 4.8, which is a dynamic building simulation software that applies equation-based modeling [63]. IDA ICE has been validated in several studies [64–68].

Electric radiators are used for SH as it is the most common space-heating system in Norwegian houses [69]. One electric radiator is placed in each room with a power equal to the nominal SH power of the room at design outdoor temperature (DOT) of −19 °C. DHW is produced in a storage tank. An electric resistance heater with a capacity of 3 kW is used for DHW heating and is installed in the lower third of the tank. IDA ICE has a one-dimensional model of a stratified water tank that accounts for the heat conduction and convection effects in the tank. The DHW storage tank is here divided into four horizontal layers to account for stratification effects. The DHW storage volume is calculated by

$$V_{DHW} \cong S \cdot 65 \cdot n_{people}^{0.7} \quad [\text{litre}] \qquad (9)$$

where S is the safety margin and n_{people} the number of occupants. S is set to 125% for a low number of people [70]. The charging of the DHW storage tank is controlled by two temperature sensors that are installed in the bottom and the top of the tank. The electric resistance heater starts heating as soon as the temperature in the upper part of the tank drops below the set-point and it continues until the set-point for the temperature sensor in the lower part of the tank is reached. The DHW start temperature is called the DHW TSP here. The DHW stop temperature is always taken 3 °C above to start temperature.

The internal heat gains from electrical appliances, occupants and lighting are defined according to the Norwegian technical standard, SN/TS 3031:2016 [71]. The daily DHW profiles are taken from the same standard. Schedules for electrical appliances are also based on SN/TS 3031:2016, whereas the schedules for occupancy and lighting are taken from prEN16798-1 and ISO/FDIS 17772-1 standards [72,73]. The internal heat gains are distributed uniformly in space. All profiles presented in Figure A1 have an hourly resolution and are applied for every day of the year.

The energy flexibility potential is evaluated for four different PRBCs by comparing them to the reference scenario that applies constant TSPs for SH (21 °C) and DHW (50 °C). The TSP for the bathroom is 24 °C. All doors are closed at all times. DR measures are applied to the common rooms only (meaning the living room, kitchen, and living room north). All cases use the weather data of 2015 (retrieved from [74]) for Trondheim, independent of the BZ considered. NordPool provides hourly day-ahead spot prices for each bidding zone [75]. They are used as an input signal for the price-based control and to calculate energy costs for heating. An electricity fee for the use of the distribution grid is not considered in the cost evaluation.

4.2. Demand Response Strategies

The reference scenario, termed BAU (for business as usual), maintains constant TSPs for SH and DHW heating at 21 °C and 50 °C respectively. Using constant TSP is the most common way to control the heating system in Norwegian residential buildings. These TSPs are varied for the DR strategies. The DHW TSP can be increased by 10 K or decreased by 5 K depending on the control signal. The limit for DHW temperature decrease is chosen with regards to Legionella protection. Regarding SH, the TSPs are increased by 3 K or decreased by 1 K. According to EN15251:2007 [76] indoor temperatures between 20 °C and 24 °C correspond to a predicted percentage dissatisfied (PPD) < 10% and −0.5 < PMV < +0.5 in residential buildings for an activity level of 1.2 MET and a clothing factor of 1.0 clo.

The control signal for the CO_2-based control is determined based on two principles. The first principle, hereafter called CSC-a, aims at operating the energy system in times of lowest $CO_{2eq.}$ intensities by increasing TSPs for SH and DHW during these periods. For Norway, this principle may in practice lead to extended periods with high TSPs because of the typical daily $CO_{2eq.}$ intensity profile (Figure 3a,b). This may lead to an unnecessary increase in annual energy use for heating. Therefore, the second principle—hereafter called CSC-b—does not aim at operating the heating system during periods with the lowest carbon intensities but rather charges the storages just before high-carbon periods in order to avoid the energy use during these critical periods. Using CSC-b, the TSPs are increased for shorter time periods compared to CSC-a, thus improving the energy efficiency.

The carbon-based PRBC uses a 24 h sliding horizon to determine a high-$CO_{2eq.}$ intensity threshold (HCT) and low-$CO_{2eq.}$ intensity threshold (LCT). At each hour, the current $CO_{2eq.}$ intensity (CI) is compared to these thresholds. Taking CI_{max} and CI_{min} as the maximum and minimum intensities for the next 24 h, LCT has been selected to CI_{min} + 0.3 (CI_{max}-CI_{min}) and HCT to CI_{min} + 0.7 (CI_{max}-CI_{min}). Regarding CSC-a, if the $CO_{2eq.}$ intensity of the current hour is below the LCT, the TSPs are increased. If the current $CO_{2eq.}$ intensity is above the HCT, the set-points are decreased to delay the start of the heating, whereas if the $CO_{2eq.}$ intensity of the current hour is between the LCT and HCT, the TSPs remain equal to the reference scenario. Regarding CSC-b, the control signal is also determined based on the three price segments, as defined for CSC-a, but, additionally, the control considers if the current $CO_{2eq.}$ intensity is increasing or decreasing with time. If the current $CO_{2eq.}$ intensity is between the LCT and HCT and the $CO_{2eq.}$ intensity increases in the next two hours, TSPs are increased. Both principles, CSC-a and CSC-b, are presented in Figure 5.

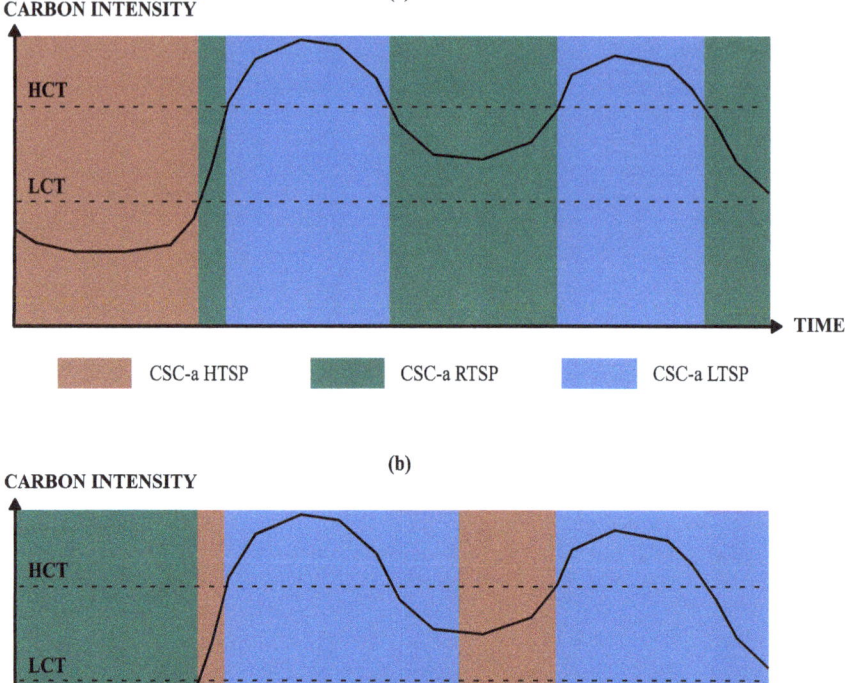

Figure 5. Principle of the determination of the carbon-based control signal according to (**a**) CSC-a and (**b**) CSC-b. (HTSP is high temperature set-points, RTSP is reference temperature set-points, LTSP is low temperature set-points).

The performance of the control with regards to emission savings is sensitive to the selection of thresholds, LCTs and HCTs. The influence of LCTs and HCTs on the number of hours per TSP-segment has been evaluated for BZ NO3. An LCT of 30% and an HCT of 70% have been chosen for calculating the control signal. The LCTs and HCTs for the other five BZs considered in the simulation study are chosen so that the number of hours in the three respective segments is equivalent to the case of NO3. An overview of the different scenarios is presented in Table 4. In fact, there is an optimum LCT and HCT for each BZ to minimize carbon emissions. However, in the case study, the two thresholds are rather chosen to have a similar number of hours in the respective TSP segments. This is done to investigate the influence of the $CO_{2eq.}$ intensity profile on the carbon emissions rather than optimizing the control principle thresholds.

Table 4. Influence of the low-carbon and high-carbon thresholds (LCT and HCT) on the number of hours per temperature set-point segment (LTSP—low-temperature set-points; RTSP—reference temperature set-points; HTSP—high temperature set-points).

	Thresholds Kept Constant						Adjusted Thresholds for Similar Segments					
	NO2	NO3	SE1	SE4	DK1	FIN	NO2	NO3	SE1	SE4	DK1	FIN
						CSC-a						
LCT [%]	30	30	30	30	30	30	22	30	27	29.5	47	46
HCT [%]	70	70	70	70	70	70	68.5	70	66	68.5	81.5	82
LTSP [h]	1812	1886	1706	1814	2584	2439	1879	1886	1884	1878	1882	1881
RTSP [h]	1822	2219	2166	2236	2731	2930	2229	2219	2223	2223	2230	2245
HTSP [h]	5126	4655	4888	4710	3445	3391	4652	4655	4653	4659	4648	4634
						CSC-b						
LCT [%]	30	30	30	30	30	30	22	30	27	29.5	21	16
HCT [%]	70	70	70	70	70	70	67	70	67.5	70	76	59
LTSP [h]	2624	2924	2758	2892	3954	3881	2912	2924	2925	2918	2933	2904
RTSP [h]	5125	4654	4487	4710	3445	3390	4651	4654	4652	4659	4648	4673
HTSP [h]	1011	1182	1115	1158	1361	1489	1191	1182	1183	1183	1179	1183

Price-based control signals, hereafter called CSP-a and CSP-b, are also determined similar to CSC-a and CSC-b, to also investigate DR measures based on the electricity spot price of each of the BZs. This means that the case study is performed for four DR signals applied to six Scandinavian BZs. In general, any form of penalty signal could be used for the control and optimization. If a proper penalty signal is chosen, a building can be controlled so that it is either energy efficient, cost efficient or CO_2 efficient. It would also be possible to select a combination of the different penalty signals for the building control [11].

4.3. Case Study Results

Figure 6 illustrates the principle of the CSC-a control during 48 h of the heating season exemplary for DK1. Figure 6a shows the $CO_{2eq.}$ intensity and both thresholds, LCT and HCT. Figure 6b presents the measured temperatures in the DHW tank and the start and stop TSPs of the DHW hysteresis control. These TSPs vary depending on the $CO_{2eq.}$ intensity signal. The same principle is shown for SH in Figure 6c. It is visible from Figure 6d that the electric radiators and the electric resistance heater for DHW heating are operated depending on the $CO_{2eq.}$ intensity signal and according to the proposed temperature hysteresis.

The energy use, $CO_{2eq.}$ emissions and costs are compared to the reference cases for each respective BZ. The relative difference in total annual emissions (Em.) for each BZ is calculated by

$$Em.(BZ_i) = \frac{Em._{BZ_i,PRBC}}{Em._{BZ_i,BAU}} * 100 - 100 \; [\%] \qquad (10)$$

Equation (10) can also be applied to determine the relative changes for annual energy use and annual costs, respectively. Results are presented in Table 5.

In general, the energy use increases for all DR cases. Regarding CSC, annual $CO_{2eq.}$ emissions decrease for NO2 while they remain rather close to the reference case for the other BZs. For NO3, the daily fluctuations in average $CO_{2eq.}$ intensity are too low to benefit from DR measures. Comparing results for NO2 and NO3 regarding overall emission savings, it demonstrates that electricity imports should be properly considered as they influence significantly the variations of $CO_{2eq.}$ intensities. These variations have shown to be important to make the $CO_{2eq.}$-based DR effective.

Regarding CSP, costs increase slightly for NO2 and NO3 while they decrease for SE4, DK1, and FIN. For SE4, DK1, and FIN, cost savings can be achieved as daily price fluctuations are sufficient. High daily fluctuations of prices do usually not occur in Norway and northern Sweden. Therefore, for NO2, NO3, and SE1, consuming during slightly decreased electricity spot prices is outbalanced by an increase in total energy use. Furthermore, using price-based control, overall emissions increase significantly for NO2, NO3, SE1, and SE4 as a result of typically high $CO_{2eq.}$ intensities at low-price periods (as shown Figure 3).

It is obvious, that reductions of $CO_{2eq.}$ emissions and costs are possible, but that they are very dependent on the characteristics of the $CO_{2eq.}$ intensity and the electricity spot price. Savings could be increased, if different thresholds (LCT and HCT) were applied, not taking NO3 as a reference regarding the number of hours per segment. Absolute savings depend on the LCT and HCT as well as the principle used to obtain the respective control signals. In addition, the maximum absolute savings should ideally be evaluated using MPC. Therefore, here, the performance of each control should be considered relative to each other rather than in absolute terms.

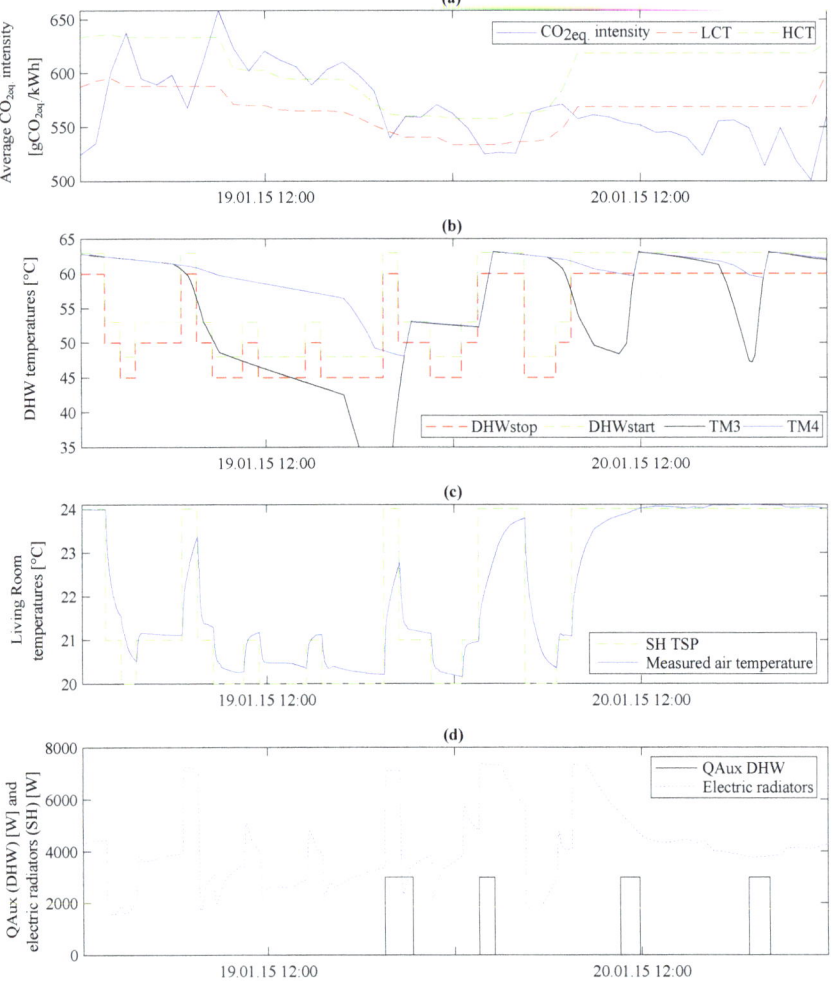

Figure 6. Illustration of the control principle for DK1 case CSC-a during a 48 h period, where (**a**) shows the $CO_{2eq.}$ intensity, (**b**) shows the DHW temperatures and hysteresis set-points, (**c**) shows the measured air temperature and the temperature set-point for space heating and (**d**) shows the power of the electric radiators and the electric auxiliary heater (DHWstart and DHWstop are the start and stop temperatures for DHW; TM is two temperature measurements in the water tank; SH TSP is space heating temperature set-point; QAux is the electric resistance heater).

Table 5. Cost and $CO_{2eq.}$ emission savings relative to the respective reference cases for all six bidding zones (E_{Use}—energy use; Em.—emissions; CSC—control strategy carbon; CSP—control strategy price).

	CSC-a			CSC-b			CSP-a			CSP-b		
	E_{Use}	Em.	Costs	E_{Use}	Em.	Costs	E_{Use}	Em.	Costs	E_{Use}	Em.	Costs
	%	%	%	%	%	%	%	%	%	%	%	%
NO2	+9	−8	+10	+3	−1	+2	+7	+21	+2	+4	+9	+1
NO3	+9	+3	+10	+3	+2	+2	+7	+13	+2	+4	+6	+1
SE1	+9	+0	+11	+3	+1	+2	+7	+14	+1	+4	+8	+1
SE4	+9	+0	+10	+3	+1	+1	+7	+11	−3	+4	+6	−1
DK1	+10	+1	+2	+3	+2	+1	+7	+1	−6	+4	+2	−1
FIN	+9	+5	+6	+3	+2	+1	+8	+8	−9	+4	+3	−3

Table 6 presents $CO_{2eq.}$ emission and cost savings for DHW heating and SH, separately. It can be seen that SH is the main contributor to total emission or cost savings because, for such a building with limited thermal insulation, the share of electricity use is much more significant for SH than for DHW heating. This will be different for buildings with better insulation levels. For example, regarding CSC-a in DK1, it can be seen that, even though DR measures for DHW heating lead to 12% emission savings, the overall emissions increase by 1% because the emissions resulting from SH increase by 3%.

Table 6. Cost and $CO_{2eq.}$ emission savings separated for DHW and SH (results are given in %).

	Emissions						Costs					
	CSC-a			CSC-b			CSP-a			CSP-b		
	Total	DHW	SH	Total	DHW	SH	Total	DHW	SH	Total	DHW	SH
NO2	−8	−17	−6	−1	+17	−3	+2	−5	+3	+1	−5	+2
NO3	+3	−3	+3	+2	+14	+0	+2	−7	+3	+1	−5	+2
SE1	+0	−3	+1	+1	+20	−1	+1	−9	+2	+1	−7	+2
SE4	+0	−7	+0	+1	+16	−1	−3	−20	+1	−1	−13	+2
DK1	+1	−12	+3	+2	−1	+2	−6	−27	−1	−1	−15	+2
FIN	+5	+0	+6	+2	+6	+2	−9	−32	−3	−3	−21	+1

Furthermore, $CO_{2eq.}$ emissions for DHW heating are decreased significantly in most of the zones using CSC-a. On the contrary, in the Norwegian and Swedish BZs, these emissions increase significantly using CSC-b. Using CSC-b, the water storage tanks are typically charged during late evenings whereas the next peak for DHW withdrawal only occurs the next morning. These peaks of DHW withdrawal happen when the $CO_{2eq.}$ intensities of the electricity mix are rather low. In other words, CSC-b shifts the operation from low $CO_{2eq.}$ intensities (meaning peaks of DHW withdrawal) to higher $CO_{2eq.}$ intensities (meaning late evenings). Regarding costs, these DR strategies applied to DHW heating are very promising to decrease the operational costs. Operational costs are decreased for all BZs for both CSP-a and CSP-b considering DHW heating. In conclusion, efficient DR measures using PRBC can be implemented for the DHW heating. However, the performance of these controls are moderate for SH. The price-based controls did not manage to decrease operational costs for SH for most cases. On the one hand, this can be due to the low daily fluctuations of the electricity spot prices. On the other hand, this can be due to the low insulation level and thermal mass of the building as well as the inherent limitations of PRBC compared to MPC.

5. Conclusions

This work consists of two distinct but complementary parts. Firstly, a generic methodology is proposed to determine the hourly average $CO_{2eq.}$ intensity of the electricity mix of a bidding zone (BZ) also considering time-varying $CO_{2eq.}$ intensities of electricity imports. Northern Europe is taken as a case. Secondly, this case study is extended to investigate the performance of demand response measures based on the evaluated $CO_{2eq.}$ intensity. These measures aim at decreasing the environmental impact of the heating system operation of a typical Norwegian residential building.

The proposed methodology to evaluate the hourly average $CO_{2eq.}$ intensity of the electricity mix is an adaptation of multi-regional input–output models (MRIO). For each electricity generation technology (EGT), the method enforces the balance between electricity generation plus imports and electricity consumption plus exports. This balance is satisfied for both EGT and BZ at each hour of the year. Regarding the CO_2 factor for a specific EGT, it is shown that they can vary significantly among references depending on their respective assumptions. Among different possible applications, the average $CO_{2eq.}$ intensity that is determined using this methodology can be used as a control signal in predictive controls for building energy systems. The average $CO_{2eq.}$ intensity can actually be predicted using the forecast on hourly electricity generation provided by ENTSO-E.

Firstly, the methodology is applied to Northern Europe using CO_2 factors per EGT from the Ecoinvent database. It enables to highlight some important characteristics of this power market. Especially in Norway and northern Sweden, the electricity generation is characterized by a high share of hydropower in the electricity mix. These plants are typically operated when the price for electricity is high which happens during periods of high electricity demand. Therefore, the average $CO_{2eq.}$ intensities are usually low at times of high electricity demand. On the contrary, for countries where the electricity generation relies more on fossil fuels, average $CO_{2eq.}$ intensities are typically high at times of high electricity demand. In this respect, it is also important to consider electricity imports and their varying $CO_{2eq.}$ intensity when evaluating the average $CO_{2eq.}$ intensity. As soon as Norway imports electricity from neighboring BZs (typically in periods with low electricity demands), the $CO_{2eq.}$ intensity can increase significantly because electricity generation is usually more carbon-intensive in Norway's neighboring countries.

Secondly, the average $CO_{2eq.}$ intensity evaluated for Northern Europe is implemented into a predictive rule-based control (PRBC) to operate the electric heating system of a typical Norwegian residential building. The demand response measures based on the average $CO_{2eq.}$ intensity aim at decreasing the environmental impact of the heating system. Results prove that carbon-based controls can achieve emission reductions if daily fluctuations of the $CO_{2eq.}$ intensity are large enough to counterbalance for the increased electricity use generated by load shifting. As an example, the potential for emission reductions is higher in NO2 compared to NO3 because the daily fluctuations in the $CO_{2eq.}$ intensity in NO2 are much larger than in NO3. As these fluctuations in NO2 are mainly generated by imports, it further confirms the need to account for these imports in demand response analysis. If the heating system is controlled according to spot prices, results also confirm that price-based controls lead to increased emissions in the Scandinavian countries as operation is usually shifted towards night-time (for the case of Norway) when cheap but carbon-intensive electricity is indirectly imported from Germany, Poland, or the Netherlands (via Denmark and Sweden).

Demand response using PRBC applied to DHW heating show a strong potential for cost and emission savings. Conclusions regarding SH are more balanced. Depending on the control strategy and BZ, the PRBC manages to decreases the SH costs or the related $CO_{2eq.}$ emissions. The case building was taken representative for the Norwegian building stock and it results to a relatively low-level of thermal insulation. Energy use for SH is thus dominant over DHW. Consequently, the overall performance of DR using PRBC for heating (i.e., SH and DHW) is also balanced. Conclusions regarding SH may be different for buildings with better insulation as they have a higher storage efficiency. In addition, PRBC relies on predefined control rules which may not always be optimal. It should be explored whether advanced controls, such as model-predictive control (MPC), would lead to significant improvements regarding overall cost and emission savings.

Author Contributions: Regarding the methodology to evaluate the hourly average $CO_{2eq.}$ intensity, conceptualization was done by L.G. and J.C.; methodology and software were handled by C.S. and J.C.; validation, formal analysis, investigation, data curation, and visualization were the work of S.S. and J.C.; writing and original draft preparation was done by J.C. Regarding the case study on demand response using the $CO_{2eq.}$

intensity as a control signal, J.C. developed the building simulation model, implemented the control, ran the simulations, analyzed the data, visualized the results, and wrote the original draft of the article. Conceptualization and methodology were done by L.G. and J.C. All authors contributed to reviewing and editing the original article. The work was supervised by L.G.

Funding: This research received no external funding.

Acknowledgments: The authors would like to acknowledge IEA EBC Annex 67 "Energy Flexible Buildings", IEA HPT Annex 49 "Design and Integration of Heat Pumps for nZEBs" as well as the Research Centre on Zero Emission Neighbourhoods in Smart Cities (FME ZEN). Sebastian Stinner gratefully acknowledges that his contribution was supported by a research grant from E.ON Stipendienfonds im Stifterverband für die Deutsche Wissenschaft (project number T0087/29896/17).

Conflicts of Interest: The authors declare no conflict of interest.

Nomenclature

BZ	Bidding zone	LCA	Life-cycle assessment
CI	Carbon intensity	LCT	Low $CO_{2eq.}$ intensity threshold
CSC	Control strategy carbon	MPC	Model-predictive control
CSP	Control strategy price	MRIO	Multi-regional input–output
DHW	Domestic hot water	n_{50}	Air changes per hour at 50 Pa pressure difference
DR	Demand response		
EGT	Electricity generation technology	PRBC	Predictive rule-based control
Em.	Emissions	SH	Space heating
ENTSO-E	European Network of TSOs for Electricity	TSO	Transmission system operator
		TSP	Temperature set-point
HCT	High $CO_{2eq.}$ intensity threshold	ZEB	Zero Emission Building
HDS	Heat distribution system	η_{HR}	Heat recovery effectiveness
HVAC	Heating, ventilation, and air-conditioning	max	Maximum
		min	Minimum

Appendix A

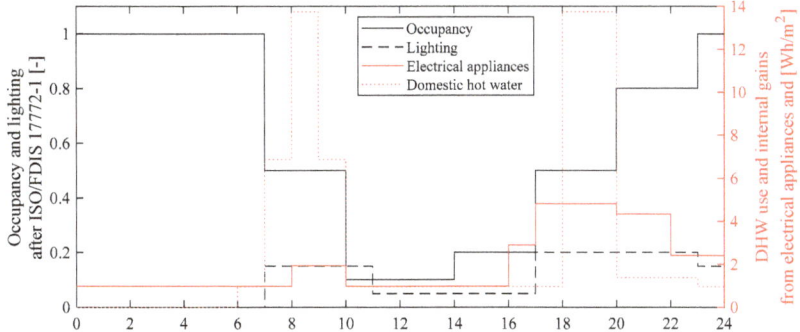

Figure A1. Daily profiles for DHW use and internal heat gains from electrical appliances, occupancy and lighting [15].

Table A1. Comparison of CO_2 factors per electricity generation technology (EGT) from different references [39].

Electricity Generation Technology (EGT)	Emission Factor [$gCO_{2eq.}/kWh_e$]		Name of EGT in Ecoinvent (for Reproduction Purposes)	Emission Factor ($gCO_{2eq.}/kWh_e$)
	IPCC	EEA		Ecoinvent (Applied Here)
Biomass	740	-	Electricity, high voltage {SE}| heat and power co-generation, wood chips, 6667 kW, state-of-the-art 2014	Alloc Rec, U	60 [1]
Fossil brown coal/Lignite	820	-	Electricity, high voltage {DE}| electricity production, lignite	Alloc Rec, U	1240
Fossil coal-derived gas	-	-	Electricity, high voltage {DE}| treatment of coal gas, in power plant	Alloc Rec, U	1667
Fossil gas	490	-	Electricity, high voltage {DK}| heat and power co-generation, natural gas, conventional power plant, 100MW electrical	Alloc Rec, U	529
Fossil hard coal	1001	-	Electricity, high voltage {DK}| heat and power co-generation, hard coal	Alloc Rec, U	1266
Fossil oil	-	-	Electricity, high voltage {DK}| heat and power co-generation, oil	Alloc Rec, U	1000
Fossil oil shale	-	-	No data in Ecoinvent (assumed value)	1000
Fossil peat	-	-	Electricity, high voltage {FI}| electricity production, peat	Alloc Rec, U	1071
Geothermal	38	-	Electricity, high voltage {DE}| electricity production, deep geothermal	Alloc Rec, U	95
Hydro pumped storage	24	-	Electricity, high voltage {NO}| electricity production, hydro, pumped storage	Alloc Rec, U	62
Hydro run-of-river and poundage	24	-	Electricity, high voltage {SE}| electricity production, hydro, run-of-river	Alloc Rec, U	5
Hydro water reservoir	24	-	Electricity, high voltage {NO}| electricity production, hydro, reservoir, alpine region	Alloc Rec, U	8
Marine	24	-	No data in Ecoinvent (assumed value - as wind offshore)	18
Nuclear	12	-	Electricity, high voltage {SE}| electricity production, nuclear, pressure water reactor	Alloc Rec, U	13
Other	-	-	No data in Ecoinvent (assumed value - avg. fossil fuels)	979
Other RES	-	-	No data in Ecoinvent (assumed value - avg. RES)	46
Solar	45	-	Electricity, low voltage {DK}| electricity production, photovoltaic, 3kWp slanted-roof installation, single-Si, panel, mounted	Alloc Rec, U	144
Waste	-	-	Electricity, for reuse in municipal waste incineration only {DK}| treatment of municipal solid waste, incineration	Alloc Rec, U	500
Wind offshore	12	-	Electricity, high voltage {DK}| electricity production, wind, 1-3MW turbine, offshore	Alloc Rec, U	18
Wind onshore	11	-	Electricity, high voltage {DK}| electricity production, wind, 1-3MW turbine, onshore	Alloc Rec, U	14
Imports from all bidding zones with calculated hourly data	-	0		0
Imports from Russia	-	-	Electricity, high voltage {RU}| market for	Alloc Rec, U	862
Imports from Estonia	-	762	Electricity, high voltage {EE}| market for	Alloc Rec, U	1179
Imports from Poland	-	671	Electricity, high voltage {PL}| market for	Alloc Rec, U	1225
Imports from Belgium	-	212	Electricity, high voltage {BE}| market for	Alloc Rec, U	365
Imports from Great Britain	-	389	Electricity, high voltage {GB}| market for	Alloc Rec, U	762

[1] Assuming that biogenic CO_2 is climate neutral; 100-year spruce rotation assumption and dynamic GWP for climate impact of burning wood; does not consider climate impact from CO_2 from wood combustion.

References

1. Aduda, K.O.; Labeodan, T.; Zeiler, W.; Boxem, G.; Zhao, Y. Demand side flexibility: Potentials and building performance implications. *Sustain. Cities Soc.* **2016**, *22*, 146–163. [CrossRef]
2. Clauß, J.; Finck, C.; Vogler-Finck, P.; Beagon, P. Control strategies for building energy systems to unlock demand side flexibility—A review. In Proceedings of the 15th International Conference of the International Building Performance Simulation Association, San Francisco, CA, USA, 7–9 August 2017.
3. International Energy Agency (IEA). *Key World Energy Statistics*; International Energy Agency: Paris, France, 2015; Volume 82, p. 6. [CrossRef]

4. Arteconi, A.; Hewitt, N.J.; Polonara, F. State of the art of thermal storage for demand-side management. *Applied Energy* **2012**, *93*, 371–389. [CrossRef]
5. Chen, Y.; Xu, P.; Gu, J.; Schmidt, F.; Li, W. Measures to improve energy demand flexibility in buildings for demand response (DR): A review. *Energy Build.* **2018**, *177*, 125–139. [CrossRef]
6. Yin, R.; Kara, E.C.; Li, Y.; DeForest, N.; Wang, K.; Yong, T.; Stadler, M. Quantifying flexibility of commercial and residential loads for demand response using setpoint changes. *Appl. Energy* **2016**, *177*, 149–164. [CrossRef]
7. Haider, H.T.; See, O.H.; Elmenreich, W. A review of residential demand response of smart grid. *Renew. Sustain. Energy Rev.* **2016**, *59*, 166–178. [CrossRef]
8. Vandermeulen, A.; van der Heijde, B.; Helsen, L. Controlling district heating and cooling networks to unlock flexibility: A review. *Energy* **2018**, *151*, 103–115. [CrossRef]
9. Vandermeulen, A.; Vandeplas, L.; Patteeuw, D.; Sourbron, M.; Helsen, L. Flexibility offered by residential floor heating in a smart grid context: The role of heat pumps and renewable energy sources in optimization towards different objectives. In Proceedings of the 12th IEA Heat Pump Conference, Rotterdam, The Netherlands, 15–18 May 2017.
10. Lopes, R.A.; Chambel, A.; Neves, J.; Aelenei, D.; Martins, J. A literature review of methodologies used to assess the energy flexibility of buildings. *Energy Procedia* **2016**, *91*, 1053–1058. [CrossRef]
11. Junker, R.G.; Azar, A.G.; Lopes, R.A.; Lindberg, K.B.; Reynders, G.; Relan, R.; Madsen, H. Characterizing the energy flexibility of buildings and districts. *Appl. Energy* **2018**, *225*, 175–182. [CrossRef]
12. Finck, C.; Li, R.; Kramer, R.; Zeiler, W. Quantifying demand flexibility of power-to-heat and thermal energy storage in the control of building heating systems. *Appl. Energy* **2017**, *209*, 409–425. [CrossRef]
13. Péan, T.; Ortiz, J.; Salom, J. Impact of Demand-Side Management on Thermal Comfort and Energy Costs in a Residential nZEB. *Buildings* **2017**, *7*, 37. [CrossRef]
14. Lizana, J.; Friedrich, D.; Renaldi, R.; Chacartegui, R. Energy flexible building through smart demand-side management and latent heat storage. *Appl. Energy* **2018**, *230*, 471–485. [CrossRef]
15. Clauß, J.; Stinner, S.; Sartori, I.; Georges, L. Predictive rule-based control to activate the energy flexibility of Norwegian residential buildings: Case of an air-source heat pump and direct electric heating. *Appl. Energy* **2019**, *237*, 500–518. [CrossRef]
16. Fischer, D.; Bernhardt, J.; Madani, H.; Wittwer, C. Comparison of control approaches for variable speed air source heat pumps considering time variable electricity prices and PV. *Appl. Energy* **2017**, *204*, 93–105. [CrossRef]
17. Alimohammadisagvand, B.; Jokisalo, J.; Kilpeläinen, S.; Ali, M.; Sirén, K. Cost-optimal thermal energy storage system for a residential building with heat pump heating and demand response control. *Appl. Energy* **2016**, *174*, 275–287. [CrossRef]
18. Alimohammadisagvand, B.; Jokisalo, J.; Sirén, K. Comparison of four rule-based demand response control algorithms in an electrically and heat pump-heated residential building. *Appl. Energy* **2018**, *209*, 167–179. [CrossRef]
19. Le Dréau, J.; Heiselberg, P. Energy flexibility of residential buildings using short term heat storage in the thermal mass. *Energy* **2016**, *111*, 991–1002. [CrossRef]
20. Dar, U.I.; Sartori, I.; Georges, L.; Novakovic, V. Advanced control of heat pumps for improved flexibility of Net-ZEB towards the grid. *Energy Build.* **2014**, *69*, 74–84. [CrossRef]
21. Vogler-Finck, P.J.C.; Wisniewski, R.; Popovski, P. Reducing the carbon footprint of house heating through model predictive control—A simulation study in Danish conditions. *Sustain. Cities Soc.* **2018**, *42*, 558–573. [CrossRef]
22. Heidmann Pedersen, T.; Hedegaard, R.E.; Petersen, S. Space heating demand response potential of retrofitted residential apartment blocks. *Energy Build.* **2017**, *141*, 158–166. [CrossRef]
23. Hedegaard, R.E.; Pedersen, T.H.; Petersen, S. Multi-market demand response using economic model predictive control of space heating in residential buildings. *Energy Build.* **2017**, *150*, 253–261. [CrossRef]
24. Dahl Knudsen, M.; Petersen, S. Demand response potential of model predictive control of space heating based on price and carbon dioxide intensity signals. *Energy Build.* **2016**, *125*, 196–204. [CrossRef]
25. Péan, T.Q.; Salom, J.; Ortiz, J. Environmental and Economic Impact of Demand Response Strategies for Energy Flexible Buildings. *Proc. BSO 2018* **2018**, 277–283.

26. Patteeuw, D.; Bruninx, K.; Arteconi, A.; Delarue, E.; D'haeseleer, W.; Helsen, L. Integrated modeling of active demand response with electric heating systems coupled to thermal energy storage systems. *Appl. Energy* **2015**, *151*, 306–319. [CrossRef]
27. De Coninck, R.; Baetens, R.; Saelens, D.; Woyte, A.; Helsen, L. Rule-based demand-side management of domestic hot water production with heat pumps in zero energy neighbourhoods. *J. Build. Perform. Simul.* **2014**, *7*, 271–288. [CrossRef]
28. Killian, M.; Kozek, M. Ten questions concerning model predictive control for energy efficient buildings. *Build. Environ.* **2016**, *105*, 403–412. [CrossRef]
29. Fischer, D.; Madani, H. On heat pumps in smart grids: A review. *Renew. Sustain. Energy Rev.* **2017**, *70*, 342–357. [CrossRef]
30. IEA; Nordic Energy Research. *Nordic Energy Technology Perspectives 2016 Cities, Flexibility and Pathways to Carbon-Neutrality*; Nordic Energy Research: Oslo, Norway, 2016.
31. NordPoolGroup. NordPool 2018. Available online: https://www.nordpoolgroup.com/message-center-container/newsroom/exchange-message-list/2018/q2/nord-pool-key-statistics--may-2018/ (accessed on 13 June 2018).
32. NordPoolGroup. NordPool-About Us 2018. Available online: https://www.nordpoolgroup.com/About-us/ (accessed on 15 October 2018).
33. NordPoolGroup. NordPool Bidding Areas 2018. Available online: https://www.nordpoolgroup.com/the-power-market/Bidding-areas/ (accessed on 28 April 2018).
34. Wiebe, K.S. The impact of renewable energy diffusion on European consumption-based emissions. *Econ. Syst. Res.* **2016**, *28*, 133–150. [CrossRef]
35. Pan, L.; Liu, P.; Li, Z.; Wang, Y. A dynamic input–output method for energy system modeling and analysis. *Chem. Eng. Res. Des.* **2018**, *131*, 183–192. [CrossRef]
36. Guevara, Z.; Domingos, T. The multi-factor energy input–output model. *Energy Econ.* **2017**, *61*, 261–269. [CrossRef]
37. Palmer, G. An input–output based net-energy assessment of an electricity supply industry. *Energy* **2017**, *141*, 1504–1516. [CrossRef]
38. Arteconi, A.; Patteeuw, D.; Bruninx, K.; Delarue, E.; D'haeseleer, W.; Helsen, L. Active demand response with electric heating systems: Impact of market penetration. *Appl. Energy* **2016**, *177*, 636–648. [CrossRef]
39. Clauß, J.; Stinner, S.; Solli, C.; Lindberg, K.B.; Madsen, H.; Georges, L. A generic methodology to evaluate hourly average CO_2eq. intensities of the electricity mix to deploy the energy flexibility potential of Norwegian buildings. In Proceedings of the 10th International Conference on System Simulation in Buildings, Liege, Belgium, 10–12 December 2018. In Proceedings of the 10th International Conference on System Simulation in Buildings, Liege, Belgium, 10–12 December 2018.
40. Energinet. Retningslinjer for Miljødeklarationen for el 2017, 1–16. Available online: https://www.google.com.hk/url?sa=t&rct=j&q=&esrc=s&source=web&cd=1&cad=rja&uact=8&ved=2ahUKEwiFtdmLn8DhAhXsyIsBHUtOChwQFjAAegQIAxAC&url=https%3A%2F%2Fenerginet.dk%2F-%2Fmedia%2FEnerginet%2FEl-RGD%2FRetningslinjer-for-miljdeklarationen-for-el.pdf%3Fla%3Dda&usg=AOvVaw3lFa3nFoF1MpOpXtSk5pRa (accessed on 3 February 2019).
41. Milovanoff, A.; Dandres, T.; Gaudreault, C.; Cheriet, M.; Samson, R. Real-time environmental assessment of electricity use: A tool for sustainable demand-side management programs. *Int. J. Life Cycle Assess.* **2018**, *23*, 1981–1994. [CrossRef]
42. Roux, C.; Schalbart, P.; Peuportier, B. Accounting for temporal variation of electricity production and consumption in the LCA of an energy-efficient house. *J. Clean. Prod.* **2016**, *113*, 532–540. [CrossRef]
43. Tomorrow. CO2-equivalent Model Explanation 2018. Available online: https://github.com/tmrowco/electricitymap-contrib/blob/master/CO2eqModelExplanation.ipynb (accessed on 4 May 2018).
44. Bettle, R.; Pout, C.H.; Hitchin, E.R. Interactions between electricity-saving measures and carbon emissions from power generation in England and Wales. *Energy Policy* **2006**, *34*, 3434–3446. [CrossRef]
45. Hawkes, A.D. Estimating marginal CO_2 emissions rates for national electricity systems. *Energy Policy* **2010**, *38*, 5977–5987. [CrossRef]
46. Corradi, O. Estimating the Marginal Carbon Intensity of Electricity with Machine Learning 2018. Available online: https://medium.com/electricitymap/using-machine-learning-to-estimate-the-hourly-marginal-carbon-intensity-of-electricity-49eade43b421 (accessed on 28 February 2019).

47. Graabak, I.; Bakken, B.H.; Feilberg, N. Zero emission building and conversion factors between electricity consumption and emissions of greenhouse gases in a long term perspective. *Environ. Clim. Technol.* **2014**, *13*, 12–19. [CrossRef]
48. Askeland, M.; Jaehnert, S.; Mo, B.; Korpas, M. Demand response with shiftable volume in an equilibrium model of the power system. In Proceedings of the 2017 IEEE Manchester PowerTech, Manchester, UK, 18–22 June 2017. [CrossRef]
49. Quoilin, S.; Hidalgo Ganzalez, I.; Zucker, A. *Modelling Future EU Power Systems Under High Shares of Renewables The Dispa-SET 2.1 Open-Source Model*; Publications Office of the European Union: Luxembourg, LUXEMBOURG, 2017. [CrossRef]
50. Tomorrow. Electricity Map Europe 2016. Available online: https://www.electricitymap.org/?wind=false&solar=false&page=country&countryCode=NO (accessed on 13 April 2018).
51. Tranberg, B. Cost Allocation and Risk Management in Renewable Electricity Networks. PhD Dissertation, Department of Engineering, Aarhus University, Danske Commodities, Aarhus, Denmark, 2019.
52. IPCC. *Climate Change 2014: Mitigation of Climate Change. Contribution of Working Group III to the Fifth Assessment Report of the Intergovernmental Panel on Climate Change*; Edenhofer, O., Pichs-Madruga, R., Sokona, Y., Farahani, E., Kadner, S., Seyboth, K., Adler, A., Baum, I., Brunner, S., Eickemeier, P., et al., Eds.; Cambridge University Press: Cambridge, UK; New York, NY, USA, 2014.
53. Schlömer, S.; Bruckner, T.; Fulton, L.; Hertwich, E.; McKinnon, A.; Perczyk, D.; Roy, J.; Schaeffer, R.; Sims, R.; Smith, P.; et al. Annex III: Technology-specific cost and performance parameters. In *Climate Change 2014: Mitigation of Climate Change. Contribution of Working Group III to the Fifth Assessment Report of the Intergovernmental Panel on Climate Change*; Edenhofer, O., Pichs-Madruga, R., Sokona, Y., Farahani, E., Kadner, S., Seyboth, K., Adler, A., Baum, I., Brunner, S., Eickemeier, P., et al., Eds.; Cambridge University Press: Cambridge, UK; New York, NY, USA, 2014.
54. Ecoinvent. 2016. Available online: www.ecoinvent.org (accessed on 9 April 2018).
55. EEA. Overview of Electricity Production and Use in Europe 2018. Available online: https://www.eea.europa.eu/data-and-maps/indicators/overview-of-the-electricity-production-2/assessment (accessed on 12 April 2018).
56. Bachmann, C.; Roorda, M.J.; Kennedy, C. Developing a multi-scale multi-region input–output model developing a multi-scale multi-region input–output model. *Econ. Syst. Res.* **2015**, *27*, 172–193. [CrossRef]
57. Kristjansdottir, T.F.; Houlihan-Wiberg, A.; Andresen, I.; Georges, L.; Heeren, N.; Good, C.S.; Brattebø, H. Is a net life cycle balance for energy and materials achievable for a zero emission single-family building in Norway? *Energy Build.* **2018**, *168*, 457–469. [CrossRef]
58. Brattebø, H.; O'Born, R.; Sartori, I.; Klinski, M.; Nørstebø, B. Typologier for Norske Boligbygg—Eksempler på Tiltak for Energieffektivisering. Available online: https://brage.bibsys.no/xmlui/handle/11250/2456621 (accessed on 3 February 2019).
59. Goia, F.; Finocchiaro, L.; Gustavsen, A. Passivhus Norden|Sustainable Cities and Buildings The ZEB Living Laboratory at the Norwegian University of Science and Technology: A Zero Emission House for Engineering and Social science Experiments. In Proceedings of the 7th Nordic Passive House Conference, Copenhagen, Denmark, 20–21 August 2015.
60. Halvgaard, R.; Poulsen, N.; Madsen, H.; Jørgensen, J. Economic model predictive control for building climate control in a smart grid. In Proceedings of the 2012 IEEE PES Innovative Smart Grid Technologies (ISGT), Washington, DC, USA, 16–20 January 2012; pp. 1–6. [CrossRef]
61. Madsen, H.; Parvizi, J.; Halvgaard, R.; Sokoler, L.E.; Jørgensen, J.B.; Hansen, L.H.; Hilger, K.B. Control of Electricity Loads in Future Electric Energy Systems. In *Handbook of Clean Energy Systems*; Wiley: Hoboken, NJ, USA, 2015.
62. Johnsen, T.; Taksdal, K.; Clauß, J.; Georges, L. Influence of thermal zoning and electric radiator control on the energy flexibility potential of Norwegian detached houses. In Proceedings of the CLIMA 2019, Bucharest, Romania, 26–29 May 2019.
63. EQUA. EQUA Simulation AB 2015. Available online: http://www.equa.se/en/ida-ice (accessed on 16 October 2015).
64. EQUA Simulation AB; EQUA Simulation Finland Oy. *Validation of IDA Indoor Climate and Energy 4.0 with Respect to CEN Standards EN 15255-2007 and EN 15265-2007*; EQUA Simulation Technology Group: Solna, Sweden, 2010.

65. EQUA Simulation AB. *Validation of IDA Indoor Climate and Energy 4.0 build 4 with Respect to ANSI/ASHRAE Standard 140-2004*; EQUA Simulation Technology Group: Solna, Sweden, 2010.
66. Achermann, M.; Zweifel, G. *RADTEST—Radiant Heating and Cooling Test Cases Supporting Documents*; HTA Luzern: Luzern, Switzerland, 2003.
67. Bring, A.; Sahlin, P.; Vuolle, M. *Models for Building Indoor Climate and Energy Simulation Models for Building Indoor Climate and Energy Simulation 1. Executive Background and Summary*; Department of Building Sciences, KTH: Stockholm, Sweden, 1999.
68. Sahlin, P. *Modelling and Simulation Methods for Modular Continuous Systems in Buildings*; Royal Institute of Technology: Stockholm, Sweden, 1996.
69. Kipping, A.; Trømborg, E. Hourly electricity consumption in Norwegian households—Assessing the impacts of different heating systems. *Energy* **2015**, *93*, 655–671. [CrossRef]
70. Fischer, D.; Lindberg, K.B.; Madani, H.; Wittwer, C. Impact of PV and variable prices on optimal system sizing for heat pumps and thermal storage. *Energy Build.* **2017**, *128*, 723–733. [CrossRef]
71. SN/TS3031:2016. *Bygningers Energiytelse, Beregning av Energibehov og Energiforsyning*; Standard Norge: Oslo, Norway, 2016.
72. Ahmed, K.; Akhondzada, A.; Kurnitski, J.; Olesen, B. Occupancy schedules for energy simulation in new prEN16798-1 and ISO/FDIS 17772-1 standards. *Sustain. Cities Soc.* **2017**, *35*, 134–144. [CrossRef]
73. ISO17772-1. *Energy Performance of Buildings—Indoor Environmental Quality—Part1: Indoor Environmental Input Parameters for the Design and Assessment of Energy Performance of Buildings*; International Organization for Standardization: Geneva, Switzerland, 2017.
74. OpenStreetMap. Shiny Weather Data 2017. Available online: https://rokka.shinyapps.io/shinyweatherdata/ (accessed on 20 May 2017).
75. NordPoolGroup. NordPool Historical Market Data. Available online: https://www.nordpoolgroup.com/historical-market-data/ (accessed on 16 April 2018).
76. Standard Norge. *NS-EN 15251:2007 Indoor Environmental Input Parameters for Design and Assessment of Energy Performance of Buildings Addressing Indoor Air Quality, Thermal Environment, Lighting and Acoustics*; Standard Norge: Oslo, Norway, 2007.

© 2019 by the authors. Licensee MDPI, Basel, Switzerland. This article is an open access article distributed under the terms and conditions of the Creative Commons Attribution (CC BY) license (http://creativecommons.org/licenses/by/4.0/).

Article

Novel Proposal for Prediction of CO_2 Course and Occupancy Recognition in Intelligent Buildings within IoT

Jan Vanus *,†,‡, Ojan M. Gorjani ‡ and Petr Bilik

Department of Cybernetics and Biomedical Engineering, Faculty of Electrical Engineering and Computer Science, VSB - Technical University of Ostrava, 70833 Ostrava-Poruba, Czech Republic; ojan.majidzadeh.gorjani@vsb.cz (O.M.G.); petr.bilik@vsb.cz (P.B.)
* Correspondence: jan.vanus@vsb.cz; Tel.: +420-597-325-856
† Current address: 17. listopadu 2172/15, 70800 Ostrava, Czech Republic.
‡ These authors contributed equally to this work.

Received: 31 October 2019; Accepted: 23 November 2019; Published: 28 November 2019

Abstract: Many direct and indirect methods, processes, and sensors available on the market today are used to monitor the occupancy of selected Intelligent Building (IB) premises and the living activities of IB residents. By recognizing the occupancy of individual spaces in IB, IB can be optimally automated in conjunction with energy savings. This article proposes a novel method of indirect occupancy monitoring using CO_2, temperature, and relative humidity measured by means of standard operating measurements using the KNX (Konnex (standard EN 50090, ISO/IEC 14543)) technology to monitor laboratory room occupancy in an intelligent building within the Internet of Things (IoT). The article further describes the design and creation of a Software (SW) tool for ensuring connectivity of the KNX technology and the IoT IBM Watson platform in real-time for storing and visualization of the values measured using a Message Queuing Telemetry Transport (MQTT) protocol and data storage into a CouchDB type database. As part of the proposed occupancy determination method, the prediction of the course of CO_2 concentration from the measured temperature and relative humidity values were performed using mathematical methods of Linear Regression, Neural Networks, and Random Tree (using IBM SPSS Modeler) with an accuracy higher than 90%. To increase the accuracy of the prediction, the application of suppression of additive noise from the CO_2 signal predicted by CO_2 using the Least mean squares (LMS) algorithm in adaptive filtering (AF) method was used within the newly designed method. In selected experiments, the prediction accuracy with LMS adaptive filtration was better than 95%.

Keywords: KNX; Neural Network (NN); Multilayer Perceptron (MLP); Random Tree (RT); Linear Regression (LR); Cloud Computing (CC); Internet of Things (IoT); LMS (Least Mean Squares) Adaptive filter (AF); gateway; monitoring; occupancy; prediction; IBM SPSS; Intelligent Buildings (IB); energy savings

---

1. Introduction

Until recently, automated smart wiring was the privilege of commercial buildings. However, more and more people want to take advantage of the smart installation options in their homes. There are already many companies on the market involved in smart installations. The prices of such systems with high level of reliability are becoming more affordable in relation to what the customer gains. One of the customer's motivations for monitoring and controlling the operational and technical conditions of the building can be tracking the events in the building. By measuring various physical quantities, for example, a window left open, unauthorized entry or the presence of people in a building can be inferred. Another motivation for space monitoring can be care for disabled people or seniors. The IoT is increasingly being used to monitor and control the operational status of a building today, which greatly facilitates remote communication with the building from anywhere in the world. This is an area that is relatively new and constantly developing. However, multinational companies such as Google, IBM, Amazon, etc. are becoming increasingly important players in the IoT world. They come with their services, which they run on their servers and offer to customers. These services are generally referred to as Cloud Computing (CC). IB and IoT are two distinct concepts that are closely related, as IB requires real-time interaction with different technological processes. The interactions take place within the system on the basis of a programmed model or directly with the users, and IoT inherently helps. This research is focused on current trends in building automation and monitoring of operational and technical conditions in IB within IoT. The authors of [1–23] focused on new approaches for home automation solutions within IoT. The authors investigated the new possibilities and trends of IB and smart home (SH) automation solutions with appropriate use of IBM IoT tools [2–6], with ensuring appropriate security in IB and SH automatization [7–9], and with the possible overlap of applications within the Smart Cities (SC) [10,11] and Smart Home Care (SHC) [12] platforms.

The above-stated articles contain technical terms such as SH, IB, IoT, communication technology, data transfer in connection with communication protocols, CC, "Cloud of Things". etc. SH is an automated or intelligent home. This is an expression for a modern living space in contrast to existing conventional buildings. SH is characterized by a high degree of automation of operational and technical processes, which are operated by people in standard households. These include lighting control, air temperature and air-conditioning control, kitchen equipment, security technology, door and window control, energy management, multimedia, and consumer electronics control, among many others. SH connects a number of automated systems and appliances with interoperability among the decentralized or centralized management technologies used. One of the main objectives is to increase the required level of user comfort while saving energy costs [24]. SH should, of course, also offer the possibility of an additional configuration of functionality.

IoT is a network of interconnected devices that can exchange information with each other and be mutually interactive. Each of these devices must be clearly identifiable and addressable. Standardized communication protocols are used for communication. The aim is to create autonomous systems that are able to perform the expected activities completely independently on the basis of data obtained from the equipment. In practice, this means that the more data are collected, the better they can be analyzed, which may be useful in IoT implementation in specific fields [25]. In his article on the introduction to the IoT, Vojacek described IoT architecture as a model consisting of three basic elements [26]. The first block includes the "things" themselves, which are all devices connected to the network, whether cable or wireless, providing completely independent data. The second block represents the network, which serves as a means of communication between the things themselves and the control system. The third block includes data processing, which can be on the cloud, i.e. a remote server on which the data are further processed. The development of the Internet of Things is now mainly driven by the rapid development of new IoT-enabled devices and their ever-decreasing price. By 2020, it is estimated that the number of connected IoT devices may exceed 30 billion [27]. However, this figure varies for different sources, but it is certain that the numbers go to tens of billions. This brings about the issue of Internet addressing, because addresses from the IPv4 address space were depleted in 2011, but, with the advent of IPv6, address depletion should not occur as there are 2^{128} addresses in this address space, which is a huge number; for your information, it is about 66 trillion IP addresses per square centimeter of the Earth's surface, including the oceans [28,29]. The Internet of Things may be used in many sectors of human activities. Smart Cities (SC) and Smart Grids (SG) can also be included in the current IoT application areas.

Obviously, it is necessary to transfer the data collected effectively. When it comes to wireless data transmissions, the most important parameters include especially range, energy intensity, the security of transmission, format, intensity of data processing, and transmission speed. Competition in this segment is relatively high [30–33]. Several communication protocols, such as MQTT, Constrained Application Protocol (CoAP), and Extensible Messaging and Presence Protocol (XMPP), are available for IoT data transmission [34]. In this work, the MQTT protocol is used. Aazam et al. described the topic of CC in the work "Cloud of Things: Integrating the Internet of Things and CC and the issues involved" [35]. This is a growing trend in IT technologies, which is also used in IoT. The characteristic feature of CC is the provision of services, software, and hardware of servers that are accessible to the users via the Internet from anywhere in the world. The cloud technology provider will enable the user to use their computing power, data storage, and the software offered. For the users, these services are appealing because they may not have the knowledge of the intrinsic functionality of the hardware and software leased. It also offers high-level data security or a high-quality, user-friendly web interface. Depending on the use of services, CC can be divided into groups [36]: IaaS (Infrastructure as a Service), SaaS (Software as a Service), PaaS (Platform as a Service), and NaaS (Network as a Service). The basic idea behind CC is to move application logic to remote servers. The individual devices in SH with an Internet connection can be connected directly to the cloud. The data received are collected and processed here. Based on the data evaluated, there is a feedback interaction with the devices in SH. It is often necessary to connect a device that does not have an Internet connection to the system. This can be performed using a gateway that is connected to the Internet. This gateway does not have to mediate communication to only one device. Often, cloud services also offer their own applications for visualization of the data processed and their storage in the database. There is a huge amount of data that can be used for other complex calculations, such as machine learning. Leading providers of these cloud technologies include IBM, which, inter alia, offers IBM Watson IoT, and Amazon with its Amazon Web Services.

Similar to this study, the obtained results in [12,18,37,38] showed that the accuracy of CO_2 prediction can possibly exceed 90%. Khazaei [39] employed a multilayer perceptron neural network for the purpose of indoor CO_2 concentration levels, where relative humidity and temperature were used as inputs. The most accurate model (based on the calculated mean-square-error (MSE)) was five steps ahead of the reference signal. On average, the difference between reference and prediction signals was less than 17 ppm. Wang [40] used a recurrent neural network (RNN) based dynamic backpropagation (BP) algorithm model with historical internal inputs. The models were developed to predict the temperature and humidity of a solar greenhouse. The obtained results demonstrate that the RNN-BP model provides reasonably good predictions. One of the benefits of the prediction of CO_2 concentration levels is low-cost indirect occupancy monitoring. Szczurek et al. proposed a method to provide occupancy determination in intermittently occupied space in real-time and with predefined temporal resolution [41]. Galda et al. discovered insignificant difference between a room with and without plants, while measuring values of CO_2 concentration, inner temperature, and relative humidity [42]. In our article, we used KNX technology for IB–IoT connectivity by means of the MQTT protocol within the IBM Watson IoT platform. The target of this article is to propose and verify a novel method for predicting CO_2 concentration (ppm) course from the temperature (T_{indoor} (°C)) and relative humidity (rH_{indoor} (%)) courses measured with operational accuracy to monitor occupancy in SH using conventional KNX operating sensors with the least investment cost. To achieve the goal, it is necessary to program the required KNX modules in the ETS 5 SW tool. Next, a software application is created to ensure connectivity between the KNX technology and the IoT IBM platform for real-time storage and visualization of the data measured. To predict the course of CO_2 concentration, it is necessary to verify and compare mathematical methods (linear regression (LR), random trees (RT), and multilayer perceptron's (MLP)), and to select the method with the best results achieved. To increase the accuracy of CO_2 prediction, it is necessary to implement the LMS AF in an application for suppressing the additive noise from the CO_2 signal predicted and to assess the advantages and disadvantages of using AF in the newly proposed method. The proposed method was verified using the measured data in springtime with day-long and week-long intervals (2–10 May 2019). The proposed method can save investment costs in the design of large office buildings. Recognition of occupancy of monitored premises in the building leads to the optimization of IB automation in connection with the reduction of operating costs of IB.

2. Materials and Methods

2.1. Building Automation—KNX Technology

KNX technology is a worldwide standard in the field of wiring of intelligent buildings and their automation. It is a decentralized solution. KNX technology provides the interconnection of KNX modules using the Twisted pair (TP) communication medium. The devices used in this work are an MTN6005-0001 sensor of CO_2, temperature, and air humidity (technical parameters: current consumption from the bus, max. 10 mA; ambient temperature, −5 to 45 °C; measuring range for CO_2, 300–9999 ppm; measuring range for temperature: 0–40 °C; and measuring range for humidity, linear 20–100%). A laboratory (EB312) on the premises of the building of the new FEI at VSB Technical University of Ostrava was used as the experimental location. This location often holds educational classes or it is visited by staff and researchers. Furthermore, in this laboratory, there are bus buttons MTN6172 and MTN6171 for controlling the lighting and switching off (disconnecting) the sockets. In the switchboard in the corridor, there is the switching actuator MTN 649204 and, for lighting control, the dimming actuator KNX/DALI gate MTN680191. Each device in the KNX installation has an identifiable individual address and, together with other devices, forms a function block using a group address. As a whole, these components have been parameterized into a functional unit in the ETS5 configuration software offered by KNX associations (Figure 1).

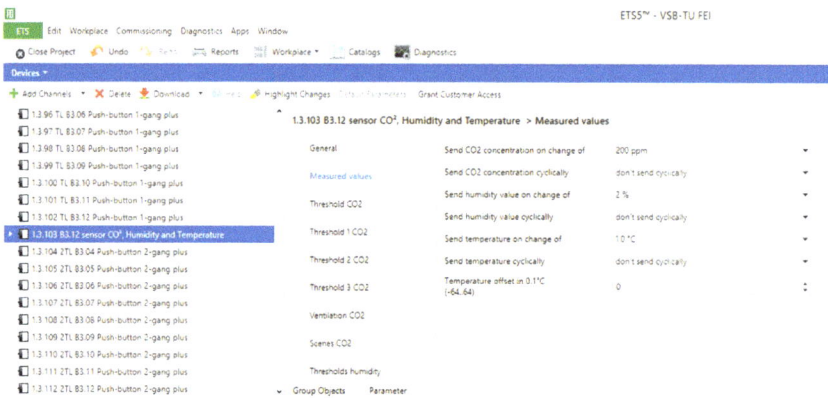

Figure 1. MTN6005-0001 sensor configuration.

As mentioned above, the measurement was performed in laboratory EB312, on the premises of the building of the new FEI at VŠB TU Ostrava. Therefore, the necessary structure of the building was created using the ETS 5 SW tool (Figure 2). The topology in the KNX installation consists of the backbone, the main line, and five secondary lines (Figure 3).

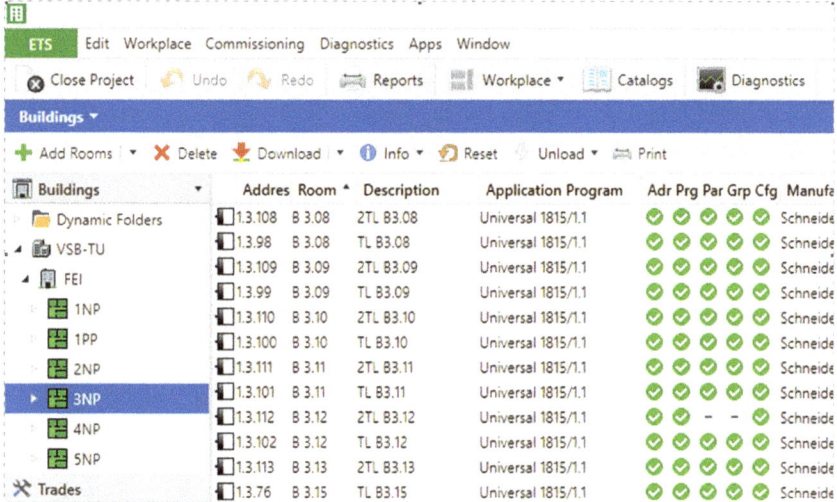

Figure 2. The structure of the building of New FEI VŠB-TU Ostrava created in the ETS 5 SW tool.

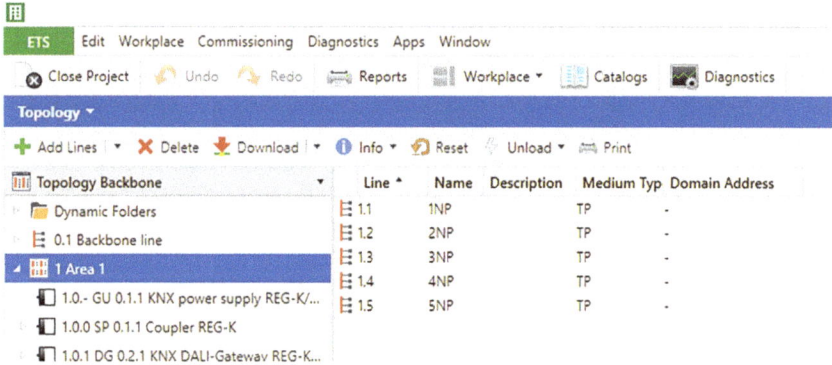

Figure 3. Building Topology of New FEI VŠB-TU Ostrava created in the ETS 5 SW tool.

The connection between the lines is provided by a bus coupler, which allows connection of up to 64 KNX modules. To ensure proper communication between the individual KNX modules by means of the KNX telegram, a group address, which uniquely determines the functionality of the individual operational and technical functions, is defined (Figure 4) [43]. The topology in the KNX installation consists of the backbone, the main line, and five secondary lines (Figure 3).

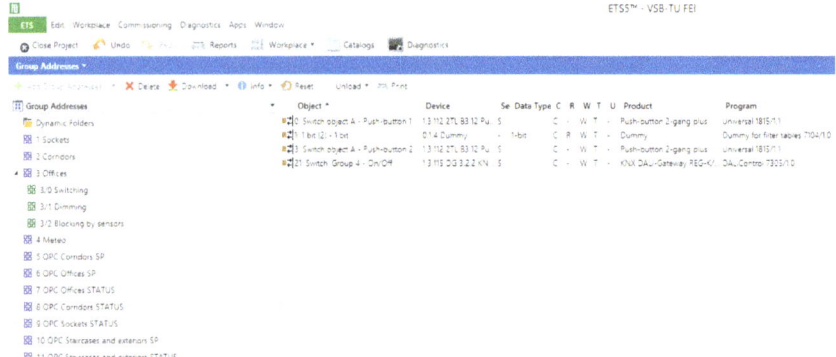

Figure 4. The structure of the group addresses of the building of New FEI VŠB-TU Ostrava created in the ETS 5 SW tool.

KNX Association, in response to the expanding IoT trend, introduced its own solution—KNX Web Services. This solution removes obstacles to accessing KNX as part of the Internet of Things. It enables reliable integration of a large installation in which multiple suppliers and manufacturers are installed. KNX Web Services focus on the existing web services, such as oBIX, OPC UA, and BACnet-WS. Web services are standalone modular software components that can be described, published, and activated via the web [44,45]. It is necessary to use a suitable interface between the KNX installation and the IP network. On the Internet side, control devices, such as mobile devices, web applications, etc., that communicate with the interface via web services can be connected. For the implementation, however, it is necessary to know the configuration of the KNX installation as it was created by the ETS program. This is conducted by means of an ETS software supplement called ETS Exporter App [46]. There are several applications on the market that are able to communicate with KNX installations. The ones that are known include KNX DashBoard, KNXWeb2 or LinKNX [47], and Home In Hand [48]. In this work, the connection of the KNX technology and the IBM cloud technology is provided using SW that we developed, which enables communication between the IBM Watson IoT service and the KNX smart installation (Figure 5). MQTT protocol is used as a communication protocol.

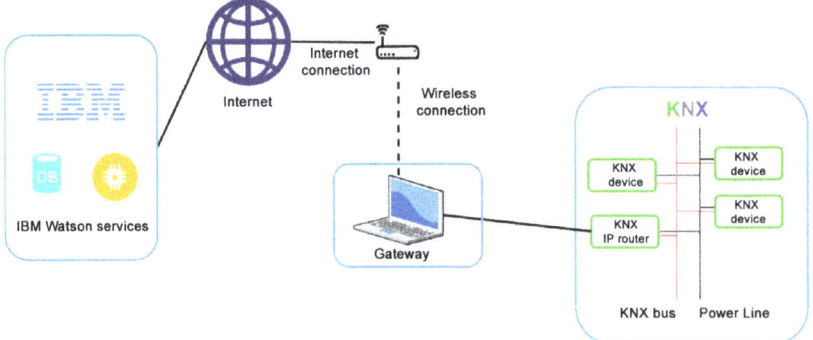

Figure 5. A block diagram of the KNX technology SW connection created in the New FEI VŠB-TU Ostrava building with the IBM IoT Watson platform.

Microsoft Visual Studio 2017 was chosen as the development environment. The .NET framework is used as the development framework and C# is used both from the KNX Association and from IBM. A personal computer (PC) with the Windows 10 operating system is used as a hardware solution for running the developed SW. The IBM Watson IoT platform web service is used to access the KNX installation within IoT. The function and the requirement of this software are to run continuously on the selected device and to maintain a connection for two-way communication, respectively. This software is implemented to serve as access for an independent application that monitors the presence of people in the room. Room EB312 located in the FEI building at VŠB TU Ostrava is monitored. Libraries for communication with the KNX bus, the IBM Watson IoT platform, the Cloudant database, and the JSON files were used as third-party libraries. The Falcon SDK library was used to establish connectivity with the KNX installations using the Manufacturer SDK version. The IBM Watson IoT-CSharp library was used to interact with the IBM Watson IoT platform. The individual KNX modules communicate with IoT IBM using a KNXnet/IP router. The data obtained from the KNX installation by the software implemented must be transferred for further processing in real-time. To do this, the IBM Watson IoT platform service controlled through the Web interface is used. This is a PaaS service. This service acts as a broker, an intermediary for real-time communication between the applications and the IoT devices using the MQTT protocol. The software implemented acts as both a publisher and a subscriber because it is a gateway to the KNX installation from the Internet and allows two-way communication. Multiple publishers and subscribers can be logged into this service. Another SW application we developed is the Console Gateway software, which is implemented as a console application (Figure 6). Its function is to maintain a continuous connection between the KNX installation and the IBM Watson IoT platform. During this connection, the application monitors the KNX bus and captures and sends the predetermined telegrams to the Watson IoT platform. The console application was created primarily for faster implementation and high reliability. The Console Gateway SW tool created was deployed on a laboratory PC that runs continuously as long as it is necessary to maintain a connection between the cloud service and the KNX installation. Upon start-up, it informs the user of the success or failure of the connection to the KNXnet/IP router and the Watson IoT cloud service. If both are OK, the user is prompted to start the data transfer. Then, the console writes data that are sent to the cloud.

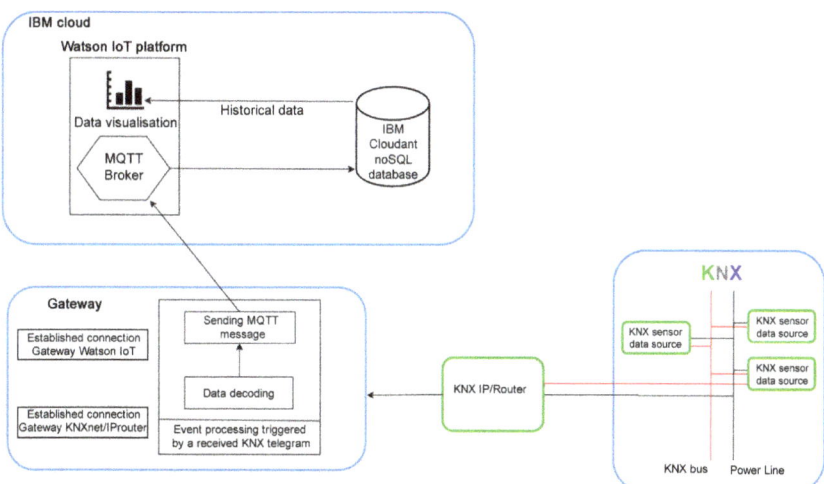

Figure 6. A diagram of the ConsoleGateway SW solution implemented for online data transfer from the KNX technology to IoT IBM Watson.

The official .Net framework Falcon library provided by the KNX association is used for communication with the KNXnet/IP router. This library allows establishing connections, sending commands to the KNX installation, monitoring communication on the KNX bus, and requesting information about KNX devices. The connection to the router is established in the tunneling mode. To establish a connection, it is necessary to configure the connection parameters, namely the IP address and the port number. When the program is closed, it is necessary to terminate the connection.

2.2. Predictive Analysis

Predictive models are based on variables that are most likely to influence the outcome. These variables are also known as predictors [49]. Predictive modeling has a wide range of applications such as weather forecasting, business, Bayesian spam filters, advertising and marketing, fraud detection, etc. The formulation of the statistical model requires the collection of data for relevant predictors. Ahmad et al. compared different methods of predictive modeling for solar thermal energy systems such as random forest, extra trees, and regression trees [50]. The predictive analysis performed in this article falls under the category of machine learning with a supervised learning strategy. In simple words, machine learning is the process of computers learning and recognizing patterns based on the data, which helps the computer program to make intelligent decisions. Supervised learning is one of the methods used in machine learning. Usually, in this method, a set of solved (labeled) examples are presented to the machine for training, which helps the machine to establish the pattern between the problem and the answer. Once the pattern is established, the machine can solve similar problems on its own [51]. Kachalsky et al. explained an application of supervised learning techniques in a computer card game that resulted in a high percentage of won games by the computer [52]. Unsupervised learning is based on similar principles such as clustering in data mining, which uses clustering to discover classes within the data. This method can find a pattern in unsolved (unlabeled) examples that may not have a semantic meaning. Nijhawan et al. used unsupervised learning for classification of land cover by taking advantage of relative positions of pixels in the image of the land [53]. Another class of machine learning techniques is semi-supervised learning, which uses labeled examples to learn class models and unlabeled examples to refine the boundaries between classes. Liu et al. demonstrated an example of the application of semi-supervised learning combined with multitask learning in landmine detection [54]. The fourth machine learning approach is active learning, which requires the active role of the user in the learning process by asking the user to label an unlabeled example. This improves the quality of analysis by taking advantage of human knowledge [51].

In [37], the authors used the radial basis functions (feedforward NN) for CO_2 prediction. They examined the performance of the developed models based on the number of neurons and interval lengths. The results indicate that the day-long interval lengths are the most suitable for training the radial basis function. This article compares the performance of three different statistical methods (linear regression (LR), random trees (RT), and multilayer perceptron (MLP)) for predictions of the CO_2 concentration level in an intelligent administrative building (New FEI Building) in VSB TU Ostrava (Figure 7). As explained above, the authors developed a software tool that is capable of obtaining and recording the values of temperature, humidity, and CO_2 concentration through the KNX Platform. The obtained data were divided into various intervals that were processed using the IBM SPSS Modeler software tool.

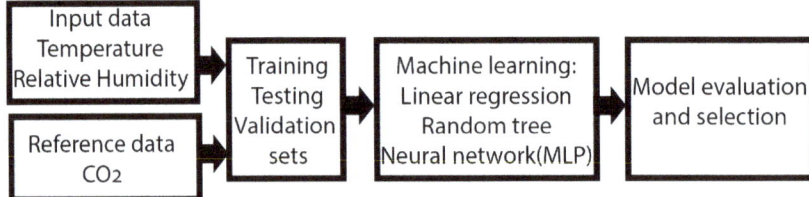

Figure 7. Predictive analysis concept.

The process of CO_2 predictions can be broken into the following parts:

1. Data collection
2. Pre-processing of the collected data
3. Predictions using IBM SPSS

4. Evaluation and comparison of the obtained results

The data collection was performed by a personal computer that was running the developed KNX-IBM gateway software. As explained above, the developed software records and transmits any message that is traveling through the KNX bus (twisted pair). The developed software creates a new database entry for every KNX bus message that contains a predefined group address. Using the developed method, the data collection rate can vary from one to ten samples per minute (approximately 7000 samples per 24 h). The obtained data intervals were normalized (feature scaling or min-max normalization) using Microsoft Excel. In the third step of the implementation, the IBM SPSS Modeler software was used for the development of predictive models. Figure 8 shows an example of a data stream in IBM SPSS Modeler 18. The data are imported using an Excel node. The filter and type nodes select relevant data and assign correct data types (continuous, categorical, etc.). Additionally, the type node can assign which parameters are used as predictors (input) and which as predictions (target). In IBM SPSS Modeler, K-fold, V-fold, N-fold, and partitioning methods are commonly used as validation methods. For optimization purposes, it is recommended to use a partitioning method for large datasets [55]. The partitioning method randomly divides (the ratio of this division can be defined) the input datasets into three parts of training, testing, and validation. In this experiment, 40% of the input data were used for training of the models and 30% of the data for testing partition, which were mainly used for selecting the most suitable model and prevention of overfitting. The validation partition (30% of the input data) was used to determine how well the models truly perform. Using this partition, the developed models perform predictions using only predictors. The obtained results were compared with the reference signal. The partitioned data were fed directly to statistical predictive models (linear regression, random tree, and neural networks (MLP)). The resulting predictions can be exported to Excel files or analyzed using built-in functions such as plots and analysis nodes.

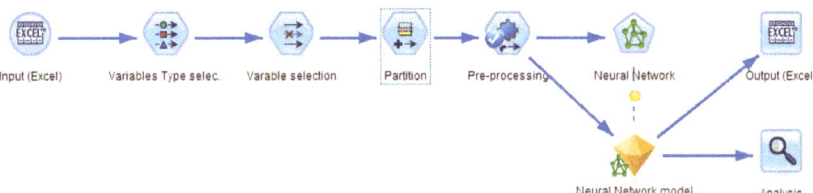

Figure 8. Representation of a data stream in IBM SPSS.

2.2.1. Linear Regression (LR)

LR is one of the oldest and most commonly used algorithms in supervised machine learning [56]. Generally, it estimates the coefficients of the linear equation. These coefficients involve one or more independent variables that can best predict the value of the dependent variable [56–61]. Equation (1) represents a simple mathematical representation of LR, where it evaluates the influence of the variable X on y (α and β are coefficients and ε is error variable). If the response y is influenced by more than one predictor variable, the regression function can be modeled with the function in Equation (2) ($\beta_0, \beta_1, \beta_2$, and β_k are regression coefficients). Equations (3)–(6) are matrix representations of y, X, β and ε, respectively [57,58].

$$y = \alpha + \beta X + \varepsilon, \tag{1}$$

$$y = \beta_0 + \beta_1 X_1 + \beta_2 X_2 + ... + \beta_K X_K + \varepsilon, \tag{2}$$

$$y = \begin{pmatrix} y_1 \\ y_2 \\ \vdots \\ y_n \end{pmatrix}, \quad (3)$$

$$X = \begin{pmatrix} 1 & x_{11} & \cdots & 1x_{1p} \\ 1 & x_{21} & 1x_{2p} \\ \vdots & \vdots & \ddots & \vdots \\ 1 & x_{n1} & \cdots & x_{np} \end{pmatrix}, \quad (4)$$

$$\beta = \begin{pmatrix} \beta_0 \\ \beta_1 \\ \vdots \\ \beta_p \end{pmatrix}, \quad (5)$$

$$\varepsilon = \begin{pmatrix} \varepsilon_1 \\ \varepsilon_2 \\ \vdots \\ \varepsilon_n \end{pmatrix}. \quad (6)$$

Few available methods can be used to create a variety of regression models from the same sets of variables. These methods specify how independent variables are entered into the analysis. A few of the common methods are enter (regression), stepwise, backward elimination, and forward selection [58–61]. In entering configuration, all variables in a block are entered in a single step. However, in the stepwise method, at each step, the independent variable with the smallest probability is entered, given that it is not in the equation and the probability is sufficiently small. The variables already in the regression equation are removed if their probability becomes sufficiently large. The stepwise method terminates if there is no need for inclusion or removal of variables [62,63].

In the backward elimination, all the variables are entered in the equation, and then sequentially removed. The variable with the smallest partial correlation with the dependent variable is considered first for removal. The process repeats for the remaining variables until there are no variables in the equation that satisfy the removal conditions [64]. The forward selection is a stepwise variable selection procedure where variables are sequentially entered into the model. The variable with the largest positive or negative correlation with the dependent variable is considered first for entry to the equation. However, this variable should satisfy the entry conditions. The process repeats for remaining variables that are not in the equation. Once there are no variables that satisfy the entry condition, the process eliminates [65]. All variables must pass the tolerance criterion to be entered in the equation, regardless of the entry method specified. All selected independent variables are added to a single regression model. However, it is possible to specify different entry methods for different subsets of variables [66].

2.2.2. Random Tree (RT)

Decision trees can be used as a predictive model to drive certain conclusions from the evaluation of data. Decision trees can be mainly divided into two types of classification and random trees. Classification trees determine which class the data belong to. Random trees are a type of decision tree that provides possibilities of a continuous variable at the output [67]. The RT is a modern and sophisticated method that is a tree-based classification (by majority voting) and regression tree methodology (by average). In general, it allows building an ensemble model that consists of multiple decision trees. This method is capable of providing supervised learning for categorical or continuous target variables. In general, it uses groups of classification or regression trees (C&R trees) and randomness to make predictions that are robust when applied to new observations. The C&R

trees are binary; each field splits the results into two branches, and the categories are grouped into two groups based on the inner splitting conditions [68,69].

The RT uses recursive partitioning to split the training records into segments with similar output field values. The algorithm initially starts by finding the best split from the available input variables. These splits are evaluated from the resulting reduction in an impurity index by the split. The binary split defines two subgroups, where each is split into two other binary subgroups and the splitting process continuous for all splits until one of the stopping conditions is satisfied [70–73]. The RT model offered by the IBM SPSS Modeler software tool uses bootstrap sampling with replacement to generate sample data [73]. The RT method is much less likely to overfit due to the use of bagging and field sampling. As the second feature in this method, at each split of the tree, only a sampling of the input fields is considered for the impurity measure. Therefore, it is often used as a robust method when dealing with large datasets and many fields are required.

2.2.3. Multilayer Perceptron (MLP) Neural Network (NN)

The neural network resembles the brain in some aspects: first, the knowledge is obtained by the learning process and, second, the interneuron connection strengths known as synaptic weights are used to store the knowledge [74]. In other words, the NN can be defined as a massively parallel distributed processor which can store experiential knowledge and make it available for use [75]. Neural networks are the preferred tool for many predictive applications because of their power, flexibility, and ease of use. Predictive NNs are particularly useful in applications where the underlying process is complex. Zarei et al. used NNs to study the effective parameters of a greenhouse on the freshwater production [76]. Moosavi et al. used an NN to predict CO_2- foam injection in a Laboratory [77]. The multilayer perceptron (MLP) is a feedforward NN. It uses a supervised learning strategy to create predictive applications, in the sense that the model-predicted results can be compared against known values of the target variables. In addition to input and output layers, the MLP can contain multiple hidden layers, each of which may contain multiple neurons (Figure 9). The IBM SPSS Modeler Algorithm [78] guide provides the following mathematical expression for the NN (MLP).

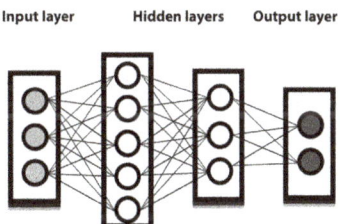

Figure 9. Representation of a Multilayer Perceptron NN.

The general architecture for MLP networks is:

Input layer: $j_0 = p$ units, $a_{0:1}, ..., a_{0:j_0}$; with

$$a_{0:j_0} = x_j,\qquad(7)$$

where j is the number of neurons in the layer and X is the input.

***i*th hidden layer:** j_i units, $a_{i:1}, ..., a_{i:j_i}$; with

$$a_{i:k} = \gamma_i(C_{i:k})\qquad(8)$$

and
$$C_{i:k} = \sum_{j=0}^{J_i-1} \omega_{I:j_1,k} a_{i-1:j},\qquad(9)$$

where $a_{i-1:0} = 1$, γ_i is activation function for layer i and $\omega_{I:j_1,k}$ is weight leading from layer $i-1$, unit j to layer i, unit k.

Output layer: $j_I = R$ units, $a_{I:1}, ..., a_{I:J_I}$; with

$$a_{I:k} = \gamma_I(C_{I:k})\qquad(10)$$

and

$$C_{I:k} = \sum_{j=0}^{J_1} \omega_{I:j_1,k} a_{i-1:j},\qquad(11)$$

where $a_{i-1:0} = 1$.

2.2.4. Evaluation Methods

For evaluation of the obtained results, the following parameters were used:

Mean Absolute Error (MAE): It measures the difference between two continuous variables, so that difference is always positive. It is given by the following mathematical expression [79]:

$$MAE = \frac{1}{n}\sum_{i-1}^{n}|y_i - \hat{y}_i|.\qquad(12)$$

Mean Square Error (MSE): Tt measures the average of the error squares between two signals. It is given by the following mathematical expression [80]:

$$MSE = \frac{1}{n}\sum_{i-1}^{n}(y_i - \hat{y}_i)^2.\qquad(13)$$

Linear Correlation (LC): It corresponds to a degree of dependence (correlation) between two variables. It may be calculated using the following mathematical expression [81]:

$$LC = \frac{n\sum_{i=1}^{n} y_i\hat{y}_i - \sum_{i=1}^{n} y_i \sum_{i=1}^{n} \hat{y}_i}{\sqrt{n\sum_{i=1}^{n} y_i^2 - (\sum_{i=1}^{n} y_i)^2}\sqrt{n\sum_{i=1}^{n} \hat{y}_i^2(\sum_{i=1}^{n}\hat{y}_i)^2}}.\qquad(14)$$

2.3. AF Theory for Smoothing the Predicted Signal

AFs are used in signal processing areas where it is not possible to pre-identify an unknown environment. AFs are also used in areas of time-varying environments where time-varying parameters are not known in advance or whose development cannot be predicted in the future. An AF is able to obtain the necessary information (estimates of individual quantities) about the environment during signal filtering. The AF should be able to respond to time changes in the environment with a certain speed and to also process signals generated by non-stationary processes [82]. AFs do not require preliminary identification of the signal source, but it is necessary to provide them with additional information in the form of a so-called test signal. The signal measured and the test signal is fed to the AF input. The test signal is closely related to the desired filter output that it somehow approximates [82–87].

In this study, an AF with the LMS algorithm, which is the most widespread in practice, was used. The wide range of applications of the LMS algorithm is attributed to its simplicity and robustness during the calculation of the signals processed [83]. In practice, the following types of LMS algorithms are used [85]: LMS algorithm, LMS algorithm with complex data values, Normalized LMS algorithm

(NLMS), Variable Step Size algorithm (VSLMS), leaky LMS algorithm, linearly constrained LMS algorithm, and LMS algorithm with self-correction of individual parameters (SCAF, Self-Correcting Adaptive Filtering). In this study, an LMS algorithm was applied to suppress the additive noise from the filtered course.

Algorithms Description

Figure 10 describes the Mth-order transversal AF. The filter input signal is denoted by $x(n)$, the required signal is denoted by $d(n)$, and the filter output signal is denoted by $y(n)$. The output sequence of the individual values of the AF signal $y(n)$ is calculated used Equation (15):

$$y(n) = \sum_{i=0}^{M-1} w_i(n)x(n-i). \tag{15}$$

The values of the weighting filtration coefficients $w_0(n), w_1(n)$, and $w_{M-1}(n)$ of the AF with the LMS algorithm are adjusted gradually so that the calculated deviation value $e(n)$ (Equation (16)) was as small as possible, in accordance with the concept of the minimum mean square error

$$e(n) = d(n) - y(n). \tag{16}$$

The individual values of $w_0(n), w_1(n)$, and $w_{M-1}(n)$ of the weighting filtration coefficients of the AF of the given order M change over time. These values of the weighting filtration coefficients are gradually calculated according to Equation (17) or Equation (23). The LMS algorithm adjusts the values of the weighting filtration coefficients of the AF by minimizing the error $e(n)$ in terms of the smallest square deviation, hence the name of the least mean square (LMS) AF algorithm. If the input signal $x(n)$ and the desired signal $d(n)$ are stationary, then the LMS converges to the optimal value of the vector of the weighting filtration coefficients w_0, which is the result of the Wiener–Hopf equation (Equation (24)) of the Wiener filter. A common LMS algorithm is a stochastic implementation of the steepest drop algorithm (Equation(17))

$$w(n+1) = w(n) - \mu \nabla e^2(n), \tag{17}$$

where $w(n) = [w_0(n), w_1(n), ...w_{M-1}(n)]^T$, μ is a parameter determining the degree of correction of algorithm coefficients, and ∇ is a gradient operator, defined as a column vector

$$\nabla = \begin{bmatrix} \frac{\delta}{\delta W_0} & \frac{\delta}{\delta W_1} & \cdots & \frac{\delta}{\delta W_{M-1}} \end{bmatrix}. \tag{18}$$

Equation (19) describes the method of expressing the ith gradient member of vector

$$\frac{\delta e^2(n)}{\delta W_i} = 2e(n)\frac{\delta e(n)}{\delta W_i}. \tag{19}$$

By substitution of Equation (16) into the last term on the right side of Equation (18) and provided that $d(n)$ is independent of w_i, Equation (20) can be described as

$$\frac{\delta e^2(n)}{\delta W_i} = -2e(n)\frac{\delta y(n)}{\delta W_i}. \tag{20}$$

Substitution of $y(n)$ from Equation (15) gives Equation (21)

$$\frac{\delta e^2(n)}{\delta W_i} = -2e(n)x(n-i). \tag{21}$$

Using Equations (18) and (21) gives

$$\nabla e^2(n) = -2e(n)x(n), \tag{22}$$

where $x(n) = [x(n), x(n-1), ..., x(n-M-1)]^T$.

Substitution of the result from Equation (22) to Equation (17) gives

$$w(n+1) = w(n) + 2\mu e(n)x(n). \tag{23}$$

The relation in Equation (23) is referred to as LMS recursion. A simple procedure for recursive adaptation of the weighting filtration coefficients after passing of each new input sample $x(n)$ and its corresponding required sample $d(n)$ are outlined. Equations (15), (16) and (23) specify three steps necessary to complete the LMS algorithm recursion. Equation (15) is referred to as filtration. After calculating Equation (15), the results are output signals $y(n)$ of the AF. Equation (16) is used to calculate the deviation estimate $e(n)$. Equation (23) is used to calculate the values of the vector of weighting filtration coefficients $w(n)$ by adaptive recursion [83]. The most advantageous feature of the LMS algorithm is its simplicity. The implementation of the LMS algorithm of the Mth order AF requires:

- $2M + 1$ multiplication;
- M multiplication for calculating the output signal $y(n)$;
- one multiplication for calculating $2\mu e(n)$;
- M multiplication for calculating the product $2\mu e(n)x(n)$; and
- $2M$ addition.

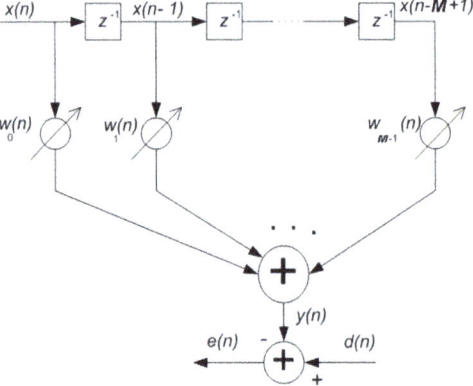

Figure 10. Mth order transversal AF [83].

Another important feature of the LMS algorithm is its stability and robustness for different signal processing conditions. The biggest disadvantage of the LMS algorithm is its slow convergence when the basic input process has a very high frequency spectrum [83]. The Wiener–Hopf equation (Equation (24)) is:

$$\mathbf{w}_o = \mathbf{R}^{-1}\mathbf{p}, \tag{24}$$

where $\mathbf{p}$ is a vector of correlation $\mathbf{x}(n)$ and $d(n)$ with the dimension of $M \times 1$.

$$\mathbf{p} = E[\mathbf{x}(n)d(n)] = [p_0, p_1 \cdots p_{M-1}]^T, \tag{25}$$

where E [.] is the probability operator,

R is the autocorrelation matrix of the input vector $\mathbf{x}(n)$ with the dimension of $M \times M$.

3. Experiments and Results

By taking advantage of the developed software tool, the data (relative humidity, temperature, and CO_2) corresponding to the interval between 2 May 2019 and 10 May 2019 were collected. The analysis was performed for four individual day-long (00:00:00–24:00:00) intervals (3, 6, 7, and 9 May 2019) and the entire recorded interval. The predictive analysis was performed using the IBM SPSS Modeler software tool. Three statistical methods of LR, RT, and NN (MLP) were examined and evaluated. The day-long interval of 3 May was chosen for the purpose of visual inspection of the reference and prediction signals. This interval contains ordinary CO_2 waveform for this type of location. The obtained results were denormalized and evaluated using LC, MAE, and MSE.

3.1. LR Prediction

The LR predictions were performed using four settings of entering, stepwise, backward, and forward. The overall trend of results indicated that stepwise, backward, and forward methods produced very similar results. In some cases, enter method produced slightly less accurate results (Table 1). The most accurate results in terms of LC (LC: 0.952, MAE: 22.413, and MSE: 5.603×10^2) was achieved in the day-long interval 6 May 2019, where all methods resulted in identical statistics. The enter method for a week-long interval of 2–10 May 2019 (LC: 0.459, MAE: 62.573, and MSE: 6.471×10^3) resulted in the least accurate predictions. Table 2 shows that the day-long interval 6 May represents the most accurate interval (LC: 0.952) where 3 and 9 May intervals performed slightly less accurate (LC: 0.929 and LC: 0.905). Due to its regularly repeated trend, the day interval of 3 May was chosen for the purpose of visual inspection of the reference and prediction signals. Figure 11 shows the prediction using the forward setting, where the overall observations of the signal point toward various inaccuracies and excessive noise in the early and late hours of the day.

Table 1. Average results of each implemented setting (LR prediction).

Method	LC	MAE	MSE
Enter	0.826	35.359	2.539×10^3
Stepwise	0.826	35.360	2.536×10^3
Backward	0.826	35.360	2.536×10^3
Forward	0.826	35.360	2.536×10^3

Table 2. Average results of each interval (LR prediction).

Interval	LC	MAE	MSE
3 May	0.929	22.413	9.368×10^2
6 May	0.952	22.413	5.604×10^2
7 May	0.884	39.969	2.890×10^2
9 May	0.905	32.859	1.825×10^3
2–10 May	0.459	62.573	6.471×10^3

To filter the predicted course of CO_2 concentration by LR, an LMS AF was applied to suppress the additive noise from the CO_2 reference signal measured (ppm) (Figure 11), while maintaining the convergence and stability with the parameters set (filter order $M = 48$, step size parameter $\mu = 5.9 \times 10^{-3}$). The implemented adaptive LMS filtration significantly improved the course of the resulting predicted course of CO_2 concentration. The calculated correlation coefficient between the courses of reference CO_2 (ppm) and the LMS AF of CO_2 (ppm) (Figure 11) reached 98.66%, which is higher than the calculated value of the correlation coefficient of 93.60% between Reference CO_2 (ppm) and LR predicted CO_2 (ppm) in Figure 11 as well as in comparison with the results indicated in Tables 1 and 2.

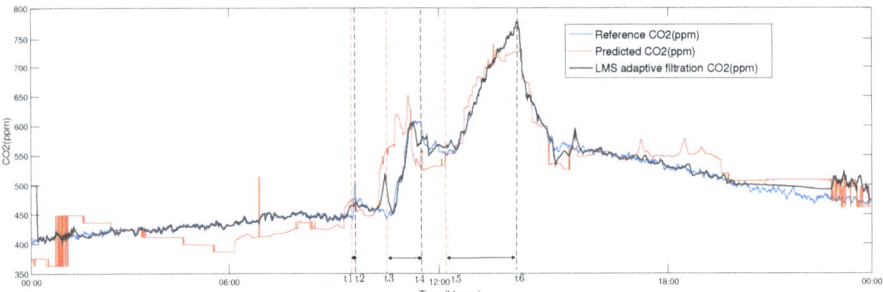

Figure 11. CO_2 (ppm) reference waveform (blue), prediction waveform using LR (Forward setting) (red), and LMS adaptive filtration of prediction waveform using LR (black), (training day-long interval 3 May 2019). **Legend** (hh:mm:ss): arrival t_1 = 09:48:52, departure t_2 = 09:53:32, and length of stay in the room EB312 Δt = 00:05:40; arrival t_3 = 10:58:55, departure t_4 = 11:37:19, and length of stay in the room EB312 Δt = 00:39:24; and arrival t_5 = 12:16:20, departure t_6 = 13:34:58, and length of stay in the room EB312 Δt = 01:18:38.

3.2. RT Prediction

The RT method was implemented using 1–20 trees. The maximum number of nodes was set to 10,000 (by default). The maximum tree depth was set to 10 and minimum child nodes (Min Bucket) size to 5. The model stopped building once the accuracy could no longer improve. The analysis performed with the day-long interval of 3 May showed increasing the number of trees results in increased accuracy of the models. However, by increasing this number any further than eight trees, the result did not imply any significant improvements. Similar behaviors were observed for the other intervals. Table 3 shows that the models with 13, 16, and 18 trees represent exactly the same average results, which are similar to the models with 10 and 11 trees, which implies the ideal range of 13–18 trees. The most accurate results were obtained from day-long interval of 3 May 2019 (LC: 0.991, MAE: 7.476, and MSE: 1.427×10^3) and the least accurate result was obtained from the day-long interval of 2–10 May 2019 and one tree model (LC: 0.562, MAE: 13.862, and MSE: 6.675×10^4). Table 4 shows the average of each time interval. It can be easily observed that day-long intervals are significantly more accurate than the week-long interval. Figure 12 demonstrates mostly accurate predictions of CO_2 concentration levels. However, between 15:00 and 24:00, various noises and glitches can be observed. This indicates a slight overfitting.

Table 3. Average of results for each implemented setting (number of trees) (RT prediction).

Number of Trees	LC	MAE	MSE
1	0.7822	37.655	4.163×10^3
3	0.8138	29.083	2.439×10^3
5	0.8464	31.453	1.465×10^3
8	0.8646	31.453	1.359×10^3
10	0.8672	28.016	1.319×10^3
11	0.8674	22.062	1.314×10^3
13	0.8676	27.944	1.085×10^3
16	0.8676	27.944	1.085×10^3
18	0.8676	27.944	1.085×10^3
20	0.8486	31.596	1.085×10^3

Table 4. Average results of each interval (RT prediction).

Interval	LC	MAE	MSE
3 May	0.975	11.356	4.187×10^2
6 May	0.699	43.181	2.176×10^3
7 May	0.879	28.543	1.601×10^3
9 May	0.895	37.9874	6.474×10^2
2–10 May	−17.5829	26.1345	3.355×10^3

To filter the predicted course of CO_2 concentration by RT (using 18 trees), an LMS AF was applied to suppress the additive noise from the CO_2 Reference signal measured (ppm) (Figure 12), while maintaining the convergence and stability with the parameters set (filter order M = 48, step size parameter $\mu = 5.9 \times 10^{-3}$). The implemented adaptive LMS filtration significantly improved the course of the resulting predicted course of CO_2 concentration. The calculated correlation coefficient between the courses of Reference CO_2 (ppm) and the LMS adaptive filtration of CO_2 (ppm) (Figure 12) reached 99.25%, which is higher than the calculated value of the correlation coefficient of 98.96% between Reference CO_2 (ppm) and RT (using 18 trees) Predicted CO_2 (ppm) in Figure 12 as well as in comparison with the results indicated in Tables 3 and 4.

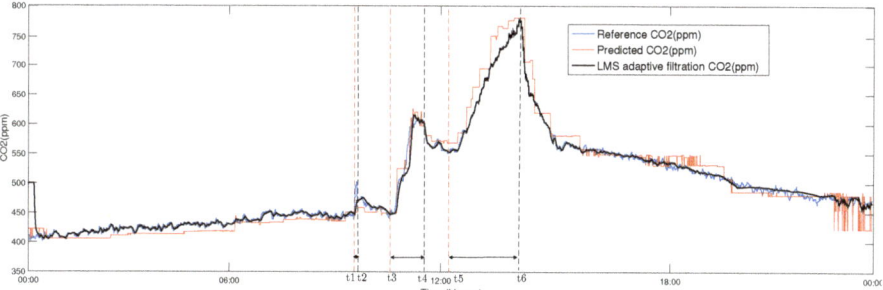

Figure 12. CO_2 reference waveform (blue), prediction waveform using RT (using 18 trees) (red), and LMS adaptive filtration of prediction waveform using RT (using 18 trees) (black), (training day-long interval 3 May 2019). **Legend** (hh:mm:ss): arrival t_1 = 09:48:52, departure t_2 = 09:53:32, and length of stay in the room EB312 Δt = 00:05:40; arrival t_3 = 10:58:55, departure t_4 = 11:37:19, and length of stay in the room EB312 Δt = 00:39:24; and arrival t_5 = 12:16:20, departure t_6 = 13:34:58, and length of stay in the room EB312 Δt = 01:18:38.

3.3. NN (MLP) Prediction

The implementation of a NN (MLP) was performed using several different settings for the number of neurons in the hidden layers. For the purpose of comparisons, these models were tested using the identical data-intervals. The model trained with data day-long interval of 6 May 2019, with settings of 500 neurons in the first hidden layer and 20 neurons in the second hidden layer provided the highest overall accuracy (LC: 0.997, MAE: 4.597, and MSE: 3.641×10^1). Meanwhile, the week-long training data interval (2–10 May 2019) combined with 100 neurons in the first hidden layer and 50 neurons in the second hidden layer provided the least accuracy (LC: 0.947, MAE: 19.764, and MSE: 8.414×10^2). By investigating the overall trend of the results, it was observed that the models which had 500 neurons in the first layer and 20 neurons in the second layer generally performed better. These results can be observed in Table 5, which demonstrates the overall trend of analysis results in terms of model settings. On the other hand, Table 6 shows that the day-long intervals hold better average accuracies. Additionally, the day-long interval of 6 May appears to be the most suitable training interval. Figure 13 shows the CO_2 reference and prediction signal using a neural network. The applied model used 500 neurons in the first and 20 neurons in the second hidden layer. The day-long interval of 3 May 2019 was used for training of this model. Although the signals demonstrate accurate overall prediction of CO_2 concentration levels, in the final hours (between 22:30 and 24:00) of the day, some noises can be observed due to minor overfitting.

Table 5. Average of results for each implemented setting (number of neurons) (NN prediction).

Number of Neurons Layer1-Layer2	LC	MAE	MSE
10–10	0.988	9.7558	2.112×10^2
25–50	0.989	9.315	1.927×10^2
50–25	0.990	8.6476	1.636×10^2
50–100	0.989	9.0496	1.803×10^2
100–50	0.990	8.9098	1.763×10^2
500–20	0.991	8.2812	1.626×10^2
20–500	0.984	10.839	2.634×10^2

Table 6. Average of each interval (NN prediction).

Interval	LC	MAE	MSE
3 May	0.990	8.658	1.364×10^2
6 May	0.996	5.024	4.662×10^1
7 May	0.993	8.818	2.018×10^2
9 May	0.989	10.442	2.15×10^2
2–10 May	0.972	14.600	4.518×10^2

To filter the predicted course of CO_2 concentration by NN (using 500 neurons in Layer 1 and 20 neurons in Layer 2), an LMS AF was applied for additive noise canceling from the CO_2 reference signal measured (ppm) (Figure 13), while maintaining the convergence and stability with the parameters set (filter order $M = 48$, step size parameter $\mu = 5.9 \times 10^{-3}$). The implemented adaptive LMS filtration significantly improved the course of the resulting predicted course of CO_2 concentration. The calculated correlation coefficient between the courses of Reference CO_2 (ppm) and the LMS adaptive filtration of CO_2 (ppm) (Figure 13) reached 99.29%, which is higher than the calculated value of the correlation coefficient of 98.96% between Reference CO_2 (ppm) and NN (using 500 neurons in Layer 1 and 20 neurons in Layer 2) Predicted CO_2 (ppm) in Figure 13 as well as in comparison with the results indicated in Tables 5 and 6.

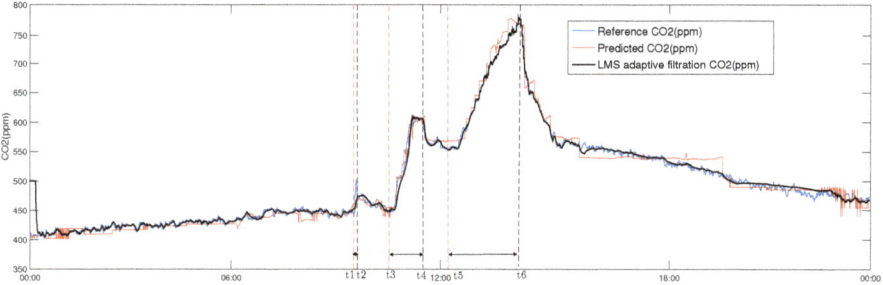

Figure 13. CO_2 reference waveform (blue), prediction waveform using NN (using neurons 500 in Layer 1 and 20 neurons in Layer 2) (Table 5) (red), and LMS adaptive filtration of prediction waveform using NN (using neurons 500 in Layer 1 and 20 neurons in Layer 2) (black) (training day-long interval 3 May 2019). **Legend** (hh:mm:ss): arrival t_1 = 09:48:52, departure t_2 = 09:53:32, and length of stay in the room EB312 Δt = 00:05:40; arrival t_3 = 10:58:55, departure t_4 = 11:37:19, and length of stay in the room EB312 Δt = 00:39:24; and arrival t_5 = 12:16:20, departure t_6 = 13:34:58, and length of stay in the room EB312 Δt = 01:18:38.

4. Discussion

In the first part of the practical work, the development of a dedicated console application for KNX installation and IBM Watson IoT platform connectivity was described. When creating this program, the emphasis was placed on simplicity and reliability. Continuous running of the program was controlled by a web browser with an Internet connection, as the Watson IoT IBM platform provides a friendly web interface for visualizing the data received. After satisfactory verification of reliability, the development of a desktop version, which provides the user with a clear user interface, was commenced. Its core function, i.e., connectivity of the KNX and IoT IBM technologies, is the same as in the case of the console application. In the second part of the practical work, based on the data obtained from the KNX installation, a visualization application allowing the user both to display the values measured, namely their historical values depending on the database size and to visualize and process the data transmitted in real-time, was being developed. Emphasis was placed on information about the current occupancy of the room monitored. The user can monitor the measured values of the course of the CO_2 sensor concentration, from the development of which the occupancy of the space monitored can be derived. When designing the new method for predicting the course of CO_2 concentration based on temperature and humidity measurements, one of the motivations for CO_2 prediction was to save the initial investment resources for the acquisition of CO_2 sensors, which are significantly more expensive compared to temperature and humidity sensors. For prediction, IBM offers the SPSS modeler desktop application or the Watson Studio cloud service. Both options were used during the implementation. When working with the SPSS modeler, the prediction was conducted based on the input data in the form of a csv file. Using Watson Studio, which also offers many features that can be used to process data for CO_2 prediction, there was also a successful connection with the Cloudant database for the access to historical data and, at the same time, there was a connection of this service with the Watson IoT platform for real-time data transfer from the KNX installation.

By reviewing the obtained results from LR, it becomes apparent that changing the algorithms of the models does not have any major impact on the results. Furthermore, day-long intervals showed significantly better accuracy. By summing up the tabulated and graphical results, it was concluded that the predictions obtained from this method would be accurate enough for the detection of presence. The prediction results included some inaccurate predictions. Therefore, it is not suitable for accurate CO_2 concentration level predictions. The RT method showed significantly improved generalizations of CO_2 concentration levels. Unlike the previous method, model settings affected the result noticeably. The general trend of the prediction pointed toward better accuracy with an increase in the number of trees. Specifically, the region between 10 and 18 trees appeared to be optimal for accurate predictions. Except for 6 May 2019, the prediction results from day-long intervals showed better overall characteristics. Further investigations found that 6 May 2019 showed short occupancy day-long intervals. Despite the fact the prediction waveform contained few minor inaccuracies and small noisy sections, it was still considered as a suitable method for both occupancy monitoring and accurate CO_2 waveform predictions. The NN (MLP) showed the best numerical characteristics. The visual investigation verified the good generalization of the trained models. All models and methods showed noisy predictions at late hours of the day. Nevertheless, these noise levels were reduced in NN (MLP). Similar to previous methods, the day-long training intervals were more accurate. Overall, the NN (MLP) proved to be the most accurate method for both occupancy monitoring and accurate CO_2 waveform predictions.

The use of an LMS AF in an application for suppressing the additive noise in the CO_2 concentration course predicted significantly improved the resulting CO_2 concentration course compared to the reference course measured for all of the above-stated LR, RT, NN (MLP) prediction methods (Figures 11–13). The best results, in terms of prediction accuracy using the LMS AF, were obtained for the NN method, where the calculated correlation coefficient between the courses of CO_2 reference (ppm) and the LMS adaptive CO_2 filtration (ppm) (Figure 13) reached 99.29% for the adaptive LMS filter order set (filter order $M = 48$, step size parameter $\mu = 5.9 \times 10^{-3}$) (Table 6). It can be compared with the correlation coefficient values calculated, as shown in Tables 5 and 6, or the with other methods in Tables 1–4. The disadvantage of the LMS AF is the initial peak (Figures 11–13) when the AF does not know the signal course processed. This is due to the vector of the weighting filtration coefficients $\mathbf{w}(n)$, which was set to zero for the initial conditions and the start-up of the predicted CO_2 course processing by the LMS AF (Equation (23)).

5. Conclusions

The article describes the implementation of a novel method for predicting CO_2 course from the temperature and relative humidity courses measured (within the operational accuracy) to monitor occupancy in IB using conventional KNX operating sensors. The article further describes the programming of the required KNX modules in the ETS 5 SW tool. The IoT-based automation of smart homes has been investigated in numerous studies [88–92]. However, this article develops a novel SW tool that ensures connectivity between the KNX technology and the IoT IBM platform for real-time storage and visualization of the data measured. Furthermore, the article describes a comparison of mathematical methods (linear regression (LR), random trees (RT), and multilayer perceptron (MLP)) in the framework of CO_2 prediction. To increase the accuracy of the CO_2 prediction, the authors then describe the implementation of the LMS AF in a novel application. It suppresses the additive noise in the predicted CO_2 waveform. There are similar works [39,41,93] focused on CO_2 modeling or indirect occupancy monitoring. The novelty of this article is within its comprehensiveness. It offers a complete solution: programming of KNX modules, development of pc based IoT gateway, predictive analysis, and filtration. The authors used the IBM SPSS Software tool for CO_2 concentration waveforms prediction using relative humidity (indoor) and temperature (indoor) values. These three important mathematical methods were investigated numerically and visually. LR showed the least accuracy with many offsets and noise output signal. NN (MLP) showed the most accurate results with minor noise and glitches. Based on the results achieved with a prediction accuracy of more than 98% for the selected experiments, it can be stated that the proposed objectives and procedures were met. In the next work, the authors will focus on the application of the proposed method in the optimization of operating costs in IB with subsequent energy savings within IB automatization.

Author Contributions: J.V. and O.M.G., methodology; J.V. and O.M.G., software; J.V. and O.M.G., validation; J.V. and O.M.G., formal analysis; J.V. and O.M.G., investigation; J.V. and O.M.G., resources; J.V., data creation; J.V. and O.M.G., writing—original draft preparation; J.V. and O.M.G., writing—review and editing; J.V., visualization; J.V., supervision; P.B., project administration; and P.B., funding acquisition

Funding: This work was supported by the Student Grant System of VSB Technical University of Ostrava, grant number SP2019/118.

Acknowledgments: This work was supported by the Student Grant System of VSB Technical University of Ostrava, grant number SP2019/118. This work was supported by the European Regional Development Fund in the Research Centre of Advanced Mechatronic Systems project, project number CZ.02.1.01/0.0/0.0/16_019/0000867 within the Operational Programme Research, Development and Education.

Conflicts of Interest: The funders had no role in the design of the study; in the collection, analyses, or interpretation of data; in the writing of the manuscript, or in the decision to publish the results.

Abbreviations

The following abbreviations are used in this manuscript:

NN	Neural Network
MLP	Multilayer Perceptron
RT	Random Tree
LR	Linear Regression
CC	Cloud Computing
IoT	Internet of Things
LMS	Least Mean Squares
AF	Adaptive filter
IB	Intelligent Buildings
SH	Smart Home
SHC	Smart Home Care
VLC	Visible Light Communication
LED	Light-Emitting Diode
KNX	Konnex (standard EN 50090, ISO/IEC 14543)
DNS	Device Name System
MQTT	Message Queuing Telemetry Transport
ETS	Engineering Tool Software
SW	Software
CO_2	Carbon dioxide
SC	Smart Cities
SG	Smart Grids
CoAP	Constrained Application Protocol
XMPP	Extensible Messaging and Presence Protocol
IBM SPSS	Statistical Package for the Social Sciences from the company IBM

References

1. Asensio, J.A.; Criado, J.; Padilla, N.; Iribarne, L. Emulating home automation installations through component-based web technology. *Future Gener. Comput. Syst.* **2017**, *93*, 777–791. [CrossRef]
2. Aggarwal, M.; Madhukar, M. IBM's Watson Analytics for Health Care: A Miracle Made True. In *Cloud Computing Systems and Applications in Healthcare*; IGI Global: Hershey, PA, USA, 2017; pp. 117–134.
3. Petnik, J.; Vanus, J. Design of smart home implementation within IoT with natural language interface. *IFAC-PapersOnLine* **2018**, *51*, 174–179. [CrossRef]
4. Lekić, M.; Gardašević, G. IoT sensor integration to Node-RED platform. In Proceedings of the 2018 IEEE 17th International Symposium INFOTEH-JAHORINA (INFOTEH), East Sarajevo, Bosnia-Herzegovina, 21–23 March 2018; pp. 1–5.
5. Martinez, A.C. Connecting Small Form-Factor Devices to the Internet of Things. In *Advances in Human Factors and System Interactions*; Springer: Cham, Switzerland, 2017; pp. 313–322.

6. Akinsiku, A.; Jadav, D. BeaSmart: A beacon enabled smarter workplace. In Proceedings of the NOMS 2016—2016 IEEE/IFIP Network Operations and Management Symposium, Istanbul, Turkey, 25–29 April 2016; pp. 1269–1272.
7. Nandi, S. Cloud-based cognitive premise security system using ibm watson and IBM internet of things (IoT). In *Advances in Electronics, Communication and Computing*; Springer: Singapore, 2018; pp. 723–731.
8. Marksteiner, S.; Jiménez, V.J.E.; Valiant, H.; Zeiner, H. An overview of wireless IoT protocol security in the smart home domain. In Proceedings of the 2017 Internet of Things Business Models, Users, and Networks, Copenhagen, Denmark, 23–24 November 2017; pp. 1–8.
9. Tanwar, S.; Patel, P.; Patel, K.; Tyagi, S.; Kumar, N.; Obaidat, M.S. An advanced Internet of Thing based security alert system for smart home. In Proceedings of the 2017 International Conference on Computer, Information and Telecommunication Systems (CITS), Dalian, China, 21–23 July 2017; pp. 25–29.
10. Perera, C.; Liu, C.H.; Jayawardena, S. The emerging internet of things marketplace from an industrial perspective: A survey. *IEEE Trans. Emerg. Top. Comput.* **2015**, *3*, 585–598. [CrossRef]
11. Bastos, D.; Shackleton, M.; El-Moussa, F. Internet of things: A survey of technologies and security risks in smart home and city environments. In Proceedings of the Living in the Internet of Things: Cybersecurity of the IoT, London, UK, 28–29 March 2018.
12. Vanus, J.; Belesova, J.; Martinek, R.; Nedoma, J.; Fajkus, M.; Bilik, P.; Zidek, J. Monitoring of the daily living activities in smart home care. *Hum.-Centric Comput. Inf. Sci.* **2017**, *7*, 30. doi:10.1186/s13673-017-0113-6. [CrossRef]
13. Holý, R.; Kalika, M.; Havlík, J.; Makarov, A. HVAC system—Communication platform. In Proceedings of the 2018 3rd International Conference on Intelligent Green Building and Smart Grid (IGBSG), Yi-Lan, Taiwan, 22–25 April 2018; pp. 1–4.
14. Chen, M.; Wan, J.; Li, F. Machine-to-machine communications: Architectures, standards and applications. *Ksii Trans. Internet Inf. Syst.* **2012**, *6*. [CrossRef]
15. Ilieva, S.; Penchev, A.; Petrova-Antonova, D. Internet of Things Framework for Smart Home Building. In Proceedings of the International Conference on Digital Transformation and Global Society, St. Petersburg, Russia, 23–24 June 2016; Springer: Cham, Switzerland, 2016; pp. 450–462.
16. Mainetti, L.; Mighali, V.; Patrono, L. An android multi-protocol application for heterogeneous building automation systems. In Proceedings of the 2014 22nd International Conference on Software, Telecommunications and Computer Networks (SoftCOM), Split, Croatia, 17–19 September 2014; pp. 121–127.
17. Bajer, M. IoT for smart buildings-long awaited revolution or lean evolution. In Proceedings of the 2018 IEEE 6th International Conference on Future Internet of Things and Cloud (FiCloud), Barcelona, Spain, 6–8 August 2018; pp. 149–154.
18. Vanus, J.; Machac, J.; Martinek, R.; Bilik, P.; Zidek, J.; Nedoma, J.; Fajkus, M. The design of an indirect method for the human presence monitoring in the intelligent building. *Hum.-Centric Comput. Inf. Sci.* **2018**, *8*, 28. doi:10.1186/s13673-018-0151-8. [CrossRef]
19. Jung, M.; Weidinger, J.; Kastner, W.; Olivieri, A. Building automation and smart cities: An integration approach based on a service-oriented architecture. In Proceedings of the 2013 27th International Conference on Advanced Information Networking and Applications Workshops, Barcelona, Spain, 25–28 March 2013; pp. 1361–1367.
20. Vanus, J.; Stratil, T.; Martinek, R.; Bilik, P.; Zidek, J. The possibility of using VLC data transfer in the smart home. *IFAC-PapersOnLine* **2016**, *49*, 176–181. [CrossRef]
21. Díaz, M.; Martín, C.; Rubio, B. State-of-the-art, challenges, and open issues in the integration of Internet of things and cloud computing. *J. Netw. Comput. Appl.* **2016**, *67*, 99–117. [CrossRef]
22. Kelly, S.D.T.; Suryadevara, N.K.; Mukhopadhyay, S.C. Towards the implementation of IoT for environmental condition monitoring in homes. *IEEE Sens. J.* **2013**, *13*, 3846–3853. [CrossRef]
23. Koo, J.; Kim, Y.G. Interoperability of device identification in heterogeneous IoT platforms. In Proceedings of the 2017 13th International Computer Engineering Conference (ICENCO), Cairo, Egypt, 27–28 December 2017; pp. 26–29.
24. Tolentino, M. 1950s Smart Homes: Future in the Past. 2014. Available online: https://siliconangle.com/2014/02/05/1950ssmart-homes-future-in-the-past/ (accessed on 20 January 2019).
25. Pohanka, P. Internet věcí. In: i2ot.eu 2019. Available online: http://i2ot.eu/internet-of-things/ (accessed on 9 April 2019).

26. Vojacek, A. Zakladní uvod do oblasti internetu veci (IoT). 2016. Available online: https://automatizace.hw.cz/zakladni-uvod-dooblasti-internetu-veci-iot.html (accessed on 21 January 2019).
27. Bouhaï, N.; Saleh, I. *Internet of Things: Evolutions and Innovations*; Wiley-ISTE: London, UK, 2017.
28. KNX asociace. knxcz.cz. Available online: https://knxcz/cz/images/clanky/KNX-IoT_en.pdf (accessed on 31 January 2019).
29. Hagen, S. *IPv6 essentials*; O'Reilly Media, Inc.: Sebastopol, CA, USA, 2006.
30. Nilsoon, R. Bluetooth Low Energy není jen nová verze standard Bluetooth. 2013. Available online: https://www.automa.cz/cz/casopis-clanky/bluetoothlow-energy-neni-jen-nova-verze-standardu-bluetooth-2013_12_0_10907/ (accessed on 31 January 2019).
31. LoRa Alliance. LoRaWAN: What Is It? 2015. Available online: https://lora-alliance.org/sites/default/files/2018-04/what-is-lorawan.pdf (accessed on 21 January 2019).
32. Farahani, S. *ZigBee Wireless Networks and Transceivers*; Newnes: Oxford, UK, 2011.
33. Paetz, C. *Z-Wave Essentials*; Createspace Independent Publishing Platform: Zwickau, Germany, 2017.
34. MALÝ, M. Protokol MQTT: komunikační standart pro IoT. 2016. Available online: https://www.root.cz/clanky/protokol-mqttkomunikacni-standard-pro-iot/ (accessed on 21 January 2019).
35. Ranjan, R.; Wang, L.; Chen, J.; Benatallah, B. *Cloud Computing: Methodology, Systems, and Applications*; CRC Press: Boca Raton, FL, USA, 2011.
36. Intesis Software S.L.U. houseinhand.com, 2016. Available online: https://www.houseinhand.com/ (accessed on 31 January 2019).
37. Vanus, J.; Kubicek, J.; Gorjani, O.M.; Koziorek, J. Using the IBM SPSS SW Tool with Wavelet Transformation for CO_2 Prediction within IoT in Smart Home Care. *Sensors* **2019**, *19*, 1407. doi:10.3390/s19061407. [CrossRef]
38. Vanus, J.; Martinek, R.; Bilik, P.; Zidek, J.; Dohnalek, P.; Gajdos, P.; Ieee. New Method for Accurate Prediction of CO_2 in the Smart Home. In Proceedings of the 2016 IEEE International Instrumentation and Measurement Technology Conference Proceedings, Taipei, Taiwan, 23–26 May 2016; IEEE: New York, NY, USA, 2016; pp. 1333–1337.
39. Khazaei, B.; Shiehbeigi, A.; Kani, A. Modeling indoor air carbon dioxide concentration using artificial neural network. *Int. J. Environ. Sci. Technol.* **2019**, *16*, 729–736. doi:10.1007/s13762-018-1642-x. [CrossRef]
40. Wang, H.K.; Li, L.; Wu, Y.; Meng, F.J.; Wang, H.H.; Sigrimis, N.A. Recurrent Neural Network Model for Prediction of Microclimate in Solar Greenhouse. *Ifac PapersOnline* **2018**, *51*, 790–795. doi:10.1016/j.ifacol.2018.08.099. [CrossRef]
41. Szczurek, A.; Maciejewska, M.; Pietrucha, T. Occupancy determination based on time series of CO2 concentration, temperature and relative humidity. *Energy Build.* **2017**, *147*, 142–154. doi:10.1016/j.enbuild.2017.04.080. [CrossRef]
42. Galda, Z.; Sipkova, V.; Labudek, J.; Gergela, P.; SGEM. Experimental Measurements of CO_2 in the Summer Months in the Passive House. In Proceedings of the 15th International Multidisciplinary Scientific Geoconference (SGEM), Albena, Bulgaria, 18–24 June 2015; International Multidisciplinary Scientific GeoConference-SGEM; Stef92 Technology Ltd.: Sofia, Bulgaria, 2015; pp. 127–132.
43. Aazam, M.; Khan, I.; Alsaffar, A.A.; Huh, E.N. Cloud of Things: Integrating Internet of Things and cloud computing and the issues involved. In Proceedings of the 2014 11th International Bhurban Conference on Applied Sciences & Technology (IBCAST), Islamabad, Pakistan, 14–18 January 2014; pp. 414–419.
44. Asociace, K. Knx-Specifications. 2014. Available online: https://my.knx.org/en/downloads/knx-specifications (accessed on 31 January 2019).
45. Asociace, K. sti.uniurb.it. Available online: http://www.sti.uniurb.it/romanell/Domotica_e_Edifici_Intelligenti/110504-Lez10a-KNX-Datapoint%20Types%20v1.5.00%20AS.pdf (accessed on 31 January 2019).
46. Michalec, L. Komunikace v KNX. 2014. Available online: https://vyvoj.hw.cz/automatizace/komunikace-v-knx.html (accessed on 31 January 2019).
47. Intesis Software S.L.U. (Application). Available online: https://play.google.com/store/apps/details?id=com.intesis.houseinhand (accessed on 31 January 2019).
48. DELIOT. Pascal. Available online: https://www.microsoft.com/cs-cz/p/knxdashboard/9wzdncrdm06f?activetab=pivot:overviewtab# (accessed on 31 January 2019).
49. Rouse, M. Predictive modeling. Available online: https://searchenterpriseai.techtarget.com/definition/predictive-modeling (accessed on 31 May 2019).

50. Ahmad, M.W.; Reynolds, J.; Rezgui, Y. Predictive modelling for solar thermal energy systems: A comparison of support vector regression, random forest, extra trees and regression trees. *J. Clean. Prod.* **2018**, *203*, 810–821. [CrossRef]
51. Han, J.; Pei, J.; Kamber, M. *Data Mining: Concepts and Techniques*; Elsevier: Waltham, MA, USA, 2011.
52. Kachalsky, I.; Zakirzyanov, I.; Ulyantsev, V. Applying reinforcement learning and supervised learning techniques to play Hearthstone. In Proceedings of the 2017 16th IEEE International Conference on Machine Learning and Applications (ICMLA), Cancun, Mexico, 18–21 December 2017; pp. 1145–1148.
53. Nijhawan, R.; Srivastava, I.; Shukla, P. Land cover classification using super-vised and unsupervised learning techniques. In Proceedings of the 2017 International Conference on Computational Intelligence in Data Science (ICCIDS), Chennai, India, 2–3 June 2017; pp. 1–6.
54. Liu, Q.; Liao, X.; Carin, L. Semi-Supervised Life-Long Learning with Application to Sensing. In Proceedings of the 2007 2nd IEEE International Workshop on Computational Advances in Multi-Sensor Adaptive Processing, St. Thomas, VI, USA, 12–14 December 2007; pp. 1–4.
55. IBM. *IBM SPSS Modeler 18.0 User's Guide*; IBM. Available online: https://searchenterpriseai.techtarget.com/definition/predictive-modeling (accessed on 31 May 2019).
56. Yan, X.; Su, X. *Linear Regression Analysis: Theory and Computing*; World Scientific: Singapore, 2009.
57. Rencher, A.C.; Christensen, W.F. Chapter 10, Multivariate regression–Section 10.1, Introduction. In *Methods of Multivariate Analysis*; Wiley Series in Probability and Statistics; John Wiley & Sons Inc.: New York, NY, USA, 2012; Volume 709, p. 19.
58. Ralston, A.; Wilf, H.S. *Mathematical Methods for Digital Computers*; Technical Report; John Wiley and Sons Ltd.: New York, NY, USA, 1960.
59. Hocking, R.R. A Biometrics invited paper. The analysis and selection of variables in linear regression. *Biometrics* **1976**, *32*, 1–49. [CrossRef]
60. Draper, N.; Smith, H. *Applied Regression Analysis*, 2nd ed.; John Willey & Sons: New York, NY, USA, 1981.
61. SAS Institute Inc. *SAS/STAT User's Guide Version 6*, 4th ed.; SAS Institute Inc.: Cary, NC, USA, 1989.
62. Knecht, W.R. *Pilot Willingness to Take Off Into Marginal Weather. Part 2. Antecedent Overfitting with Forward Stepwise Logistic Regression*; Technical Report; Federal Aviation Administration Oklahoma City Ok Civil Aeromedical Inst: Oklahoma City, OK, USA, 2005.
63. Flom, P.L.; Cassell, D.L. Stopping Stepwise: Why Stepwise and Similar Selection Methods Are Bad, and What You Should Use. In Proceedings of the NorthEast SAS Users Group (NESUG) 2007: Statistics and Data Analysis, Baltimore, MD, USA, 11–14 November 2007.
64. Myers, R.H.; Myers, R.H. *Classical and Modern Regression with Applications*; Duxbury press: Belmont, CA, USA, 1990; Volume 2.
65. Bendel, R.B.; Afifi, A.A. Comparison of stopping rules in forward "stepwise" regression. *J. Am. Stat. Assoc.* **1977**, *72*, 46–53.
66. Kubinyi, H. Evolutionary variable selection in regression and PLS analyses. *J. Chemom.* **1996**, *10*, 119–133. [CrossRef]
67. Quinlan, J.R. Simplifying decision trees. *Int. J. Man-Mach. Stud.* **1987**, *27*, 221–234. [CrossRef]
68. Liaw, A.; Wiener, M. Classification and regression by randomForest. *R News* **2002**, *2*, 18–22.
69. Breiman, L. *Classification and Regression Trees*; Routledge: Boca Raton, FL, USA, 2017.
70. Meir, A.; Moon, J.W. On the altitude of nodes in random trees. *Can. J. Math.* **1978**, *30*, 997–1015. [CrossRef]
71. Duquesne, T.; Le Gall, J.F. *Random Trees, Lévy Processes and Spatial Branching Processes*; Société mathématique de France: Paris, France, 2002; Volume 281.
72. Le Gall, J.F. Random trees and applications. *Probab. Surv.* **2005**, *2*, 245–311. [CrossRef]
73. Pittel, B. Note on the heights of random recursive trees and random m-ary search trees. *Random Struct. Algorithms* **1994**, *5*, 337–347. [CrossRef]
74. Ripley, B.D.; Hjort, N. *Pattern Recognition and Neural Networks*; Cambridge University Press: Cambridge, UK, 1996.
75. Haykin, S. *Neural Networks: A Comprehensive Foundation*; Prentice Hall PTR: Upper Saddle River, NJ, USA, 1994.
76. Zarei, T.; Behyad, R. Predicting the water production of a solar seawater greenhouse desalination unit using multi-layer perceptron model. *Sol. Energy* **2019**, *177*, 595–603. [CrossRef]

77. Moosavi, S.R.; Wood, D.A.; Ahmadi, M.A.; Choubineh, A. ANN-Based Prediction of Laboratory-Scale Performance of CO_2-Foam Flooding for Improving Oil Recovery. *Nat. Resour. Res.* **2019**, *28*, 1619–1637. [CrossRef]
78. IBM. *IBM SPSS Modeler 18 Algorithms Guide*; IBM. Available online: ftp://public.dhe.ibm.com/software/analytics/spss/documentation/modeler/18.0/en/AlgorithmsGuide.pdf (accessed on 31 January 2019).
79. Willmott, C.J.; Matsuura, K. Advantages of the mean absolute error (MAE) over the root mean square error (RMSE) in assessing average model performance. *Clim. Res.* **2005**, *30*, 79–82. [CrossRef]
80. Lehmann, E.L.; Casella, G. *Theory of Point Estimation*; Springer Science & Business Media: New York, NY, USA, 2006.
81. Ijiri, Y. The linear aggregation coefficient as the dual of the linear correlation coefficient. *Econom. J. Econom. Soc.* **1968**, *36*, 252–259. [CrossRef]
82. Jan, J. *Cislicova Filtrace, Analyza a Restaurace Signalu*; Vutium: Brno, Czech Republic, 2002.
83. Boroujeny, B.F. *Adaptive Filters: Theory and Applications*; John Wiley & Sons: New York, NY, USA, 2013.
84. Poularikas, A.D.; Ramadan, Z.M. *Adaptive Filtering Primer with MATLAB*; CRC Press: Boca Raton, FL, USA, 2017.
85. Haykin, S.S. *Modern Filters*; Macmillan Coll Division: Upper Saddle River, NJ, USA, 1989.
86. Haykin, S.S.; Widrow, B. *Least-Mean-Square Adaptive Filters*; Wiley Online Library: New York, NY, USA, 2003; Volume 31.
87. Haykin, S.S. *Adaptive Filter Theory*; Pearson Education India: Bengaluru, India, 2005.
88. Kodali, R.K.; Jain, V.; Bose, S.; Boppana, L. IoT based smart security and home automation system. In Proceedings of the 2016 International Conference on Computing, Communication and Automation (ICCCA), Noida, India, 29–30 April 2016; pp. 1286–1289.
89. Pirbhulal, S.; Zhang, H.; E Alahi, M.; Ghayvat, H.; Mukhopadhyay, S.; Zhang, Y.T.; Wu, W. A novel secure IoT-based smart home automation system using a wireless sensor network. *Sensors* **2017**, *17*, 69. [CrossRef]
90. Pavithra, D.; Balakrishnan, R. IoT based monitoring and control system for home automation. In Proceedings of the 2015 Global Conference on Communication Technologies (GCCT), Thuckalay, India, 23–24 April 2015; pp. 169–173.
91. Wang, M.; Zhang, G.; Zhang, C.; Zhang, J.; Li, C. An IoT-based appliance control system for smart homes. In Proceedings of the 2013 Fourth International Conference on Intelligent Control and Information Processing (ICICIP), Beijing, China, 9–11 June 2013; pp. 744–747.
92. Lee, W.S.; Hong, S.H. Implementation of a KNX-ZigBee gateway for home automation. In Proceedings of the 2009 IEEE 13th International Symposium on Consumer Electronics, Kyoto, Japan, 25–28 May 2009; pp. 545–549.
93. Skön, J.; Johansson, M.; Raatikainen, M.; Leiviskä, K.; Kolehmainen, M. Modelling indoor air carbon dioxide (CO_2) concentration using neural network. *Methods* **2012**, *14*, 16.

© 2019 by the authors. Licensee MDPI, Basel, Switzerland. This article is an open access article distributed under the terms and conditions of the Creative Commons Attribution (CC BY) license (http://creativecommons.org/licenses/by/4.0/).

Article

Intelligent Energy Management Strategy for Automated Office Buildings

Simplice Igor Noubissie Tientcheu *, Shyama P. Chowdhury and Thomas O. Olwal

Department of Electrical Engineering, Tshwane University of Technology, Pretoria 0001, South Africa; spchowdhury2010@gmail.com (S.P.C.); OlwalTO@tut.ac.za (T.O.O.)
* Correspondence: simplice.co@gmail.com; Tel.: +27-78-333-2635

Received: 29 September 2019; Accepted: 30 October 2019; Published: 13 November 2019

Abstract: The increasing demand to reduce the high consumption of end-use energy in office buildings framed the objective of this work, which was to design an intelligent system management that could be utilized to minimize office buildings' energy consumption from the national electricity grid. Heating, Ventilation and Air Conditioning (HVAC) and lighting are the two main consumers of electricity in office buildings. Advanced automation and control systems for buildings and their components have been developed by researchers to achieve low energy consumption in office buildings without considering integrating the load consumed and the Photovoltaic system (PV) input to the controller. This study investigated the use of PV to power the HVAC and lighting equipped with a suitable control strategy to improve energy saving within a building, especially in office buildings where there are reports of high misuse of electricity. The intelligent system was modelled using occupant activities, weather condition changes, load consumed and PV energy changes, as input to the control system of lighting and HVAC. The model was verified and tested using specialized simulation tools (Simulink®) and was subsequently used to investigate the impact of an integrated system on energy consumption, based on three scenarios. In addition, the direct impact on reduced energy cost was also analysed. The first scenario was tested in simulation of four offices building in a civil building in South Africa of a single occupant's activities, weather conditions, temperature and the simulation resulted in savings of HVAC energy and lighting energy of 13% and 29%, respectively. In the second scenario, the four offices were tested in simulation due to the loads' management plus temperature and occupancy and it resulted in a saving of 20% of HVAC energy and 29% of lighting electrical energy. The third scenario, which tested integrating PV energy (thus, the approach utilized) with the above-mentioned scenarios, resulted in, respectively, 64% and 73% of HVAC energy and lighting electrical energy saved. This saving was greater than that of the first two scenarios. The results of the system developed demonstrated that the loads' control and the PV integration combined with the occupancy, weather and temperature control, could lead to a significant saving of energy within office buildings.

Keywords: end-use energy consumption; heating; ventilation and air conditioning (HVAC); intelligent system management; lighting electrical energy; national electricity grid; office building; Photovoltaic system; simulation; Simulink®

1. Introduction

Due to the functional and operational requirements of the offices, in a normal office building, energy consumed is affected by the heating, ventilation, air conditioning and artificial lighting. In particular, the International Energy Agency [1] has focused attention on that artificial lighting consumes a large amount of energy than heating and cooling system do. The issue of developing energy saving strategies has certain challenges. In office buildings, some units require considerable energy, such as lighting and the Heating, Ventilating and Air Conditioning (HVAC) systems. These represent an important part of the total energy consumption in buildings. According to energy efficiency, building in South Africa consumes 17% of the total electric energy [2]. Thus, the need to optimally control and save energy has motivated interest in the energy consumption of buildings. Shweta Jain et al. developed a technique which uses participatory sensing to compute real-time occupant discomfort, forcing occupants to use portable foot heaters, distribution map within the office building and enable energy saving without additional hardware [3]. Vangelis et al. [4] presented a control tool that took into account the monitoring of energy consumption in building sector based on real time. Zanoli [5] developed a model predictive control for energy savings in building automation. The proposed system integrates energy consumption for heat and light power supply, with a green energy-supplying source.

Due to the lack of an important save of electric energy, this study investigates the use of photovoltaic (PV) to power the HVAC and lighting equipped with a suitable control strategy to improve energy saving within a building, especially in office buildings where there are reports of high misuse of electricity.

Multiples control system and algorithms have been d developed for HVAC and lighting system control in office buildings in order to reduce the energy consumption. An energy-saving lighting control system has been developed, in which the system detects the employee's present position in the office and provides the corresponding brightness to the person [6]. An intervention study was done on energy saving which investigated the benefit of controlling ceiling lighting based on occupant presence information obtained at each desk [7]. Lighting energy saved due to the manual control relies on occupant awareness and attitude on energy efficiency. [8]. In his research, Tiller [9] for lighting control, used an intelligent system that gathering information from an occupancy detector. However, it was concluded that disturbances such as huge sensitivity detector and delay may reduce the savings target.

The presence of solar irradiation or daylight has a greater impact in the assessment of building energy consumption [10]. Luigi Martirano [11] took advantage of BAS (Building Automation System) and solar energy to validate a fuzzy logic approach to optimize the level of energy performance and comfort in an office space. Based on daylight and lighting of unoccupied spaces, the lighting control plays an important role by reducing the energy wasted in unoccupied hours in the building and automatically adjusts the electrical light level according to daylight based on the façade ([12,13]). In the above research the weather condition can be a problem for the use of the day light such as—sky covered, the season of the year particularly in winter.

HVAC system is assembled of distinct components that can be controlled isolatedly from one another. Control systems reveal that humidity, ventilation and zone temperature are the most significant element in an HVAC system. To solve problem that affect the HVAC electric energy consumed, such as occupancy, loads, temperature, some researchers examined the application of control algorithm based on occupancy because it has been find out that an HVAC system consumes energy even in an unoccupied space, inducing an important waste of energy. Benezeth et al. [14] designed an algorithm control system using a camera to track the presence of a human in an area. Lin Qiu et al. [15] used the control modules of America Automated Logic Corporation to establish applicable algorithmic modules to overcome the issue of bad indoor air quality which affects the energy saving.

Some of the researchers conducted work on the HVAC especially on its internal control algorithm but did not consider the occupancy into the control loop. Therefore, associating the motion sensor with the internal control of the HVAC would improve energy efficiency, which has been achieved by Reference [10]. Other researchers did not take the changes in electricity price and the ambient temperature into account but did consider the time delay and occupancy. Therefore, a control algorithm designed on any change in occupancy detection and weather conditions joined with the integration of a PV solar system will produce a significant result in terms of energy saving within an office building.

Integrated lighting control and HVAC control system have also been used to save the energy consumption within building. Martirano et al. [16] developed a fuzzy control technique for energy saving and for users' comfort where the lighting system and the heating, ventilation and air conditioning systems were integrated. In their design, the amount of daylight and the comfort limit were managed by a shading system. This shading system, allows an amount of solar radiation to penetrate inside a room which will minimize the electric load, adjusting the HVAC and lighting system to match temperature and illuminance requirements ([17–19]).

In the above-mentioned technique, the weather conditions could be a user problem; this isbecause if there is not daylight, the lights and the HVAC will consume more electricity from the national grid. In this study, it was found that removing the daylight control system and combining it with a PV sensor produced a better result in terms of energy saving.

The reason for undertaking a passive technique prior to any HVAC or electrical installation design is viewed as a good approach to make sure that an underground building design is as efficient as possible to reduce the demand for additional heating or cooling or the reliance on too much artificial lighting, thereby reducing the energy requirement of the building from the start. The energy efficiency of a building can be enhanced by using active and passive design techniques. Active design techniques focus on using different equipment inside the building such as energy efficient HVAC systems and electrical installations such as artificial lighting control systems. An energy efficiency of Building layout, shape and building envelope have been improved due to a passive design technique [20]. The material used in the building envelope has a major effect on thermal comfort, energy used by the HVAC systems and lighting levels and therefore, special attention ought to be given to insulation materials, fenestration, shading devices and the walls and roofs of buildings [21]. In this regard, the building envelope system of a building and the insulation materials used can reduce the electrical energy consumption of an HVAC heating system in the warmer months of the year by up to 20%. ([22,23]). Many buildings have been constructed without taking their shape or envelope into consideration. In such buildings, an active design technique must be used to enable the HVAC control system and lighting control system to be based on an intelligent system that will assist in saving electric energy.

Energy management is very important in term of energy reduction as well as in its contribution to reducing fossil fuel consumption and earth warming [24]. There is a difference in energy management in the office building from energy management in a residential house as energy management in a public building does not give the worker an opportunity to take decisions in this regard [25]. Energy management implementation in the office can affect the budget so the decision is one that needs to be taken by top management.

Joelle et al. [26] proposed a method of energy management in public buildings, automatic advice based on algorithm is advised, their proposed model allows the Building Energy Management System (BEMS) to use the appropriate co-design of the system by following the enumerated procedures—the identification of key stakeholders, the pre-analysis of key stakeholder, face-to-face qualitative semi-directed interviews, workshop with the stakeholders, pilot of the Building Energy Management System in the building. To reduce the energy around the office building, some researchers have opted for smart meters [27] stated that having feedback from a smart meter can reduce energy consumption. Immediate feedback allows the user to be more efficient usage in term of energy reduction. A smart meter showing the energy consumed in terms of cost allows the consumers to see their efforts in energy efficiency and the environment.

The study reviewed above requires that the consumer be aware of the energy consumed or that the user has to act in saving energy. Due to the activities that may be demanded of workers in the workplace and the stress that can be experienced by workers in an office building, employees may be too distracted to engage in energy saving. This paper suggests that the most beneficial strategy is intelligent system management, where the components that have the highest consumption rate, such as HVAC light, are controlled by considering the temperature, the workers' activities, the weather and the load itself, combined or integrated with a PV system.

2. Materials and Methods

This section covers the mathematical model or the system modelling, the material used and the simulation of the proposed intelligent system used to reduce electrical energy around the office building where each office has a length of 3.81 m (12.5 feet) and a width of 4.26 m (14 feet) with a sliding window of 2.5 m² see figure.

The mathematical relationship between energy consumption as a dependent variable and occupant activity, temperature and load consumption as the independent variable were modelled and the simulation using MATLAB Simulink was presented. The section also presents a developed intelligent integrated automatic switching system in term of occupant activities, temperature, weather condition, loads consumption and the PV system as an independent variable in order to further reduce energy consumption in the office building. The material used to collect real data (light energy consumption, HVAC energy consumption) is described.

2.1. Office HVAC Energy Model Based on Temperature

Based on the fundamental principles of thermodynamics and heat transfer, a control volume analysis of the HVAC was conducted as shown in Figure 1.

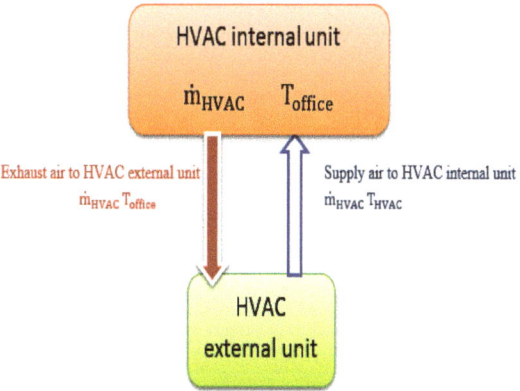

Figure 1. Schematic of the control volume.

The HVAC energy consumption is divided into two—the energy consumed by the fan when the compressor is off and the power consumed by the entire system when the compressor and the fan are ON. In the above schematic diagram, the electric energy consumed by the fan E_f can be computed as follows [28]

$$E_f = \frac{\dot{V}_{HVAC} P_{tot}}{\eta} \quad (1)$$

where η is the fan efficiency and $\dot{V}_{HVAC}$ is the air volume flow rate of the fan P_{tot} the total pressure of the fan.

2.2. Office HVAC Energy Consumption Model during Winter and Summer

2.2.1. During Winter

Based on the heat transfer principle, the heat energy is defined as follow [29]

$$Q_H = \dot{m}_{HVAC} C_p (T_{HVAC} - T_{office}) \quad (2)$$

In Equation (2), Q_H is the heat energy transfer, = $\dot{m}_{HVAC}$ is the mass flow rate, C_p is the air heat capacity and T_{HVAC} and T_{office} are the HVAC temperature and the office temperature, respectively. The energy efficiency ratio (EER) of the air conditioner in heating mode is the ratio of the heating capacity to the input power and the coefficient of performance is the ratio of the heat energy output (Q_{OUT}) to the input energy (E_H).

$$Q_{OUT} = Q_H + E_f \tag{3}$$

$$COP_{heating} = \frac{Q_{OUT}}{E_H} \tag{4}$$

Due to Equations (2)–(4), the energy consumed by the office during winter is determined as

$$E_H = \frac{\dot{m}_{HVAC} C_p (T_{HVAC} - T_{office}) + E_f}{COP_{heating}} \tag{5}$$

$$COP_{heating} = EER * 0.29307 \tag{6}$$

Figure 2 is a schematic representation of the HVAC energy consumed in winter as simulated in MATLAB Simulink® so that different variables were constructed and represented as the block in Simulink in order to simulate the HVAC energy in winter.

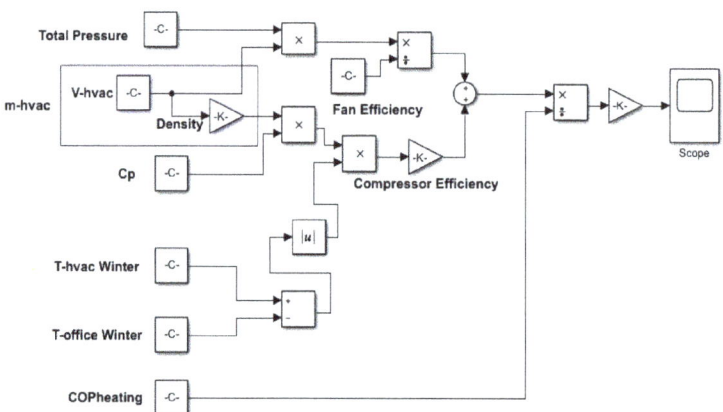

Figure 2. Heating Ventilation and Air Conditioning (HVAC) energy consumed during winter, modelled in MATLAB Simulink®.

2.2.2. During Summer

Based on the split HVAC, the energy efficiency ratio (EER) of the air conditioner in cooling mode is the ratio of the cooling capacity to the input power and the coefficient of performance is the ratio of the cooling energy output (Q_{OUT}) to the input energy (E_c).

$$COP_{cooling} = \frac{Q_{OUT}}{E_c} \tag{7}$$

From (2), (3) and (7) the cooling energy E_c consumption (see Figure 2) during summer is modelled as:

$$E_c = \frac{\dot{m}_{HVAC} C_p (T_{HVAC} - T_{office}) + E_f}{COP_{cooling} * SHF} \tag{8}$$

$$COP_{cooling} = EER * 0.29307$$

where EER is the Energy Efficient Ratio which in this case is the ratio of cooling capacity (Btu) of a required HVAC to it input power. Where SHR is the sensible heat factor which is the ratio between sensible heat load and the total heat load of the office room

The simulated model was built in Simulink (Figure 3) where independent variables were built as block and the result was shown by the scope.

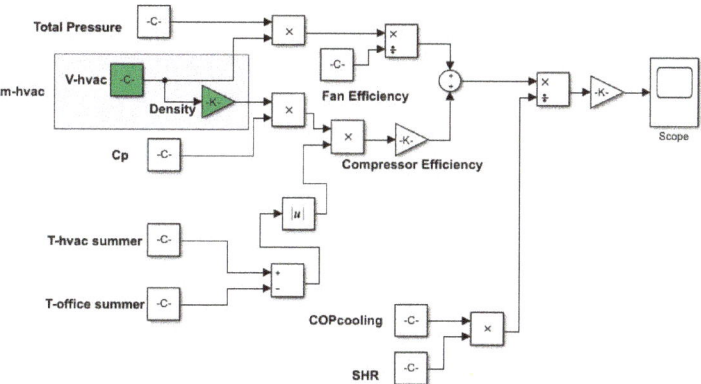

Figure 3. HVAC energy consumed simulated during summer, modelled in MATLAB Simulink.

2.3. Office HVAC Energy Efficient Based on Occupancy Control, Temperature, Loads Control

Based on the above equation regarding the HVAC office energy, consumption varies due to the office temperature changing and the occupancy behaviour.

2.3.1. Occupancy Control Strategy

To save the HVAC energy consumption in the office, the occupancy sensor is installed in the office in order to alert if the room is occupied (Figure 4) or when it is unoccupied (Figure 5). Therefore, let $k \in \{0, 1\}$ be defined as the state variable for the occupancy sensor and the HVAC by:

$$k_t = \begin{cases} 1 & if\ office = occupied \\ 0 & if\ office = unoccupied \end{cases} \qquad (9)$$

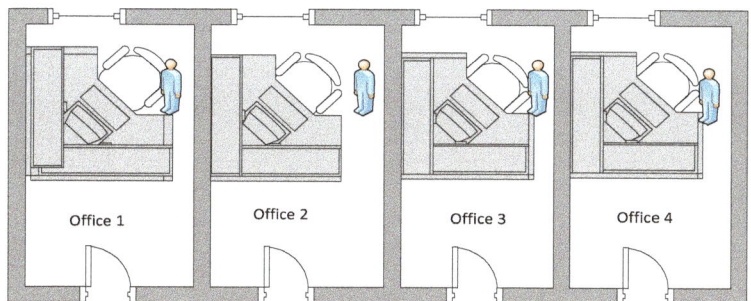

Figure 4. Four occupied offices.

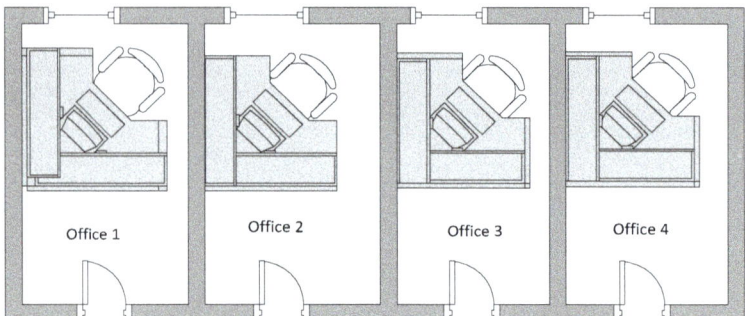

Figure 5. Unoccupied offices.

The presence of this sensor permits a schedule based on human activity in an office and the working hours to be drawn. The thermostat control strategy must be designed based on the season or weather condition, the occupancy sensor and the loads used.

2.3.2. HVAC Temperature, Occupancy, Weather, Loads Control Strategy in Summer and the Impact on the Energy Consumption

In summer, most office environments need to be cooled in order to reduce the heat. Thus, the temperature must be set in the cooling mode. In order to reduce the energy consumption, the temperature ($T_{office} \in \{T_{cold}, T_{md}, T_{hot}\}$) needs to be changed based on the occupancy sensor and working hours and loads consumed; let E_c be defined as the energy consumed during summer and **w** the energy consumed threshold. The control strategy can be expressed by:

$$T_{office} = \begin{cases} T_{cold} & if\ k_t = 1, \quad E_c < \mathbf{w} \\ T_{md} & if\ k_t = 1, \quad E_c\ otherwise \\ T_{hot} & if\ k_t = 0, \quad E_c < \mathbf{w} \end{cases} \qquad (10)$$

Due to the above condition, during summer the temperature rises, resulting in the environment being hot. However, the employee in the office needs to be comfortable. Based on the occupancy detector, the HVAC must be ON in the T_{cold} mode if the room is occupied—T_{cold} is the cooling temperature. When the office is not occupied, the HVAC should be in the T_{hot} (Figure 6) mode and this last mode is the technique to save the HVAC energy consumption during winter.

In case the HVAC electricity used is greater than the threshold value **w**, the HVAC must be ON in the T_{md} (medium temperature) mode (see Figure 7).

During summer, the HVAC energy consumed in an office when that office is occupied and when it is during, the working hour is derived as:

$$E_c = \frac{\dot{m}_{HVAC} C_p (T_{HVAC} - T_{cold}) + E_f}{COP_{cooling} * SHF}, \quad k_t = 1, \quad E_c < \mathbf{w} \qquad (11)$$

When the HVAC energy consumed in an office during summer is equal or above the threshold, the energy can be formulated as follows:

$$E_c = \frac{\dot{m}_{HVAC} C_p (T_{HVAC} - T_{md}) + E_f}{COP_{cooling} * SHF}, \quad k_t = 1, \quad E_c \geq \mathbf{w} \qquad (12)$$

The electric energy consumed in an office during summer when the office is not occupied and when it is out the working period can be obtained from the following equation:

$$E_c = \frac{\dot{m}_{HVAC}C_p(T_{HVAC} - T_{hot}) + E_f}{COP_{cooling}*SHF}, \quad k_t = 0, \quad E_c < w \tag{13}$$

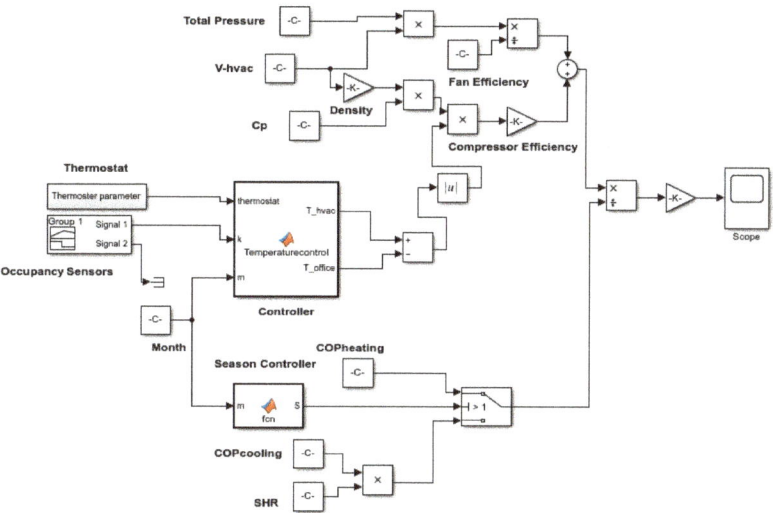

Figure 6. HVAC office energy control based on temperature, occupancy activities and weather, using MATLAB Simulink.

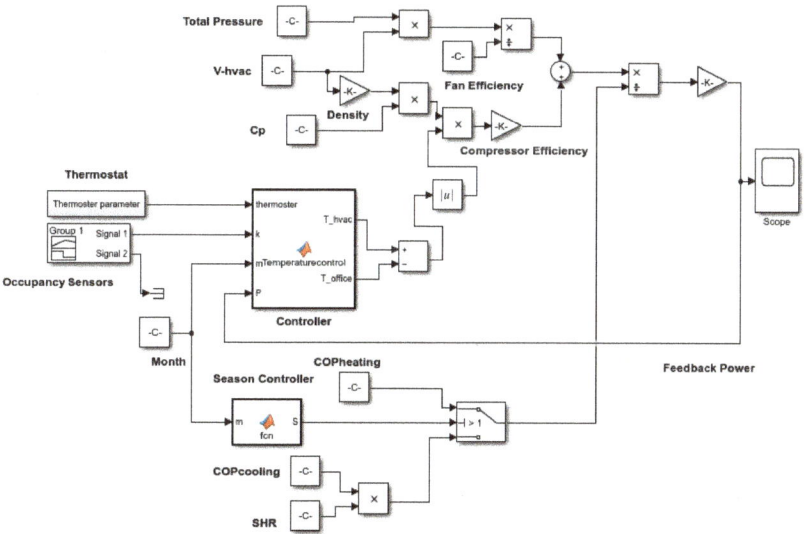

Figure 7. HVAC proposed energy control strategy using MATLAB Simulink®.

The schematic diagram above (Figure 6) was built and simulated in Simulink. It represents the HVAC energy controlled by occupancy, temperature and weather condition. The input variables were

2.3.3. HVAC Temperature, Occupancy, Weather, Loads Control Strategy in Winter and the Impact on the Energy Consumption

In winter, several offices' environments need to be heated; in order to increase the heating, the temperature has to be set in the heating mode. In order to reduce the energy consumption, that temperature ($T_{office} \in \{T_{cold}, T_{md}, T_{hot}\}$) needs to changed based on the occupancy sensor and working hours load consumed; let E_c be defined as the energy consumed during summer and **w** the energy consumed threshold. The control strategy can be expressed by:

$$T_{office} = \begin{cases} T_{hot} & if\ k_t = 1, \quad E_h < \mathbf{w} \\ T_{md} & if\ k_t = 1, \quad E_h\ otherwise \\ T_{cold} & if\ k_t = 0, \quad E_h < \mathbf{w} \end{cases} \quad (14)$$

Offices need to be kept warm in winter in order to satisfy employee health requirements, so the occupancy sensor, the HVAC, must go ON in the T_{hot} mode if the office is occupied, otherwise the HVAC must go ON in the T_{cold} mode, which is the energy saving mode (see Figure 7). Thus, during winter when an office is occupied and when it is in the working interval time and the energy is less than the threshold, the HVAC energy consumed is derived as:

$$E_h = \frac{\dot{m}_{HVAC} C_p (T_{HVAC} - T_{hot}) + E_f}{COP_{heating}} \quad for\ k_t = 1,\ and\ E_h < \mathbf{w} \quad (15)$$

And this can also be shown as follows:

$$E_h = \frac{\dot{m}_{HVAC} C_p (T_{HVAC} - T_{hot}) + E_f}{COP_{heating}} \quad (16)$$

When the HVAC energy consumed in an office during winter is equal to or above the threshold **w** (Figure 6), the energy consumed can be formulated as follow:

$$E_h = \frac{\dot{m}_{HVAC} C_p (T_{HVAC} - T_{md}) + E_f}{COP_{heating}}, \quad k_t = 1,\ E_h \geq \mathbf{w} \quad (17)$$

In this situation, the output power is used as feedback to allow the controller to take a new decision, see Figure 8.

Figure 8. The HEXING CIU EV-KP. Source: Photograph by Simplice I, T Noubissie.

The electric energy consumed in an office during winter when the office is not occupied and when it is out the working interval can be obtained from the following equation:

$$E_h = \frac{\dot{m}_{HVAC} C_p (T_{HVAC} - T_{cold}) + E_f}{COP_{heating}}, \text{ for } k_t = 0, \text{ and } E_h < w \tag{18}$$

The total HVAC electric energy used for cooling the office in summer for a day is

$$E_T = E_c + E_h \tag{19}$$

Figure 7 represents the controlled HVAC energy consumed in winter or in summer simulated in MATLAB Simulink®. In this simulation, the weather season has been considered and the occupancy sensor, the temperatures have also been used as input and the loads consumed were also used as the feedback to the control.

2.4. HVAC Energy Saved due to Temperature and Occupancy Control and Loads Control

Based on the occupancy, loads and temperature control, the amount of electric energy saved in offices during a working day in summer can be summarized and is obtained from the following equations:

$$E_{c\_save} = n * \frac{\dot{m}_{HVAC} C_p (|T_{cold} - T_{hot}|)}{COP_{cooling} * SHF}, \text{ for } E_c < w \tag{20}$$

Or

$$E_{c\_save} = n * \frac{\dot{m}_{HVAC} C_p (|T_{cold} - T_{md}|)}{COP_{cooling} * SHF}, \text{ for } E_c \geq w \tag{21}$$

$$n = \text{number of offices}$$

The amount of electric energy saved in offices during a working day in winter can be obtained by subtracting the energy consumed in winter when the office is not occupied from the energy consumed when the office is occupied; thus, the following equation is determined:

$$E_{h\_save} = n * [\text{Equation } (15) - \text{Equation } (18)] \tag{22}$$

$$E_{h\_save} = n * \frac{\dot{m}_{HVAC} C_p (|T_{cold} - T_{hot}|)}{COP_{heating}}, \text{ for } E_h < w \tag{23}$$

Or

$$E_{h\_save} = n * [\text{Equation } (15) - \text{Equation } (17)] \tag{24}$$

$$E_{h\_save} = n * \frac{\dot{m}_{HVAC} C_p (|T_{cold} - T_{med}|)}{COP_{heating}}, \text{ for } E_h \geq w \tag{25}$$

2.5. Data Collection

2.5.1. Light Energy Consumption Data Collection

In order to simulate the design, real data is needed, so the HEXING CIU EV-KP smart meter was used to collect daily data. The HEXING CIU EV-KP, (Figure 8) is a customer interface unit with a keypad for credit charging. It communicates with the metering unit by M-BUS (meter bus) for power energy consumption and credit balance monitoring, credit limitation and credit charging. This data was collected for one month from 5 fluorescent tubes in one of four offices in the building known as *Lampropoulos Heights*, in Pretoria.

As mentioned above, the graph in Figure 9 simply illustrates the lighting energy data collected.

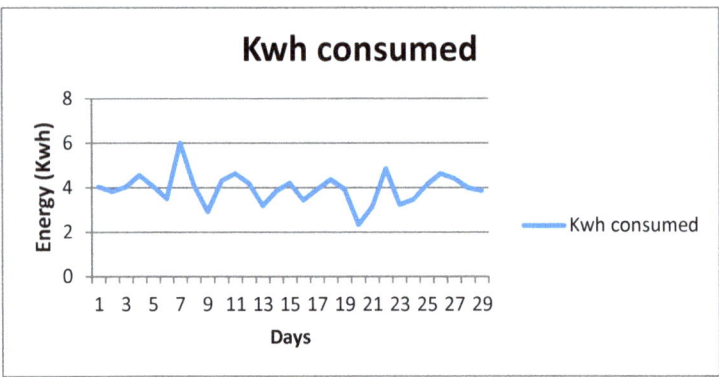

Figure 9. Fluorescent tubes' energy consumed in an office.

2.5.2. Solar Irradiation and Temperature Data Collection for PV Power Generated

To know what the performance of the PV is or what the maximum power that the PV can generate on a specific site or area is, we need to know the solar irradiation and the ambient temperature of the location. Twelve months of data were collected from the Eskom website in Aggeneis (Northern Cape region) Location: 29°17.721′ S 18°48.319′ E, Where the irradiation and temperature were collected each hour [30].

2.6. Lighting Energy Control Strategy

To establish a lighting energy control strategy in the office, workers movements or activities in and out the office was considered due to their lack of awareness on energy efficiency, they may not remember to switch off the lamp when they move out of the office. That lack of awareness has a distinct impact on energy usage.

Let P_t be the power consumed by the lamp at time t, to alert if the room is occupied or unoccupied; let $k \in \{0, 1\}$ be defined as the state variable for the occupancy sensor and the HVAC by:

$$k_t = \begin{cases} 1 & if\ office = occupied \\ 0 & if\ office = Non\ occupied \end{cases} \quad (26)$$

From the above state variable, the power consumed by the lamp can be determined as follows:

$$P_L = \begin{cases} P_t & if\ k_t = 1 \\ 0 & if\ k_t = 0 \end{cases} \quad (27)$$

So, the office power lamp consumed is expressed as

$$P_L = P_t * k_t \quad (28)$$

The total energy consumed by the offices when the offices' lamps are on, is expressed as

$$E_l = n \sum_{i=1}^{t} [P_{i,1}, P_{i,2}, P_{i,3}, \ldots \ldots P_{i,n}] * k_i * i \quad (29)$$

where n is the number of offices.

2.7. Load (HVAC, Light) Management Strategy Using PV and Grid

In this section, the PV system is modelled and simulated to further save electric energy around the office building and the fuzzy logic control is designed to manage the controlled loads (HVAC, light) by giving priority to the PV system to feed the electric loads (HVAC, light) coming from four different offices. In case the electric loads are less than the PV energy supplied, the grid will be offline. However, if the loads' energy rises and become more than the PV supply, in this case the grid together with the PV plant, will feed the loads. The all system is illustrated in Appendix A (Figure A1)

2.7.1. PV System

A grid-tied system was designed. The mathematical model of the solar panel was developed in MATLAB Simulink. The PV designed here was a 4000 W array with 5 series (n_s = 5) modules per string of 198.25 W each and individual module had 72 cells, those 5 series modules was thereafter connected in 4 parallel string (n_p = 4). Each module has consisted of voltage and current at the maximum power point of 30.5 V and 6.5 A. Therefore, p = 30.5 * 6.5 * 5 * 4 = 3.97 KW which was approximately equal to 4 KW. The specification used for the mathematical model of the solar array see Appendix C (Table 2). Figure 10 is the PV array block used in MATLAB Simulink® and Figure 11 represent the block diagram of the all PV power system.

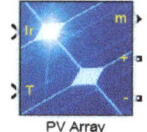

Figure 10. Photovoltaic (PV) array from MATLAB Simulink.

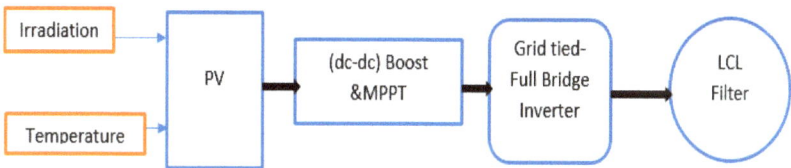

Figure 11. PV system Block diagram.

2.7.2. Fuzzy Logic Control

Fuzzy logic is the set of mathematical principles designed based on human knowledge of the system or on the degree of membership. This fuzzy logic consists of four components—fuzzification, fuzzy rules, fuzzy inference system and defuzzification.

In this part of the research we need to define the fuzzy control input and output values where this input goes through the fuzzy control system which controls the grid based on the input values of the load (HVAC and light) consumed and the PV power generated see Figure 12.

The first input is the sum of the power consumed (PL) see Figure 13 in all four offices and the order input is the power generated by the PV system (PV) see Figure 14. These inputs are used by the fuzzy logic control to provide a signal to the grid system to synchronize with the solar system in order to feed the loads when it is needed and to bring the grid off when no needed see Figure 16.

The fuzzy inference system (FIS) used in this research is a direct method called the Mamdani' fuzzy inference method (mentioned earlier). It is used in a fuzzy rule (IF-THEN) to determine the rule outcome from the given rule input and output information. This fuzzy inference system is the most used because of its simple structure and is more suitable for the system design.

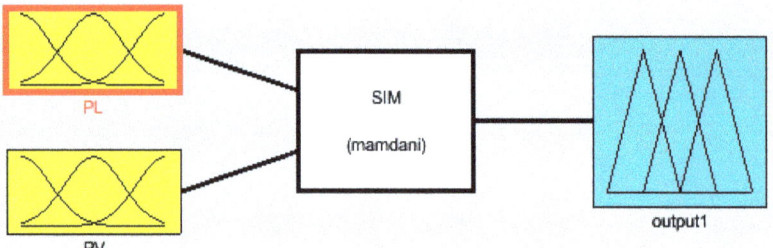

Figure 12. Fuzzy Designer.

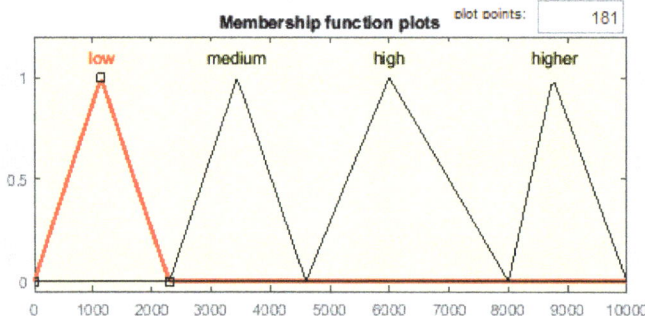

Figure 13. Input variable Power consumed by the Loads (HVAC & light power used).

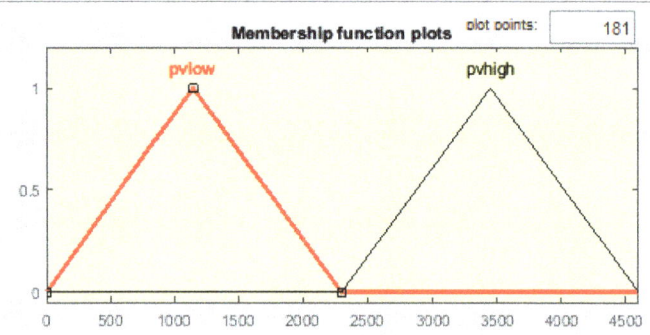

Figure 14. PV Input variable.

To allow our fuzzy to control the switching system of the grid, the defuzzification technique was applied in this research. The centroid defuzzifier technique was used.

The centroid defuzzifier technique was used to convert fuzzy sets or the membership of output linguistic variables into real numbers for the grid control output (Y) see Figure 15.

The block diagram in Figure 16 below illustrates the proposed fuzzy logic system articulated above.

A discussion on the impact of temperature and human activities on HVAC and light energy consumption in four close offices is presented, as well as the influence of the temperature control, human control and the loads' control. The results of the use of fuzzy logic control for a switching system combined with a PV system to further decrease the consumption of energy in the offices, are provided.

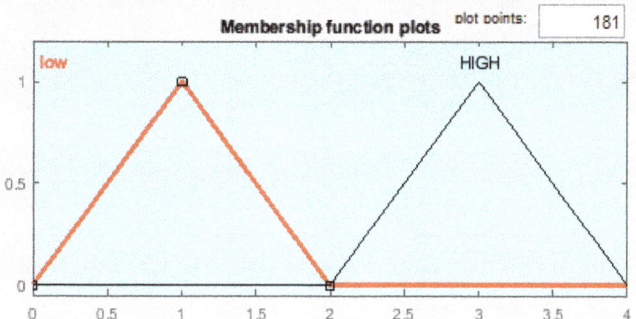

Figure 15. Fuzzy Output variable.

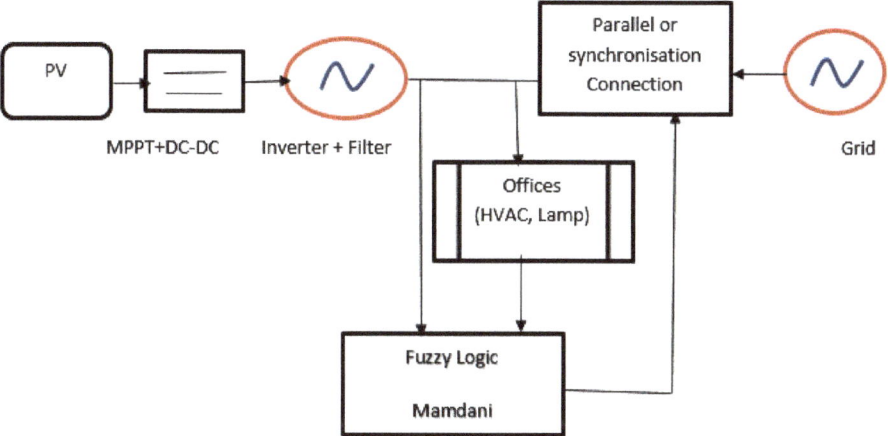

Figure 16. Block diagram of proposed fuzzy logic.

3. Results

3.1. Occupancy Sensor Results from Four Offices

To simulate the occupancy activities in offices, the data used in the investigation revealed the following results and are represented in Table 1 below and the occupancy sensor output in offices 1 and 2 is graphically provided in Figure 17.

Table 1. Office 1 and office 2 working hours' schedule.

Time	8:00–10:30	10:30–11:30	11:30–12:30	12:30–14:00	14:00–16:00	16:00–16:30	16:30–17:00
Occupancy state	1	0	1	0	1	0	1

As Table 2 shows, the graph in Figure 17 was is based on the human activities in office 1 and 2 during 9 h of working, so when the office is occupied, the sensor registers 1 and when it is not occupied, the sensor registers 0. Table 2 below, indicates the schedule of working hours in offices 3 and 4.

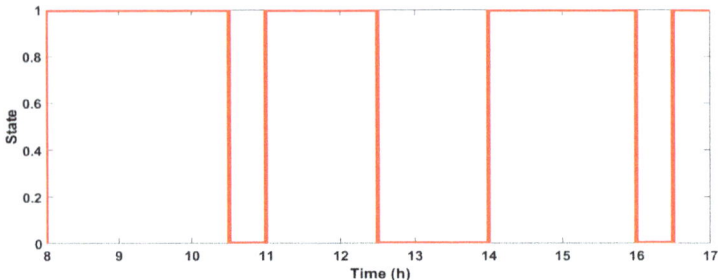

Figure 17. Occupancy sensor output in offices 1 and 2.

Table 2. Office 3 and office 4 working hours schedule.

Time	8:00–8:30	8:30–9:00	9:00–11:00	11:00–11:20	11:20–12:00	12:00–13:30	13:30–16:00	16:00–16:30	16:30–17:00
Occupancy state	1	0	1	0	1	0	1	0	1

Figure 18 depicts the different states of the employees in offices 3 and 4; as mentioned earlier the state is represented by logic 1 when the office is occupied and 0 when it is not occupied.

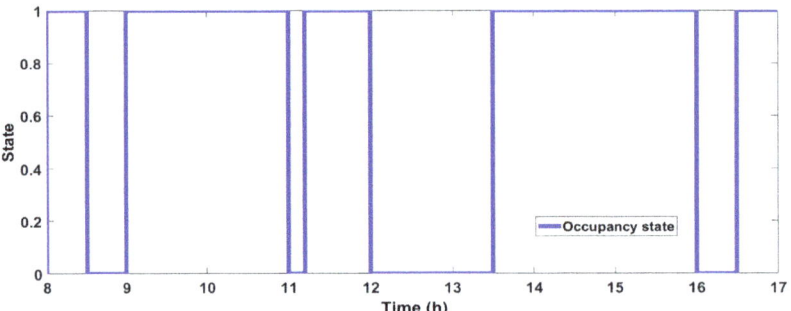

Figure 18. Occupancy sensor output in offices 3 and 4.

Both graphs (Figures 17 and 18) indicated that all employees started their work at 8 h00 but had different breaks, according to their varying needs.

3.2. Office HVAC Energy Consumed Results for the Assumed Model in Normal Conditions Without any Control

Office HVAC Energy Consumed during Summer and Winter

For the following result to be tabulated, the parameters and values were considered based on a normal 9000 Btu/ HVAC—see Appendix B (Table 1).

The graphs below, (Figures 19 and 20) represent the electric power used to cool the office during summer and to heat during winter by the office HVAC and it is based on the model developed. We realized that for a specific office the desired temperature choice is 24 °C, while the power consumed is constant (1.658 KW) during summer and 1.235 KW and it is constant during the working hours.

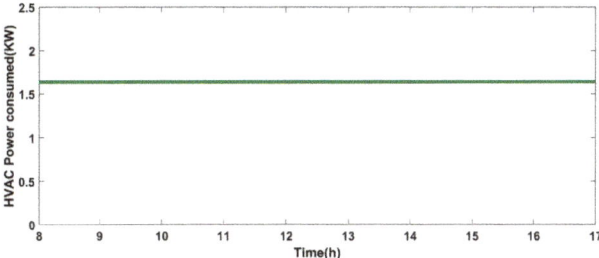

Figure 19. Power consumed by HVAC office based on model assumed during summer.

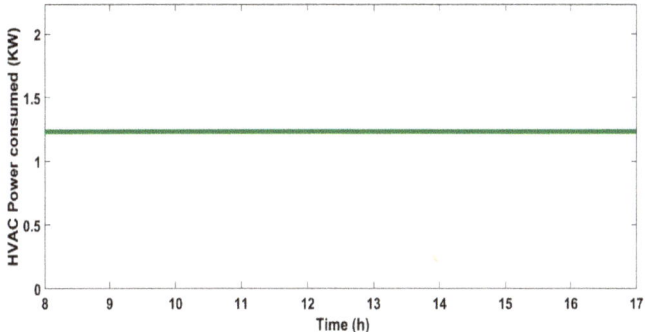

Figure 20. Power consumed by HVAC office based on model assumed during winter.

3.3. HVAC Power Consumed Based on Load, Temperature and Occupancy Control for Office

3.3.1. During Summer and Winter

The graphs in Figures 21 and 22 below are simulated results of the HVAC power consumed after the loads control were combined with the temperature and occupancy control. When the threshold power is reached for offices 1 and 2 and offices 3 and 4, when they are occupied, the electricity utilized in summer is 1.51 KW and in winter 1.143 KW and when the offices are not occupied, the electricity utilized is 0.9188 KW and 0.68 KW respectively.

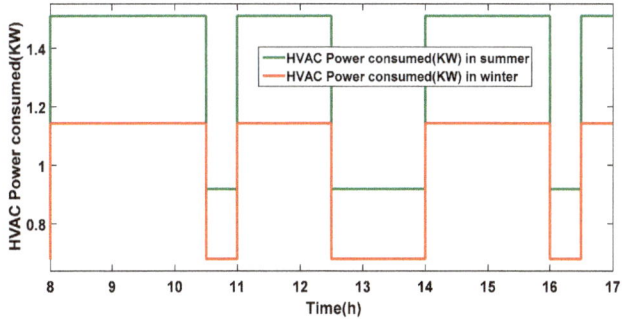

Figure 21. HVAC Power consumed based on temperature, occupancy and load control for office 1 during summer and winter.

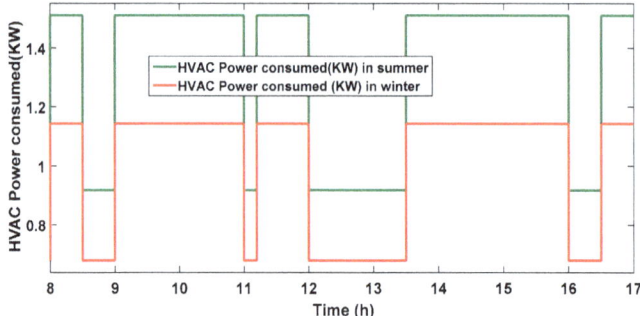

Figure 22. HVAC Power consumed based on load, temperature, occupancy and control for office 3 during summer and winter.

3.3.2. Room Temperature Behaviour

Due to equations (medium summer and winter) the energy power consumed by offices automatically reduces when the energy threshold set is reached in summer as in winter; the inside temperature which is the reference temperature (24 °C) in summer during the occupied period increased by one degree (25 °C) and in winter that temperature drops to 23 °C as indicated in Figure 23 This increase and drop of temperature have a positive impact on HVAC energy saved.

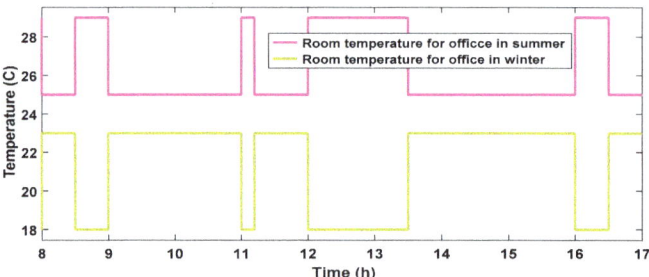

Figure 23. Room temperature for offices in summer and winter.

3.3.3. Total Energy Saved for Summer and Winter

Table 3 below lists the electric energy consumed and saved for the four offices in one year. Additionally, it shows an important saved of electric energy in summer for office 1 and 2 (18.23%) and in winter for office (26.11%). The final electric energy saved for the listed table shows that with the loads control added to the temperature and occupancy control contribute to further save energy.

3.4. Offices' Light Energy Consumed Results with Occupancy and During Summer after Simulation

The data for 5 fluorescent tubes was collected twice a day in September 2017, in the morning at 8:00 and in the afternoon at 17:00. This time was chosen based on the worker timetable. After a month (20) days we collected the data illustrated as a graph in Figure 24. This graph represents the energy consumed.

Table 3. HVAC energy used in offices 1&2 and offices 3&4 in one year after load, temperature, occupancy control.

Offices	Electric Energy Used (KWh)	Electric Energy Saved (KWh)	Savings [R] Kwh = R1.2	Percentage (%)
Office 1&2 Summer period	3875.840	899.20	1079.04	18.83%
Office 1&2 Winter period	1460.75	317.680	381.216	17.86%
Office 3&4 Summer period	3812.566	962.24	1154.688	20.15%
Office 3&4 Winter period	1314.027	464.372	557.247	26.11%
Total	10,463.183	2643.492	3172.191	20.16%

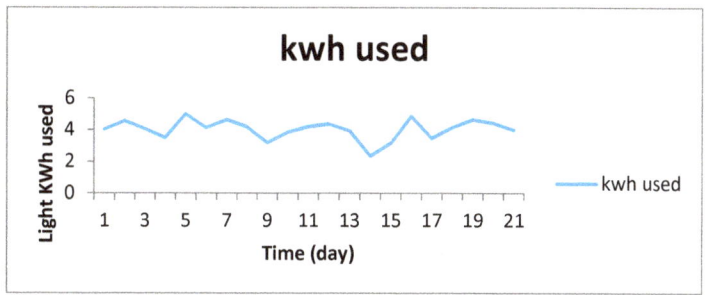

Figure 24. Five Fluorescent tubes Energy consumed in on office.

Offices (1–4) Light Energy Consumed Results with Occupancy Control

To be able to simulate the graph below, light energy data was collected in a specific office per day. To obtain the energy consumed per hour, the total energy consumed per day was divided by 9 since the work starts at 8 h and finish at 17 h. This hourly energy consumed is then used based on the occupancy sensor1 and sensor2 (Figures 16 and 17) to obtain the light energy shown in Figure 25.

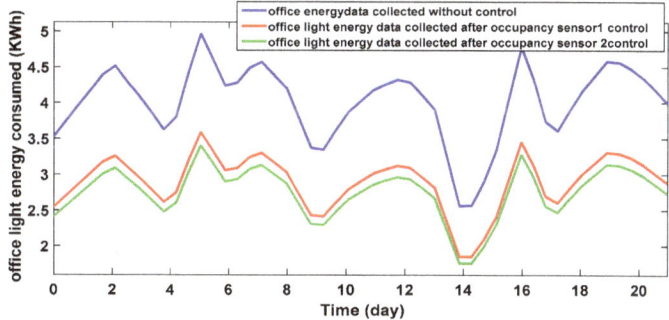

Figure 25. Office light energy consumed under occupancy sensor1 and 2.

Based on those different occupancy sensors the light energy consumed was diminished for sensor 1 and further diminished for sensor 2, thus the table below, Table 4, records the comparison between those scenarios.

In Table 4 above, we see that the occupancy sensor control allows the initial light energy used without any control to drop from 27.8% from office 1 and 2 and 31% from office 3 and 4.

Table 4. Office light total energy consumed comparison.

Office	Electric Energy Used (KWh) (Month)	Electric Energy Saved (KWh) (month)	Savings [R] Kwh = R1.2	Percentage (%)
Four Offices without sensor	84.58	-	-	-
Office1&2 with sensor1	61.08	23.5	28.67	27.8%
Office3&4 with sensor2	57.95	26.63	32.488	31%

3.5. Load (HVAC, Light) Management Strategy Results Using PV and Grid

3.5.1. Fuzzy Logic Result

Due to the intelligence systems of the fuzzy logic, the grid system was not required to support the PV system to feed the load. The results shown in Figure 26 represent two samples of the fuzzy logic rule viewer where rule number 4 was chosen among eight (8) rules to show how it operates. The first column of Figure 26 shows that the membership of input PL is at a higher range, meaning that the offices' power consumed is at 8.58 KWh. The second column also shows that the PV power generated 4.04 KWh is at a medium range as defined in the figure below. The centroid number is enumerated in the third column showing that the output is high, therefore the grid system, in this case, will be connected in series with the PV system.

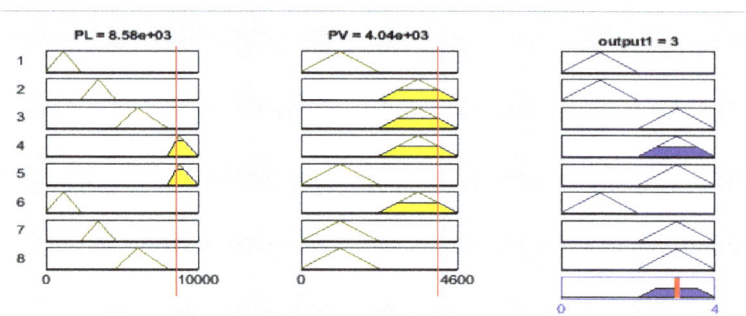

Figure 26. Rule number 4 viewer and the crisp values.

Due to the fuzzy logic, all simulations obtained in March give different state results, as illustrated in the graph below (Figure 27). The graph shows when the grid was solicited to support the PV system (Grid control state is at 1) and when it was not (Grid control state is at 0).

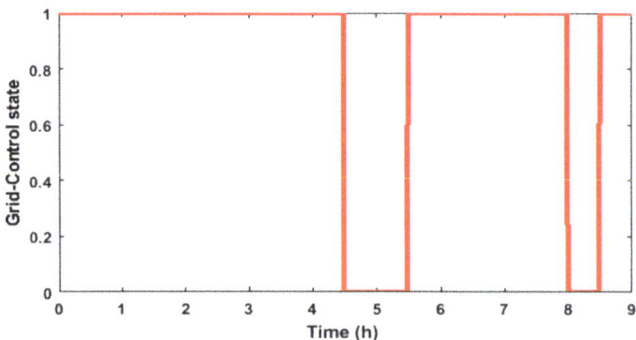

Figure 27. Grid control state due to fuzzy logic.

3.5.2. Office Load Management Results Due to the Intelligent Control of the PV and Grid

Simulated Results in Winter

The irradiation and temperature in Figure 28 are a one-day data collection and Figure 29 is a one-day simulation result in winter. The irradiation and temperature have an impact on the energy produced by the PV system (Figure 29 third graph). The first graph in Figure 29 shows how the offices load are fed by the PV system and the grid. Due to others results in the offices HVAC, lighting energy consumed in winter based on the occupancy control and temperature are lesser than other period of the year.

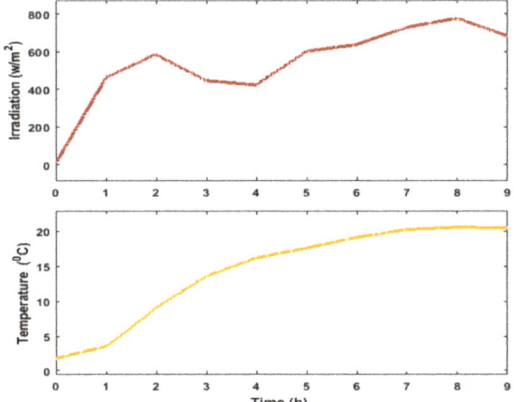

Figure 28. A day (9 h) irradiation and temperature curve in winter.

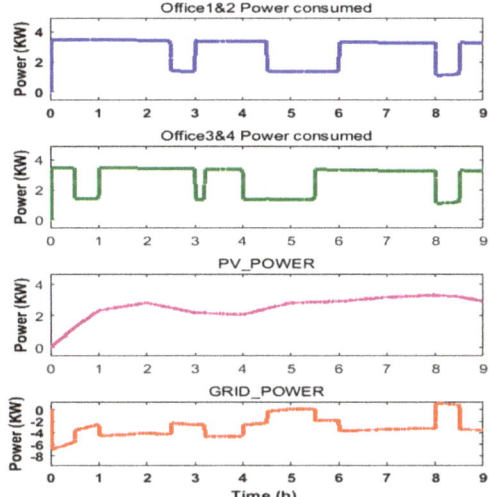

Figure 29. PV system and Grid feeding the offices load (HVAC and lighting) in winter.

We can see in Figure 29 that during the working hours and when the offices are occupied, the PV system and the grid both supplied the room but priority is given to the PV system to further save energy or reduce the offices' bills. So, at 8 h which is represented by 0 in these graphs which is when the workers start, at this time the irradiation and the temperature is at the lower values meaning that the PV system generates less power than the loads demand, therefore the grid system, in order to meet the demand, come into the feeding procedure. As soon as the PV become available (to be able to produce the necessary power), the grid system went off (0 W).

The table below (Table 5), resumed the PV energy generated, the total offices HVAC and light consumed, the grid energy supplied in one day due to the management strategy in the winter term.

Table 5. Offices load (HVAC & light), PV and Grid total electric energy comparison in one day.

Offices	Electric Energy Used (KWh)	Electric Energy Generated (KWh)	Savings [R] Kwh = R1.2	Savings Percentage
Office 1&2	25.137	-	-	-
Office 3&4	24.711	-	-	-
PV	-	23.126	27.751	46.394%
GRID	-	26.721	-	-
Total	49.848	49.847	27.51	46.394%

Due to the high number of data (irradiation and temperature), MATLAB Simulink could not run a month or three months data on a one-hour basis. Hence, based on the one-day simulated result, an extrapolation technique was conducted to reach a three-month result. Therefore, the figure shown above (Figure 30), represents the PV system's energy generated during the winter period.

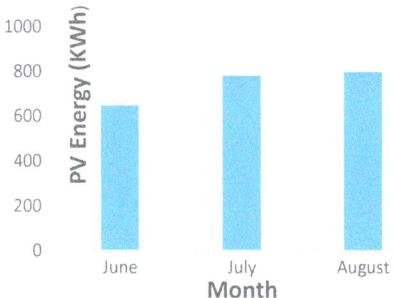

Figure 30. PV Energy generated during winter period.

The same method and technique were used to know the loads energy consumed from the PV and from the grid in others term of the year (Spring, Summer, Autumn).

3.6. Comparative Results for the Three Strategies Control Scenarios

The table below (Table 6), summarized the total energy saved in the three scenarios mentioned in Section 2.

Table 6. Total energy saved in the enumerated three scenarios mentioned in Section 1.

Strategy and Scenarios	HVAC Energy Saved %	LIGHT Energy Saved%
Occupancy, temperature, weather control	13.23%	29.3%
Loads, Occupancy, temperature, weather control	20.16%	29.3%
PV system	44%	
PV plus, Occupancy, temperature, weather control	57.23%	73.3%
PV + Loads, Occupancy, temperature, weather control	64.16%	73.3%

We can see from the table below that the energy saved around the office building is more reliable for the last strategy that was initially proposed where the PV system integrated with the loads and occupancy temperature weather control has a saving percentage of 64% for HVAC and 74% for light.

4. Discussion

This research work was conducted to further minimize the energy consumption around the automated office building. Four offices were used to test the proposed energy management strategy. Different strategies were explored.

The HVAC model developed was based on the different seasons (summer and winter) and the energy consumption due to the temperature, occupancy. The load control strategy was explored, the temperature used to heat a room in winter when the office is occupied is different from when the office is non-occupied. The office loads, which were used as a control variable, also participated in the electric energy saved.

Due to the light data collected in offices, the occupancy control strategy was implemented in order to reduce the lamp energy consumed. Therefore, in this case, the employees' working schedule was used as a sensor to track the presence of the employee in an office.

The integration strategy of the PV system with the temperature, occupancy, loads' control was conducted to further reduce the energy consumed in office. A Grid-tied inverter technique was used. The irradiations and temperature data were collected as these two parameters were different during the year, so different terms or seasons was considered to extract the maximum solar energy. Temperature and irradiation had an impact on the energy generated by the PV system but the Perturb and MPPT technique were used to continuously track the maximum power point.

Another accomplishment in this work was the design of the fuzzy logic algorithm which was used to manage the office's loads consumption by deciding exactly when the PV system would be supported by the grid to feed the needed office energy. This loads' management allows the system to use the energy generated by the PV to its maximum.

In this work, an energy audit for offices was also conducted in order to determine the energy used by the office (HVAC, light) without any strategy management. The energy saved from the grid after implementing one of the intelligent strategies was also audited.

The results obtained in this research also demonstrated how intelligent strategy of load management in offices contribute to further reducing the electrical energy consumed over the period of a year and showed how the employee activities control or diminish or have an impact on the energy used. However, this research has some limitation. The HVAC mathematical modelled proposed in this paper based on fundamental principles of heat transfer and thermodynamics has limited input variable such as the characteristic of the HVAC and the physical properties of the office buildings. The HVAC energy consumed varies due to the uncontrolled outdoor temperature. The occupancy pattern drawn from this research could not include the information due to the changing dynamics of occupancy in individual offices due to the lack of a real motion detection installed. The amount of power varies with irradiation and temperature since they are not uniform through the day and varies with the season as sunlight change. The limited space allocated to pave the solar panel influenced the amount of power generated.

In conclusion, the entire results obtained in this research project confirm that to save a significant amount of energy around an automated office building, the occupancy activity, the temperature, the weather condition, the loads control should be combined with a PV system. This was the best intelligent strategy for loads management and should be implemented; in four offices 64% energy was saved for the operation of HVAC and 74% for the light.

5. Conclusions

This paper presented the research findings, discussions and analysis of the different results for the intelligent control strategy management in four offices where the office HVAC energy consumed model without control results were found. The energy consumed by the HVAC and light due to the temperature, occupancy sensor and season strategy control results were found.

The HVAC and light energy consumed' result based on the loads control, added to others control such as temperature, occupancy and season control were also found. However, the office electric energy consumed' results from the PV-system associated with the control variables such as temperature, occupancy, loads itself, season were found. The results showed that the energy saved for one year in four automated offices was greater when the energy consumed by the loads (HVAC and light) was controlled compared to when it was not controlled energy consumed by the loads was controlled, was greater compared to when it was not controlled (20.16% and 13.23%). This paper also shows that the lighting energy saved in offices due to the occupancy activities was 29.3% and the electric energy saved due to the PV system was 44%. Therefore, the total HVAC energy saved in four offices from the intelligent control strategy articulated above, with loads' control and the PV assistance was 64.16% and the lighting 73.3%. Based on this proposed intelligent energy management strategy, the obtained energy saving in lightning is 48.3% higher than that found in the literature ([6,8]) (25% and 30%) and 23.34% higher than the energy saved in Reference [7]. This proposed intelligent system also allowed a considerable saved on the HVAC energy used which are 27.16% and 49.16 respectively higher than the one found in ([3,15]). However, due to the complexity of nature, improvement needs to be done.

Therefore, a recommendation for future work should consider the heat generated by employees, bulbs and computers into a control system to properly regulate the environment temperature which has an impact on the HVAC energy consumption and consequently on the office building. Develop a control system that will take into consideration the electricity price per unit to further reduce the electrical energy utilized. The HVAC using in office must be assigned to cover that specific office space, the inverter HVAC should be installed because of their abilities to save energy. Additionally, the LED lamp which is an energy efficient lighting technology must be installed in offices in order to reduce the amount of electricity used because LED wastes less heat energy compared to long tube lamp. The type of materials used and the architecture for office building need to be different from areas. The material and the architecture used must be favourable for the employee in term of energy saving and comfort.

Author Contributions: Conceptualization, S.P.C., S.I.N.T. and T.O.O.; methodology, S.I.N.T., S.P.C. and T.O.O.; software, S.I.N.T.; validation, S.I.N.T., S.P.C. and T.O.O.; formal analysis, S.I.N.T.; investigation, S.I.N.T.; resources, S.P.C and S.I.N.T.; data curation, S.I.N.T.; writing—original draft preparation, S.I.N.T.; writing—review and editing, S.I.N.T, T.O.O. and S.P.C.; visualization, S.P.C., S.I.N.T. and T.O.O.; supervision, S.P.C. and T.O.O.; project administration, S.P.C., T.O.O. and S.I.N.T.; funding acquisition, S.P.C. and T.O.O.

Funding: This research received no external funding.

Acknowledgments: Our gratitude and appreciation go to all members and colleagues of the Tshwane University of Technology (TUT) especially the Department of Electrical Engineering, for providing the facilities and material to conduct this research.

Conflicts of Interest: The authors declare no conflict of interest.

Appendix A

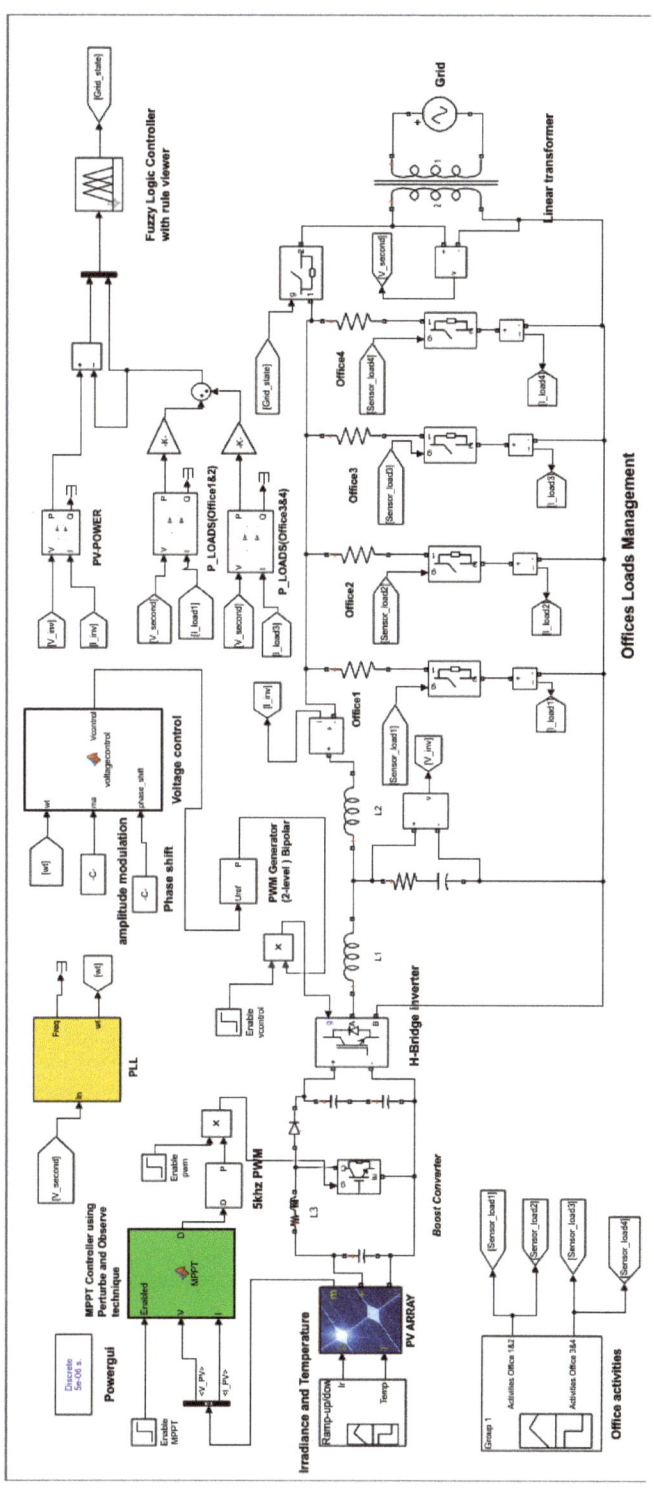

Figure A1. PV-Grid and Office Loads (HVAC, light) Complete Circuit.

B.

Table 1. HVAC Components specifications.

Parameter	Values	Unit
$\dot{V}_{HVAC}$	27	m^3/min
$COP_{heating}$	3.8	-
$COP_{cooling}$	3.4	-
P_{tot}	50	Bar
C_p	1.00	KJ/Kg
T_{HVAC} (summer)	35	°C
T_{HVAC} (winter)	11	°C
T_{cold}	24	°C
Summer T_{md}	25	°C
Winter T_{md}	23	°C
T_{hot}	29	°C
SHR	0.7	-

C.

Table 2. PV system specifications.

Specification	Value
Maximum Power	4000 W
Voltage at maximum power point (Vmp)	30.5 V
Current at maximum power point	6.5 A
Open circuit voltage Voc	48.2 V
Short circuit current Isc	10 A
Series connected array n_s	5
Parallel strings n_p	4

References

1. International Energy Agency. *Key World Energy Statistics*; International Energy Agency: Paris, France, 2014; Available online: www.iea.org (accessed on 25 October 2019).
2. Milford, D.R. Perspective on Energy Efficiency Building Regulations: A South African Perspective. Cidb. Available online: http://www.cidb.org.za (accessed on 25 October 2019).
3. Jain, M.N.S. Indoor occupancy counting to enable smart energy efficient office buildings. In Proceedings of the IEEE International Conferences on Big Data and Cloud Computing (BDCloud), Social Computing and Networking (SocialCom), Sustainable Computing and Communications (SustainCom), Atlana, GA, USA, 8–10 October 2016.
4. Marinakis, V.; Doukas, H.; Karakosta, C.; Psarras, J. An integrated system for buildings' energy-efficient automation: Application in the tertiary sector. *Appl. Energy* 2013, *101*, 6–14. [CrossRef]
5. Zanoli, S.M.; Pepe, C.; Orlietti, L.; Barchiesi, D. A model predictive control strategy for energy saving and users comfort features in building automation. In Proceedings of the 19th International Conference on System Theory, Control and Computing (ICSTCC), Cheile Gradistei, Romania, 14–16 October 2015.
6. Kaneko, Y.; Matsushita, M.; Kitagami, S.; Kiyohara, R. An energy-saving office lighting control system linked to employee's entry/exist. In Proceedings of the IEEE 2nd Global Conference on Consumer Electronics (GCCE), Tokyo, Japan, 1–4 October 2013.
7. Gonzalez, L.I.L.; Großekathöfer, U.; Amft, O. An intervention study on automated lighting control to save energy in open space offices. In Proceedings of the 2015 IEEE International Conference on Pervasive Computing and Communication Workshops (PerCom Workshops), St. Louis, MO, USA, 23–27 March 2015.

8. Yun, G.Y.; Kim, H.; Kim, J.T. Effects of occupancy and lighting use patterns on lighting energy consumption. *Energy Build.* **2012**, *46*, 152–158. [CrossRef]
9. Tiller, D.K.; Guo, X.; Henze, G.P.; Waters, C.E. Validating the application of occupancy sensor networks for lighting control. *Lighting Res. Technol.* **2010**, *42*, 399–414. [CrossRef]
10. Yu, X.; Su, Y. Daylight availability assessment and its potential energy saving estimation—A literature review. *Renew. Sustain. Energy Rev.* **2015**, *52*, 494–503. [CrossRef]
11. Martirano, L.; Parise, G.; Parise, L.; Manganelli, M. Simulation and sensitivity analysis of a fuzzy-based building automation control system. In Proceedings of the 2014 IEEE Industry Application Society Annual Meeting, Vancouver, BC, Canada, 5–9 October 2014.
12. Cziker, A.; Chindris, M.; Miron, A. Fuzzy controller for a shaded daylighting system. In Proceedings of the 2008 11th International Conference on Optimization of Electrical and Electronic Equipment, Brasov, Romania, 22–24 May 2008.
13. Cziker, A.; Chindris, M.; Miron, A. Implementation of fuzzy logic in daylighting control. In Proceedings of the 2007 11th International Conference on Intelligent Engineering Systems, Budapest, Hungary, 29 June–2 July 2007.
14. Benezeth, Y.; Laurent, H.; Emile, B.; Rosenberger, C. Towards a sensor for detecting human presence and characterizing activity. *Energy Build.* **2011**, *43*, 305–314. [CrossRef]
15. Qiu, L.; Wei, L.; Li, C. Applying research of the control strategy and algorithm system on intelligent building. In Proceedings of the Internationl Conference on Measuring Technology and Mechatronics Automation, Changsha, China, 13–14 March 2010.
16. Martirano, L.; Manganelli, M.; Parise, L.; Sbordone, D.A. Design of a fuzzy-based control system for energy saving and users comfort. In Proceedings of the 14th International Conference on Environment and Electrical Engineering (EEEIC), Krakow, Poland, 10–12 May 2014.
17. Parise, G.; Martirano, L. Impact of building automation, controls and building management on energy performance of lighting systems. In Proceedings of the Industrial & Commercial Power Systems Technical Conference—Conference Record 2009 IEEE, Calgary, AB, Canada, 3–7 May 2009.
18. Martirano, L. A smart lighting control to save energy. In Proceedings of the Intelligent Data Acquisition and Advanced Computing Systems (IDAACS), 2011 IEEE 6th International Conference, Prague, Czech Republic, 15–17 September 2011.
19. Martirano, L. Lighting systems to save energy in educational classrooms. In Proceedings of the 2011 10th International Conference on Environment and Electrical Engineering (EEEIC), Rome, Italy, 8–11 May 2011.
20. Chen, X.; Yang, H.; Lu, L. A comprehensive review on passive design approaches in green building rating tools. *Renew. Sustain. Energy Rev.* **2015**, *50*, 1425–1436. [CrossRef]
21. Jiang, Z.; Eichi, H.R. Design, modeling and simulation of a green building energy system. In Proceedings of the 2009 IEEE Power & Energy Society General Meeting, Calgary, AB, Canada, 26–30 July 2009.
22. Calcedo, J.G.S.; Rodriguez, F.L. Analysis of the performance of a high efficiency administrative building in Spain. *Int. J. Green Energy* **2017**, *14*, 55–62. [CrossRef]
23. Coma, J.; Perez, G.; Sole, C.; Castell, A.; Cabeza, L.F. Thermal assessment of extensive green roofs as passive tool for energy savings in buildings. *Renew. Energy* **2015**, *85*, 1106–1115. [CrossRef]
24. Mpelogianni, V.; Groumpos, P.; Tsipianitis, D.; Mantas, P.; Michos, S. Fuzzy inference tool for the achievement of sustainable energy solutions. In Proceedings of the 2015 6th International Conference on Information, Intelligence, Systems and Applications (IISA), Corfu, Greece, 6–8 July 2015.
25. Weber, L. *Energy Relevan Decisions in Organizations within Office Buildings*; Swiss Federation Institute of Technology: Zurich, Swiss, 2000.
26. Mastelic, J.; Emery, L.; Previdoli, D.; Papilloud, L.; Cimmino, F.; Genoud, S. Energy management in public building: A case study co-designing the building energy management system. In Proceedings of the 2017 International Conference on Engineering, Technology and Innovation (ICE/ITMC), Funchal, Portugal, 27–29 June 2017.
27. Darby, S. *The Effectiveness of Feedback on Energy*; Environmental Change Institute: Oxford, UK, 2006.
28. The Engineering Toolbox. 25 October 2018. Available online: https://www.engineeringtoolbox.com/fans-efficiency-power-consumption-d_197.html (accessed on 13 October 2019).

29. Lienhard, J.H. (Ed.) The general problem of heat transfert. In *A Heat Transfert Texbook*, 3rd ed.; 2001; pp. 7–8. Available online: http://www.mie.uth.gr/labs/ltte/grk/pubs/ahtt.pdf (accessed on 13 October 2019).
30. Eskom Solar and Met Data, 2009–2010. Available online: http://www.eskom.co.za/AboutElectricity/RenewableEnergy/Pages/Solar_Information.aspx (accessed on 13 October 2019).

 © 2019 by the authors. Licensee MDPI, Basel, Switzerland. This article is an open access article distributed under the terms and conditions of the Creative Commons Attribution (CC BY) license (http://creativecommons.org/licenses/by/4.0/).

Article

Investigation of Thermal Comfort Responses with Fuzzy Logic

József Menyhárt [1] and Ferenc Kalmár [2,*]

1. Department of Mechanical Engineering, Faculty of Debrecen, Ótemető street 2-4, 4028 Debrecen, Hungary; jozsef.menyhart@eng.unideb.hu
2. Department of Building Services and Building Engineering; Faculty of Debrecen, Ótemető street 2-4, 4028 Debrecen, Hungary
* Correspondence: fkalmar@eng.unideb.hu

Received: 20 April 2019; Accepted: 8 May 2019; Published: 11 May 2019

Abstract: In order to reduce the energy consumption of buildings a series of new heating, ventilation and air conditioning strategies, methods, and equipment are developed. The architectural trends show that office and educational buildings have large glazed areas, so the thermal comfort is influenced both by internal and external factors and discomfort parameters may affect the overall thermal sensation of occupants. Different studies have shown that the predictive mean vote (PMV)—predictive percentage of dissatisfied (PPD) model poorly evaluates the thermal comfort in real buildings. At the University of Debrecen a new personalized ventilation system (ALTAIR) was developed. A series of measurements were carried out in order to test ALTAIR involving 40 subjects, out of which 20 female (10 young and 10 elderly) and 20 male (10 young and 10 elderly) persons. Based on the responses of subjects related to indoor environment quality, a new comfort index was determined using fuzzy logic. Taking into consideration the responses related to thermal comfort sensation and perception of odor intensity a new the fuzzy comfort index was 5.85 on a scale from 1–10.

Keywords: indoor environment quality; thermal comfort; personalized ventilation; fuzzy logic

1. Introduction

1.1. Energy Use of Buildings

Nowadays, reduction of energy use in buildings is one of the main goals in European countries. According to the Energy Performance Directive, Member States shall ensure that, by 31 December 2020, all new buildings are nearly zero-energy buildings [1]. Vorsatz et al. in their paper described the trends of thermal energy use of buildings on a global and regional basis [2]. They found that in the residential building sector the final energy use was almost constant for the last 30 years. Even though the energy sources were diversified, the efficiency of technologies and services were significantly enhanced in the analyzed regions. In the modelling of the future energy use and trends, climate change has to be taken into account. Isaac and van Vurren assess the potential development of energy use for future residential heating and air conditioning in the context of climate change [3]. Based on their study they stated that the energy use for air conditioning in buildings will increase rapidly over the 21st century, mostly driven by income growth. Levesque et al. show that, without further climate policies, global final energy demand of buildings could be tripled by 2100 [4]. Improving the thermal performances of the building envelope important energy saving can be obtained. The effectiveness and the time dependence of the thermal properties of built-in insulation materials have to be known in order to optimize the insulation of the buildings' envelope. Water content and humidity has a negative influence on the thermal conductivity of the commonly used insulation materials [5]. Position in the building element and structure of the insulation material are to be taken into account when building energy simulations are carried out [6]. Improving the structure of traditional insulation material (polystyrene) the insulation effectiveness is increased and further energy savings may be obtained [7]. However, the increment of effectiveness is fragile and the construction process should be of high standard in order to obtain the presumed energy saving. Vacuum technologies may further improve the efficiency of insulation materials. However, the stability of insulation properties of these materials should be analyzed appropriately before to be used on a large scale [8]. In order to identify the optimal shape and structure of the building envelopes complex energy and cost simulation should be conducted. Baglivo et al. presented in their study the interrelation between operative temperature and configuration of the buildings' envelope in warm climates [9]. Utilization of renewable energy sources may further contribute to the reduction of fossil fuel use in the building sector. However, to integrate efficiently the renewable energy sources in the buildings energy system, complex energy-economic analysis has to be done. Molinari et al. analyzed in their study the performance of ground source heat pumps depending on the building thermal insulation, boreholes number and spacing [10]. Their methodology allows quantifying the impact of different design configurations on the need for end-use energy. Brown et al. developed a method for assessing renovation packages drawn up with the goal of increasing energy efficiency in buildings [11]. The method includes calculation of bought energy demand, life-cycle cost (LCC) analysis, and assessment of the building according to the Swedish environmental rating tool.

1.2. Indoor Environment Quality

In addition to energy saving endeavors designers have to emphasize proper thermal comfort and appropriate indoor air quality assurance in buildings. Al horr et al. in their review presented a series of findings enlighten the effects of the indoor environment quality on occupant well-being and comfort [12]. Björnsson et al. discussed in their study the sick building syndrome and its relation with bronchial hyperresponsiveness [13]. They found that sick building symptoms are common in the general population and among women, while atopy and anxiety increase the risk of reporting such symptoms. Wargocki and Wyon draw attention to the relation between quality of indoor environment and productivity [14]. They revealed that labor costs in buildings exceed energy costs by two orders of magnitude. Moreover, they claim that thermal and air quality conditions that the majority of building occupants currently even accept can be shown to reduce performance by 5–10% for adults and by 15–30% for children. Kang et al. conducted a study analyzing the impact of indoor environmental quality on work productivity in university open-plan research offices [15]. Their results proved that different age groups can have different sensitivities to artificial lighting, natural lighting and office noise. Furthermore, different genders can have different sensitivities to ventilation and temperature. Schaudienst and Vogdt demonstrated that elderly people and women prefer a higher ambient temperature [16]. Cheung et al. analyzed the accuracy on PMV–PPD model using the ASHRAE Global Thermal Comfort Database II [17]. According to their study, the PMV model accuracy in predicting the observed thermal comfort was only 34%, meaning that the thermal sensation is incorrectly predicted two out of three times. Their findings emphasized the need of development of new thermal comfort models. Geng et al. investigated through experiments with objective environmental measurements, subjective surveys, and productivity tests, the impact of thermal environment on occupant IEQ perception and productivity [18]. They found that when thermal environment was unsatisfactory, it weakened the "comfort expectation" of other IEQ factors, which accordingly resulted in the less dissatisfaction with other IEQ factors. Furthermore, the optimal productivity was obtained when people felt "neutral" or "slightly cool", and the increase of thermal satisfaction had a positive effect on productivity.

1.3. Comfort Indices

The evaluation of the thermal comfort in closed spaces is based on the heat exchange of the human body with surrounding environment. Based on the heat balance of the human body Fanger developed the predictive mean vote (PMV) and predictive percentage of dissatisfied (PPD) [19]. The model works well in well-controlled environments. However, there are plenty of non-air conditioned buildings. For free-running buildings Nicol and Humphreys proposed to use sustainable comfort standards based on an adaptive algorithm [20], which stand for variable indoor temperatures during the year. Analyzing a large database Brager and de Dear introduced the adaptive thermal comfort model in naturally ventilated buildings [21]. They proposed the indoor operative temperature as a function of prevailing outdoor air temperature. Fanger and Toftum provided in their study the extended PMV model, which can be used in non-air conditioned buildings [22]. They concluded that in warm climates the PMV model overestimates the warmth sensation of occupants in free-running buildings. Yao et al. developed a theoretical adaptive PMV(aPMV) model for thermal comfort [23]. The aPMV index proposed in their study takes into account the both physiological and behavioral adaptation of occupants. The wet bulb globe temperature (WBGT) is used both in closed and open spaces and takes into account the dry bulb air temperature, the wet bulb temperature, and the globe temperature [24]. The WBGT is used widely mainly to evaluate heat stress. Sakoi et al. tried to expand the WBGT index into a rational thermal comfort index of the human body [25]. Local thermal discomfort sensation caused by asymmetric radiation, draught, vertical temperature difference, and floor temperature are taken into account in the ISO 7730 standard [26]. The air diffusion performance index (ADPI) takes into account the mean and local temperatures and air velocities. ADPI is useful to determine the critical points of a closed space from draught sensation point of view. Ng et al. used computational fluid dynamics and response surface methodology in order to predict ADPI in the case of a displacement ventilated office [27].

As a response to energy saving and indoor environment quality endeavors, at the University of Debrecen advanced personalized ventilation (PV) equipment was developed (ALTAIR). The main idea of this PV equipment was the changing of air jet direction around the chest and head of the occupants. A series of measurements were carried out in order to analyze the thermal comfort sensation of young and elderly female and male subjects [28]. The results show significant differences between young women and men and young and elderly women, respectively. Analyzing the obtained responses related to indoor environment quality, using fuzzy logic, a new comfort index was developed. The method and results are presented in this paper.

2. The Basics of Fuzzy Logic

In the context of traditional logical sense, there are only two values; 0 (false) and 1 (true) [29–31]. It has been discovered several times in engineering that this statement is not always true and subjective opinions and statements are often used for some purpose. For perspective, it may occur during reading different measuring instruments that values are read incorrectly, generating inaccuracy [32]. It may be declared that the problem 'gets fuzzied'. Inaccuracy is illustrated with a membership function. The process in fuzzy systems is: fuzzification, interpretation, summary, and defuzzification [33,34]. The first step is fuzzification, which means the input of crisp values into the system. To perform fuzzification, categories used in modelling and the related membership functions are necessary to be specified. Thus, the core determining factors need to be considered. The number of categories is a crucial factor. By increasing the number of categories, we obtain a more accurate picture of the system under consideration, but increasing the number of categories complicates the study [33].

A membership function is necessary to be defined for each of the categories. Several ways exist to define them. The membership function of μ $(x;a)$ specifies the extent to which property 'x' belongs to set 'a'. Crisp scale determination is an important step, which needs to be selected from the scales of 0–10, 1–10, 0–100, and 1–100. The aim is to compare things to be studied as simply as possible. Rules have to be set in the interpretation phase based on categories, that is to say, a rule base for the fuzzy model is established. Results other than zero, obtained in the interpretation phase, are integrated with the properties of the controlled process (fuzzy process). As a result, the fuzzy set is established. It is a primary or preliminary conclusion; therefore, results are then interpreted in defuzzification.

Defuzzification is the last step in the process. Afterwards, specific values need to be chosen based on fuzzy conclusions that best describe the particular fuzzy set depending on the application and model [33,34].

Depending on thy type of application, the meaning of fuzzy set may be varied, therefore, different defuzzification methods are available to achieve results, such as [33]:

- Center of gravity (COG);
- Center of area (COA); and
- Weighted average method.

3. Artificial Intelligence in HVAC Systems

Artificial intelligence is gaining a wider ground these days. According to studies, artificial intelligence will have existence in several new fields by 2030, of which safety and security systems are among the most significant [35]. Not only should future developments be taken into account, as artificial intelligence is found in our daily equipment; personal assistants may be the most popular, including Siri or Cortana [36,37].

Artificial neural networks (A.N.N.) and fuzzy logic applications have more significant role in building management systems, where their main responsibilities include energy saving. Heating, ventilation, and air-conditioning (HVAC) systems are among the key sources of energy consumption of buildings [38]. Significant progress has been achieved in AI developed for HVAC in recent years [39]. The systems are able to perform control, management, optimization, fault detection, and diagnosis. These applications are the basis of the modern smart home business. According to studies, 41% of homes in the USA had programmable thermostats in 2017 [38,39].

The first electric thermostat was developed by W.S. Johnson in 1883. The 130 years since then have been full of innovations with respect to thermostats. During these years, thermostats underwent significant changes, and integrated modern IT solutions, such as sensors, actuators, hardware, software, and cloud solutions [39]. More than 18 AI solutions are available for HVAC systems developed during the last 20 years, but only three of them have become especially popular: weather forecasting, optimization, and predictive controls [40]. Most of the technical building systems today do not work as efficiently as those with artificial intelligence. This is due to the fact that existing sensors are not able to meet the requirements of AI [39,40]

Reducing energy consumption is only one task in today's environmentally conscious development. Systems with AI are expected to give feedback on the state of the internal environment, as well. These indices may help optimize HVAC systems, such as the Harris Index, but this does not characterize the internal environment from a comfort theory point of view [40–42]

4. Experiments with an Advanced Personalized Ventilation System

At the Department of Building Services and Building Engineering, University of Debrecen, a new PV system was developed [43–46]. The main novelty of the system (ALTAIR), is that the air jet is blown on the head and chest of the occupant sitting at a desk alternatively from three different directions (Figure 1).

Using the ALTAIR PV equipment a series of measurements were carried out in order to investigate to thermal comfort sensation and indoor environment quality (IEQ) perception of 10 young and 10 elderly, 10 male and 10 female subjects. The experiments were performed in controlled environment setting the mean radiant and air temperature to 30 °C. The subjects were not informed about the values of the indoor air parameters, neither about the schedule of the experiments. Fresh air was assured continuously during the measurement (50 m^3 h^{-1}), by displacement air distribution mode at a temperature of 30 °C. The speed of the built-in fan of the advanced ALTAIR PV equipment can be chosen and set by the user through a touchscreen. At higher fan speed values the air flow blown on the occupants head and chest and the air velocity are higher. Taking into account the elevated air- and mean radiant temperatures in the test room the air flow circulated by ALTAIR was set to 20 m^3·h^{-1}. Assuming the head of the occupant in the center of a circle, the air terminal devices are placed on the circle with 0.6 m radius. With these boundary parameters, the air velocity was 0.48 m·s^{-1} around the head and chest of subjects. During the experiments the background ventilation provides the required fresh air for one person sitting in the test room, so the ALTAIR PV was not connected to the ventilation system, it circulates the air in the room only. The measurement was two hours long, which was split into four different periods of 30 min each.

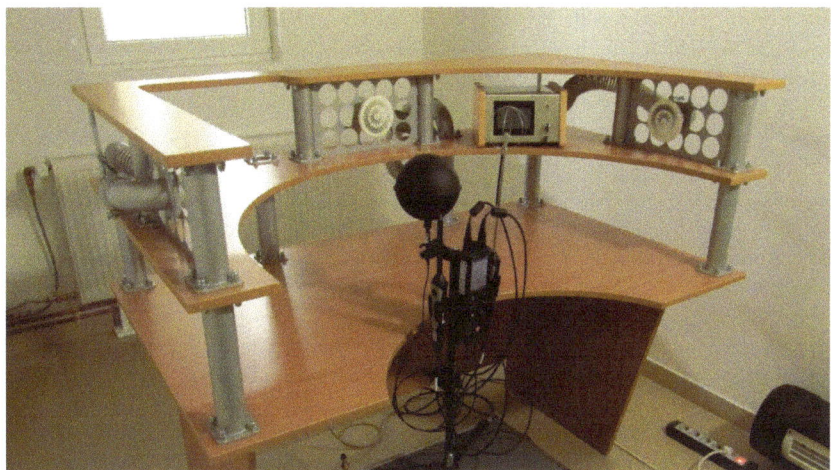

Figure 1. ALTAIR PV equipment.

During the first 30 min subjects were sitting at the desk, but the ALTAIR's built in fan was not in operation. Practically, these 30 min were considered as an accommodation period to the elevated temperatures. After the first 30 min, the ALTAIR's own fan was switched on. The air flow circulated by the ALTAIR was constant during the next periods of the experiments. In the second period of 30 min, the air jet direction blown on the subjects was changed every 30 min. Having a 30 °C air temperature and a 0.48 m·s^{-1} air velocity, the turbulence intensity measured with a TESTO 480 instrument was 18.8%. During the third period of the experiments, the time step of changing the air jet direction was set to 20 s. The measured air turbulence was 19.1%. In the last period of the measurements the time step of changing the direction of the air jet was reduced to 10 s. The turbulence intensity in this period increased to 20.6%. First and foremost, measurements were performed without subjects and all indoor air parameters were registered with a TESTO 480 instrument. According to this measurement, the PMV shown by the instrument was 1.44 in the first period (ALTAIR switched off) and 0.84 in the 2nd, 3rd, and 4th periods of the experiment (ALTAIR in switched on). The time step of air jet direction changing had no effect on the measured PMV. During the experiments, subjects sitting at the desk could learn or read a book. Eating or changing the position or clothes was prohibited. Taking into account the elevated temperatures drinking water was allowed. Every 10 min, subjects were asked to fill in a questionnaire [28]. First, they had to evaluate on the seven-point scale about the thermal environment. On this scale 0 means neutral (the optimal thermal comfort sensation). The other values are −3 (cold), −2 (cool), −1 (slightly cool), +1 (slightly warm), +2 (warm), and +3 (hot). Then, subjects had to evaluate the odor in the indoor air (no special odor was used during the experiments) on a six point scale. On this scale 0 means that no odor is perceived. Other values are: 1 (slight odor), 2 (moderate odor), 3 (strong odor), 4 (very strong odor) and 5 (overpowering odor). The acceptability of the indoor environments was evaluated by subjects on a three-point scale: +1 (clearly acceptable), 0 (just acceptable), and −1 (clearly unacceptable). Other question was related to the velocity of the indoor air. Subjects were asked if they can accept, or not, the air velocity. In the case of a negative response, they had to clarify whether they want to increase or decrease the air velocity. Draft may lead to thermal discomfort, even in properly designed environments, so subjects were asked about draft perception. In the case of positive responses, they had to identify the body segment, where they perceived the draft (head, neck, arms, back, legs, and ankles). Air freshness is other important question, related to the indoor environment quality. Subjects were asked whether they were content or not with the air freshness. The last question was related to the surface temperatures. Subjects were asked whether they were content or not with the surface temperatures. In the case of negative responses, subjects were asked to identify the building element which had an unacceptable surface temperature for them, and they had to clarify what to do with the surface temperature (increase or decrease).

Based on the results of this investigation presented in [28], using fuzzy logic, a fuzzy comfort index was developed.

5. Fuzzy Decision System

Approaching comfort theory with fuzzy logic is based on the 'tipping' fuzzy example. In the course of the research, the authors of the article distinguished nine parameters that could describe the inner environment of a building. Of the nine attributes, only two were selected to set up the mathematical model for simpler proofing. Two properties were chosen to characterize the quality of the internal environment. Odor intensity was evaluated on a scale of 0–5, while thermal sensation was evaluated on a −3 to 3 scale. The internal environment is ranked by these values and then it is calculated using these numbers by fuzzy inferences which index number is used to characterize the internal environment. The fuzzy rule system consists of:

1. Odor Intensity is categorized with six fuzzy membership functions. According to the classification the food may be *no odor, slight odor, moderate odor, strong odor, very strong, overpowering* (Figure 2).

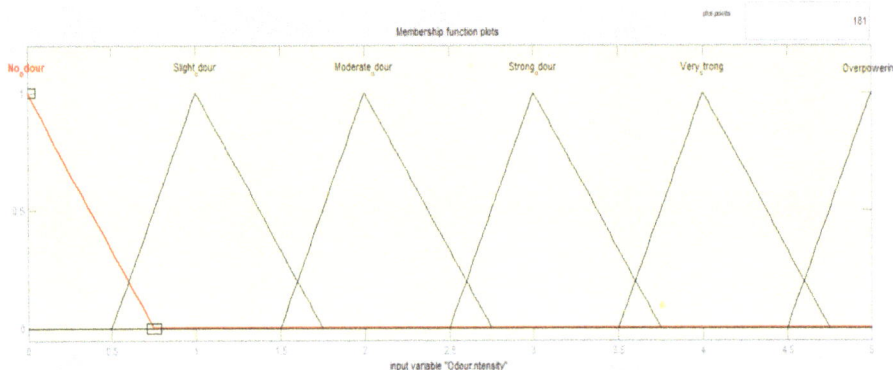

Figure 2. Membership functions of odor intensity.

2. Thermal sensation is divided into seven fuzzy sets: *cold, cool, slightly cool, neutral, slightly warm, warm, hot* (Figure 3).

$$\mu(no\_odour) = \begin{cases} 0, & if \quad x \geq 0.75 \\ 1, & if \quad x = 0 \\ \frac{0.75-x}{0.75}, & if \quad 0 \leq x \leq 0.75 \end{cases} \quad (1)$$

$$\mu(Slight\_odour) = \begin{cases} 0, & if \quad x \leq 0.5 \quad or \quad x \geq 1.75 \\ \frac{x-0.5}{0.5}, & if \quad 0.5 \leq x < 1 \\ 1, & if \quad x = 1 \\ \frac{1.75-x}{0.75}, & if \quad 1 < x \leq 1.75 \end{cases} \quad (2)$$

$$\mu(Moderate\_odour) = \begin{cases} 0, & if \quad x \leq 1.5 \quad or \quad x \geq 2.75 \\ \frac{x-1.5}{0.5}, & if \quad 1.5 \leq x < 2 \\ 1, & if \quad x = 2 \\ \frac{2.75-x}{0.75}, & if \quad 2 < x \leq 2.75 \end{cases} \quad (3)$$

$$\mu(Strong\_odour) = \begin{cases} 0, & if \quad x \leq 2.5 \quad or \quad x \geq 3.75 \\ \frac{x-2.5}{0.5}, & if \quad 2.5 \leq x < 3 \\ 1, & if \quad x = 3 \\ \frac{3.75-x}{0.75}, & if \quad 3 < x \leq 3.75 \end{cases} \quad (4)$$

$$\mu(Very\_strong) = \begin{cases} 0, & if \quad x \leq 3.5 \quad or \quad x \geq 4.75 \\ \frac{x-3.5}{0.5}, & if \quad 3.5 \leq x < 4 \\ 1, & if \quad x = 4 \\ \frac{4.75-x}{0.75}, & if \quad 4 < x \leq 4.75 \end{cases} \quad (5)$$

$$\mu(Overpowering) = \begin{cases} 0, & if \quad x \leq 4.5 \\ \frac{x-4.5}{0.5}, & if \quad 4.5 \leq x < 5 \\ 1, & if \quad x = 5 \end{cases} \quad (6)$$

The formulas above describe the logical connections between each opinion. The "odor intensity" contains three sets: unpleasant, acceptable, and excellent. These scales are not developed by a real questionnaire, and it is an important part of follow-up research. Figure 2 shows the graphical view of the functions, where the scale is between 0 and 10. It is very easy to see the common parts of the sets.

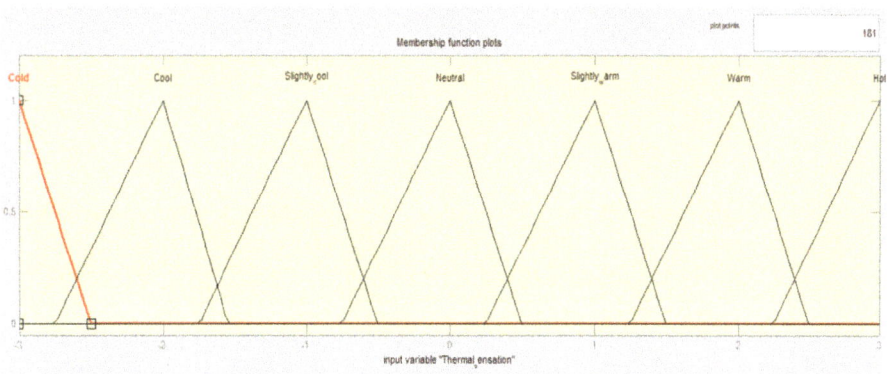

Figure 3. Membership functions of thermal sensation.

Thermal sensation is the second part of the research. The formulas work on the similar way like in the previous case. It is containing three parts: unpleasant, acceptable and excellent. The scale is, again, between 0 and 10.

Figure 3 shows the functions. When you identify internal comfort, you consider odor intensity and the thermal sensation between rate 0 and 10. We can use this to identify the internal comfort.

$$\mu(Cold) = \begin{cases} 0, & if \quad x \geq -2.5 \\ 1, & if \quad x = -3 \\ \frac{-2.5-x}{0.5}, & if \quad -3 < x \leq -2.5 \end{cases} \quad (7)$$

$$\mu(Cool) = \begin{cases} 0, & if \quad x \leq -3 \quad or \quad x \geq -2.75 \\ \frac{x+2.75}{0.75}, & if \quad -2.75 \leq x < -2 \\ 1, & if \quad x = -2 \\ \frac{-1.5-x}{0.5}, & if \quad -2 < x \leq -1.5 \end{cases} \quad (8)$$

$$\mu(Slightly_{cool}) = \begin{cases} 0, & if \quad x \leq -1.75 \quad or \quad x \geq -0.5 \\ \frac{x+1.75}{0.75}, & if \quad -1.75 \leq x < -1 \\ 1, & if \quad x = -1 \\ \frac{-0.5-x}{0.5}, & if \quad -1 < x \leq -0.5 \end{cases} \quad (9)$$

$$\mu(Neutral) = \begin{cases} 0, & if \quad x \leq -0.5 \quad or \quad x \geq 0.5 \\ \frac{x-0}{0.75}, & if \quad -0.75 \leq x < 0 \\ 1, & if \quad x = 0 \\ \frac{0.5-x}{0.5}, & if \quad 0 < x \leq 0.5 \end{cases} \quad (10)$$

$$\mu(Slightly\_warm) = \begin{cases} 0, & if \quad x \leq 0.25 \quad or \quad x \geq 0.5 \\ \frac{x-0.25}{0.75}, & if \quad 0.25 \leq x < 1 \\ 1, & if \quad x = 1 \\ \frac{1.5-x}{0.5}, & if \quad 1 < x \leq 1.5 \end{cases} \quad (11)$$

$$\mu(Warm) = \begin{cases} 0, & if \quad x \leq 1.25 \quad or \quad x \geq 2.5 \\ \frac{x-2}{0.75}, & if \quad 1.25 \leq x < 2 \\ 1, & if \quad x = 2 \\ 1 \\ \frac{2.5-x}{0.5}, & if \quad 2 < x \leq 2,5 \end{cases} \quad (12)$$

$$\mu(Hot) = \begin{cases} 0, & if \quad x \leq 2.25 \\ \frac{x-2.25}{0.75}, & if \quad 2.25 \leq x < 3 \\ 1, & if \quad x = 3 \end{cases} \quad (13)$$

Inputs
Odor intensity

- Universe (i.e., crisp value range): To what extent the smell was deemed good?

Thermal sensation

- How the temperature was deemed?

Outputs
Internal comfort

- Fuzzy set: malaise, acceptable well-being, excellent well-being

6. Possibilities of Fuzzy Comfort Index

Correlation between fuzzy membership functions of input data (Odor Intensity and Thermal Sensation) and output data (Fuzzy Comfort Index) is developed by fuzzy rules. This particular task included 30 fuzzy rules (Figure 4).

```
1. If (Odour_Intensity is Slight_odour) then (Fuzzy_Comfort_Index_-_FCI is Excellent_well-being) (1)
2. If (Odour_Intensity is Moderate_odour) then (Fuzzy_Comfort_Index_-_FCI is Excellent_well-being) (1)
3. If (Odour_Intensity is Strong_odour) then (Fuzzy_Comfort_Index_-_FCI is Acceptable_well-being) (1)
4. If (Odour_Intensity is Very_strong) then (Fuzzy_Comfort_Index_-_FCI is Malaise) (1)
5. If (Odour_Intensity is Overpowering) then (Fuzzy_Comfort_Index_-_FCI is Malaise) (1)
6. If (Odour_Intensity is Slight_odour) and (Thermal_sensation is Cold) then (Fuzzy_Comfort_Index_-_FCI is Malaise) (1)
7. If (Odour_Intensity is Slight_odour) and (Thermal_sensation is Cool) then (Fuzzy_Comfort_Index_-_FCI is Malaise) (1)
8. If (Odour_Intensity is Slight_odour) and (Thermal_sensation is Slightly_cool) then (Fuzzy_Comfort_Index_-_FCI is Acceptable_well-being) (1)
9. If (Odour_Intensity is Slight_odour) and (Thermal_sensation is Neutral) then (Fuzzy_Comfort_Index_-_FCI is Excellent_well-being) (1)
10. If (Odour_Intensity is Slight_odour) and (Thermal_sensation is Slightly_warm) then (Fuzzy_Comfort_Index_-_FCI is Excellent_well-being) (1)
11. If (Odour_Intensity is Slight_odour) and (Thermal_sensation is Warm) then (Fuzzy_Comfort_Index_-_FCI is Acceptable_well-being) (1)
12. If (Odour_Intensity is Slight_odour) and (Thermal_sensation is Hot) then (Fuzzy_Comfort_Index_-_FCI is Malaise) (1)
13. If (Odour_Intensity is Moderate_odour) and (Thermal_sensation is Cold) then (Fuzzy_Comfort_Index_-_FCI is Malaise) (1)
14. If (Odour_Intensity is Moderate_odour) and (Thermal_sensation is Cool) then (Fuzzy_Comfort_Index_-_FCI is Malaise) (1)
15. If (Odour_Intensity is Moderate_odour) and (Thermal_sensation is Slightly_cool) then (Fuzzy_Comfort_Index_-_FCI is Acceptable_well-being) (1)
16. If (Odour_Intensity is Moderate_odour) and (Thermal_sensation is Neutral) then (Fuzzy_Comfort_Index_-_FCI is Acceptable_well-being) (1)
17. If (Odour_Intensity is Moderate_odour) and (Thermal_sensation is Slightly_warm) then (Fuzzy_Comfort_Index_-_FCI is Excellent_well-being) (1)
18. If (Odour_Intensity is Moderate_odour) and (Thermal_sensation is Warm) then (Fuzzy_Comfort_Index_-_FCI is Malaise) (1)
19. If (Odour_Intensity is Moderate_odour) and (Thermal_sensation is Hot) then (Fuzzy_Comfort_Index_-_FCI is Malaise) (1)
```

Figure 4. Fuzzy rules.

Membership functions and rules are shown in the following (Figure 5) graph surface.

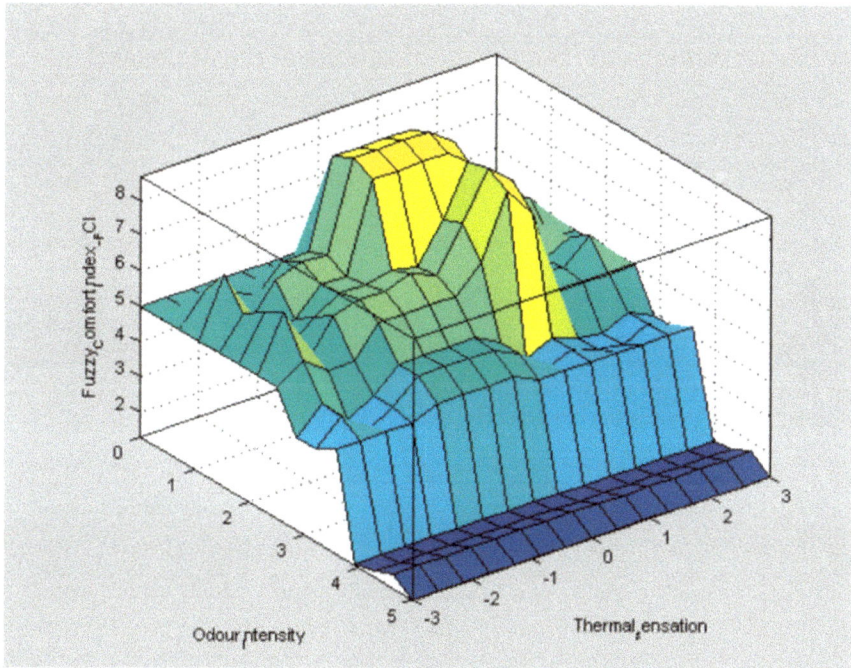

Figure 5. 3D surface.

The surface shown in Figure 5 clearly defines parameter intervals which determine the quality of the internal environment. According to the dark blue surfaces, comfort is not satisfactory, while in the case of the citrus yellow surface, the comfort of the participants tested is particularly satisfactory.

The rule system can show the result for us (Figure 6).

Figure 6. Fuzzy rules.

On the right side of Figure 8 we can see the Fuzzy Comfort Index (FCI). The index result is 5.85 (Figure 7).

Figure 7. Fuzzy Comfort Index.

This index may be a specific number for the internal space. The nine sets and a special rule system can support or give a special fuzzy scale which can absolutely describe the room regarding comfort. The authors need to study how the problem can be solved with nine properties.

This paper suggests that the Fuzzy Logic can describe the comfort property of a room, but the future work needs more experiments and calculations. Three types of fuzzy classes characterize the internal environment: (Fuzzy Comfort Index, FCI) (Figure 8), which may be subdivided into malaise, acceptable well-being, and excellent well-being.

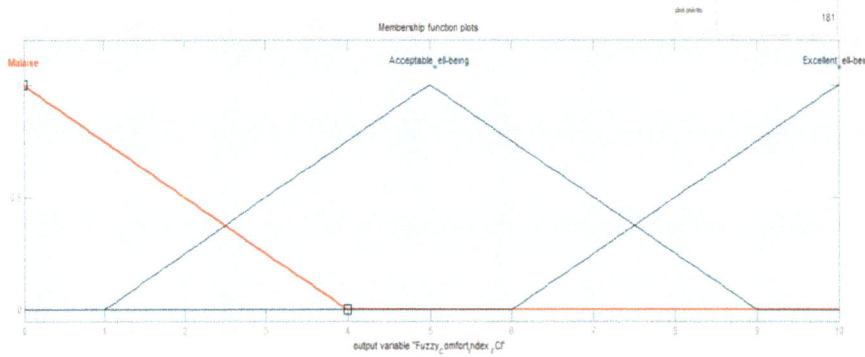

Figure 8. Fuzzy Comfort Index sets.

Figure 8 represents the FCI index sets. The results can be between 0 and 10 on this line. The results above (5.85) are in the middle range of this system. These sets are made by the authors, which depend on the investigated person's subjective opinion and experience.

7. Conclusions

The study presents a possible comfort theory indicator with fuzzy logic and MATLAB. Graphs allow determining an index number that describes a closed space, a room/indoor environment. Out of the possible nine parameters, the article examines only two; the future goal is the extension of the mathematical model to the nine possible properties and the processing of real time data, which takes the dynamic changes in the examined room into account. The entire process described above, getting to the fuzzy output from the numerical inputs by input fuzzy membership functions and the fuzzy rules, is defuzzificated with the center-of-gravity method, it is the so-called Mamdani's inference. The fuzzy system may have much more than two inputs. If we assign 3–4 or more fuzzy membership functions to each input, thousands of fuzzy rule units are required in the simulations, and this may make it very difficult to calculate both the fuzzy output and the full defuzzification process. Another, commonly used solution to avoid the complicated defuzzification and fuzzy output calculation is to teach the behavior of a complete fuzzy system to an artificial neural network (ANN). The neural network requires learning data system to learn. These are created by determining the output of the complicated fuzzy system for several different inputs. These outputs will be the expected values during the teaching of the neural network when the same inputs are received by the neural network as the fuzzy system. When the neural network is able to provide the same output as the fuzzy system for learning data, it will be able to model further fuzzy system operation on additional test data. The output will only be approximately the same as the fuzzy system, but this output may be calculated much faster, which means a higher efficiency. Such neural network modelled fuzzy systems are called neuro-fuzzy systems. Researchers intend to implement these points in their future research on the quality of the internal environment [47].

The created index allows to describe an internal room or to characterize an existing internal environmental quality, which may even be recorded by legislation.

8. Limitations

Measurements were performed in the laboratories of the Faculty of Engineering, University of Debrecen. The size of the test room, the building materials and surfaces of building elements of the test room, and the type and location of air terminal devices of mixing and displacement ventilation were considered as given boundary conditions. In the development of the new fuzzy comfort index only two parameters were taken into consideration.

Author Contributions: Conceptualization: F.K. and J.M.; methodology: J.M.; investigation: J.M. and F.K.; resources: F.K.; data curation: J.M.; writing—original draft preparation: F.K. (CH. 1, 4, 7) and J.M. (ch. 2, 3, 5,6); writing—review and editing: F.K.; visualization: J.M. and F.K.; supervision: F.K.; project administration: F.K

Funding: The research was financed by the Higher Education Institutional Excellence Programme of the Ministry of Human Capacities in Hungary, within the framework of the Energetics Thematic progTramme of the University of Debrecen.

Conflicts of Interest: The authors declare no conflict of interest.

References

1. Directive 2010/31/EU of the European Parliament and of The Council of 19 May 2010 on the energy performance of buildings. Available online: https://eur-lex.europa.eu/legal-content/EN/TXT/?uri=CELEX%3A32010L0031. (accessed on 10 March 2019).
2. Ürge-Vorsatz, D.; Cabeza, L.F.; Serrano, S.; Barreneche, C.; Petrichenko, K. Heating and cooling energy trends and drivers in buildings. *Renew. Sustain. Energy Rev.* **2015**, *41*, 85–98. [CrossRef]

3. Isaac, M.; van Vuuren, D.P. Modeling global residential sector energy demand for heating and air conditioning in the context of climate change. *Energy Policy* **2009**, *37*, 507–521. [CrossRef]
4. Levesque, A.; Pietzcker, R.C.; Baumstark, L.; de Stercke, S.; Grübler, A.; Luderer, G. How much energy will buildings consume in 2100? A global perspective within a scenario framework. *Energy* **2018**, *148*, 514–527. [CrossRef]
5. Szodrai, F.; Lakatos, Á. Effect of wetting time in the sorption and in the thermal conductivity of the most commonly used structural materials. *Build. Serv. Eng. Res. Technol.* **2017**, *38*, 475–489. [CrossRef]
6. Szodrai, F.; Lakatos, Á. Effect of the air motion on the heat transport behaviour of wall structures. *Int. Rev. Appl. Sci. Eng.* **2017**, *8*, 67–73. [CrossRef]
7. Ákos, L.; István, D.; Umberto, B. Thermal characterization of different graphite polystyrene. *Int. Rev. Appl. Sci. Eng.* **2018**, *9*, 163–168.
8. Lakatos, Á. Stability investigations of the thermal insulating performance of aerogel blanket. *Energy Build.* **2019**, *185*, 103–111. [CrossRef]
9. Baglivo, C.; Congedo, P.M.; Di Cataldo, M.; Colucci, L.D.; D'Agostino, D. Envelope Design Optimization by Thermal Modelling of a Building in a Warm Climate. *Energies* **2017**, *10*, 1808. [CrossRef]
10. Molinari, M.; Lazzarotto, A.F.B. The application of the parametric analysis for improved energy design of a ground source heat pump for residential buildings. *Energy Build.* **2013**, *63*, 119–128. [CrossRef]
11. Brown, N.W.O.; Malmqvist, T.; Bai, W.; Molinari, M. Sustainability assessment of renovation packages for increased energy efficiency for multi-family buildings in Sweden. *Build. Environ.* **2013**, *61*, 140–148. [CrossRef]
12. Al horr, Y.; Arif, M.; Katafygiotou, M.; Mazroei, A.; Kaushik, A.; Elsarrag, E. Impact of indoor environmental quality on occupant well-being and comfort: A review of the literature. *Int. J. Sustain. Built Environ.* **2016**, *5*, 1–11. [CrossRef]
13. Björnsson, E.; Janson, C.D.; Norbäck, G.B. Symptoms related to the Sick Building Syndrome in a general population sample: associations with atopy, bronchial hyper-responsiveness and anxiety. *Int. J. Tuberc. Lung. Dis.* **1998**, *2*, 1023–1028.
14. Wargocki, P.; Wyon, D.P. Ten questions concerning thermal and indoor air quality effects on the performance of office work and schoolwork, Ten questions concerning thermal and indoor air quality effects on the performance of office work and schoolwork. *Build. Environ.* **2017**, *112*, 359–366. [CrossRef]
15. Kang, S.; Ou, D.; Mak, C.M. The impact of indoor environmental quality on work productivity in university open-plan research offices. *Build. Environ.* **2017**, *124*, 78–89. [CrossRef]
16. Schaudienst, F.; Vogdt, F.U. Fanger's model of thermal comfort: A model suitable just for men? *Energy Procedia* **2017**, *132*, 129–134. [CrossRef]
17. Cheung, T.; Schiavon, S.; Parkinson, T.; Lib, P.; Brager, G. Analysis of the accuracy on PMV–PPD model using the ASHRAE Global Thermal Comfort Database II. *Build. Environ.* **2019**, *153*, 205–217. [CrossRef]
18. Geng, Y.; Ji, W.; Lin, B.; Zhu, Y. The impact of thermal environment on occupant IEQ perception and productivity. *Build. Environ.* **2017**, *121*, 158–167. [CrossRef]
19. Fanger, P.O.; Comfort, T. *Analysis and Applications in Environmental Engineering*; Danish technical Press: Copenghagen, Denmark, 1970; pp. 114–131.
20. Nicol, J.F.; Humphreys, M.A. Adaptive thermal comfort and sustainable thermal standards for buildings. *Energy Build.* **2002**, *34*, 563–572. [CrossRef]
21. de Dear, R.; Brager, G.S. The adaptive model of thermal comfort and energy conservation in the built environment. *Int. J. Biometeorol.* **2001**, *45*, 100–108. [CrossRef]
22. Fanger, P.O.; Toftum, J. Extension of the PMV model to non air conditioned buildings in warm climates. *Energy Build.* **2002**, *34*, 533–536. [CrossRef]
23. Budd, G.M. Wet-bulb globe temperature (WBGT)—Its history and its limitations. *J. Sci. Med. Sport* **2008**, *11*, 20–32. [CrossRef] [PubMed]
24. Yao, R.; Li, B.; Liu, J. A theoretical adaptive model of thermal comfort—Adaptive Predicted Mean Vote (aPMV). *Build. Environ.* **2009**, *44*, 2089–2096. [CrossRef]
25. Sakoi, T.; Mochida, T.; Kurazumi, Y.; Kuwabara, K.; Horiba, Y.; Sawada, S. Heat balance model for a human body in the form of wet bulb globe temperature indices. *J. Therm. Biol.* **2018**, *71*, 1–9. [CrossRef]

26. ISO 7730. *ISO 7730:2005, Ergonomics of the thermal environment-Analytical determination and interpretation of thermal comfort using calculation of the PMV and PPD indices and local thermal comfort criteria*; International Organization for Standardization: Geneva, Switzerland, 2005.
27. Ng, K.C.; Kadirgama, K.; Ng, E.Y.K. Response surface models for CFD predictions of air diffusion performance index in a displacement ventilated office. *Energy Build.* **2008**, *40*, 774–781. [CrossRef]
28. Kalmár, F. An indoor environment evaluation by gender and age using an advanced personalized ventilation system. *Build. Serv. Eng. Res. Technol.* **2017**, *38*, 505–521. [CrossRef]
29. Kocsis, I. Sets, University of Debrecen, Faculty of Engineering. Available online: http://users.atw.hu/dejegyzet/download.php?fname=./szem1/DiszkretMatek/halmaz.pdf (accessed on 14 January 2019). (In Hungarian)
30. Obádovics, J.G. *Mathematics*, 19th ed.; Scolar Kiadó: Budapest, Hungary, 2012; ISBN 978-963-244-330-0. (In Hungarian)
31. Benesóczky, Z. *Boole Logic, Basic Logics*; Budapest University of Technology and Economics: Budapest, Hungary, 2004. (In Hungarian)
32. Subways.net, Sendai Subway. Available online: http://www.subways.net/japan/sendai.htm (accessed on 3 December 2016).
33. Menyhárt, J.; Pokorádi, L. Batteries Fuzzy Based Estimation. *Hadmérnök-Katonai Műszaki Tudományok Online, IX* **2014**, *2*, 48–55. (In Hungarian)
34. László, T.K. Tikk Domonkos: Fuzzy Systems. Available online: http://www.typotex.hu/download/Fuzzy/output.xml (accessed on 23 September 2016).
35. Artificial Intelligence and Life in 2030: One Hundred Years Study on Artificial Intelligence, Report of the 2015 Study Panel. September 2016. Available online: https://ai100.stanford.edu/sites/g/files/sbiybj8861/f/ai_100_report_0906fnlc_single.pdf (accessed on 10 March 2019).
36. Apple: Siri. Available online: https://www.apple.com/siri/ (accessed on 10 March 2019).
37. Microsoft: Cortana. Available online: https://www.microsoft.com/en-us/cortana (accessed on 3 October 2019).
38. Dutta, N.N. DAT T. Artificial intelligence techniques for energy efficient H.V.A.C. system design. In Proceedings of the International Conference on Emerging Technologies for Sustainable and Intelligent HVAC&R Systems, Kolkata, India, 27–28 July 2018.
39. IoT for All: The Future of HVAC Lies in AI and IoT. Available online: https://www.iotforall.com/future-hvac-ai-iot/ (accessed on 3 October 2019).
40. Cheng, C.-C.; Lee, D. Artificial Intelligence-Assisted Heating Ventilation and Air Conditioning Control and the Unmet Demand for Sensors: Part 1. Problem Formulation and the Hypothesis. *Sensors* **2019**, *19*, 1131. [CrossRef]
41. Shukla, A.; Sharma, A. *Sustainability through Energy-Efficient Buildings*; CRC Press: Boca Raton, FL, USA, 2018; ISBN 9781138066755.
42. Yang, L.; Yan, H.; Lam, J.C. Thermal comfort and building energy consumption implications—A review. *Appl. Energy* **2014**, *115*, 164–173. [CrossRef]
43. Kalmár, F. Innovative method and equipment for personalized ventilation. *Indoor Air* **2015**, *25*, 297–306. [CrossRef]
44. Kalmár, F.; Kalmár, T. Alternative personalized ventilation. *Energy Build.* **2013**, *65*, 37–44. [CrossRef]
45. Kalmár, F. Impact of elevated air velocity on subjective thermal comfort sensation under asymmetric radiation and variable airflow direction. *J. Build. Phys.* **2017**, *42*, 173–193. [CrossRef]
46. Kalmár, F.; Kalmár, T. Study of human response in conditions of surface heating, asymmetric radiation and variable air jet direction. *Energy Build.* **2018**, *179*, 133–143. [CrossRef]
47. Benyó, B.; Benyó, Z.; Paláncz, B.; Szilágyi, L.; Ferenci, T. Theory of technical and biology systems. Available online: https://www.tankonyvtar.hu/hu/tartalom/tamop412A/2011_0079_benyo_muszaki_es_biologiai_rendszerek/ch09s03.html (accessed on 26 April 2019). (In Hungarian)

© 2019 by the authors. Licensee MDPI, Basel, Switzerland. This article is an open access article distributed under the terms and conditions of the Creative Commons Attribution (CC BY) license (http://creativecommons.org/licenses/by/4.0/).

Article

Literature Review of Net Zero and Resilience Research of the Urban Environment: A Citation Analysis Using Big Data

Ming Hu [1,*] and Mitchell Pavao-Zuckerman [2]

1. School of Architecture, Planning and Preservation, University of Maryland, College Park, MD 20742, USA
2. Department of Environmental Science and technology, University of Maryland, College Park, MD 20742, USA; mpzucker@umd.edu
* Correspondence: mhu2008@umd.edu

Received: 25 March 2019; Accepted: 22 April 2019; Published: 24 April 2019

Abstract: According to the fifth Intergovernmental Panel on Climate Change (IPCC) assessment report, the urban environment is responsible for between 71% and 76% of carbon emissions from global final energy use and between 67% and 76% of global energy use. Two important and trending domains in urban environment are "resilience" and "net zero" associated with high-performance design, both of which have their origins in ecology. The ultimate goal of net zero energy has become the ultimate "high-performance" standard for buildings. Another emerging index is the measurement and improvement of the resilience of buildings. Despite the richness of research on net zero energy and resilience in the urban environment, literature that compares net zero energy and resilience is very limited. This paper provides an overview of research activities in those two research domains in the past 40 years. The purpose of this review is to (1) explore the shared ecological roots of the two domains, (2) identify the main research areas/clusters within each, (3) gain insight into the size of the different research topics, and (4) identify any research gaps. Finally, conclusions about the review focus on the major difference between the net zero movement and resilience theory in the urban environment and their respective relations to their ecological origins.

1. Motivation and Background

According to the Fifth Assessment Report (AR5) of the UN Intergovernmental Panel on Climate Change (IPCC), the urban environment is responsible for between 71% and 76% of carbon emissions from global final energy use and between 67% and 76% of global energy use [1]. Within the urban environment, buildings represent the greatest unmet energy savings and carbon emission reduction potential because existing and future buildings will determine a large portion of global energy demand [2]. As developing countries keep building and maintaining their standard of living by providing housing and infrastructure, the total energy use in urban environments, particularly use related to buildings, could triple by mid-century [3].

To date, primary research about the urban environment related to climate change has focused on energy efficiency or resilience. For energy efficiency, the ultimate goal is to realize the net zero energy (NZE) goal. NZE and resilience have overlapping origins in systems ecology [4,5] but have developed independently. Investigating how they differ allows us to see how they can inform each other's future development and potentially be integrated to create a more holistic framework for evaluating sustainable development. In 2010, buildings accounted for 32% of total global final energy use and 19% of energy-related greenhouse gas emissions [2]. The energy-centric approach is easy to understand and easily implemented in the building code. Large amounts of research have quantified building energy performance, and robust methodologies have been developed and established. The number of net zero buildings increased by over 700% between 2012 and 2018 in the United States [6], and the steep upward curve can be expected to continue. On the other hand, resilience has a broad range of implications in the urban environment and includes the comprehensive measurement of performance rather than just energy performance. Methods have been implemented and tested to quantify the resilience of a single building, multiple buildings, or an entire city. For example, the city resilience index developed by the Rockefeller Foundation is used to understand and measure the resilience of a city. However, since resilience covers a wider range of issues than NZE, resilience research on the urban environment still lacks a consistent framework or definition that would prove helpful for communication among researchers and practitioners.

This paper's review of state-of-the-art studies regarding the NZE movement and resilience concept in the urban environment was based on a systematic screening of peer-reviewed articles by titles, keywords, and abstracts. The purpose of the review is to (1) identify the main research areas within each domain, (2) gain insight into the size of the different research focal points, and (3) identify any research gaps and trends. The paper is organized as follows. First, the intellectual origin of NZE and resilience is introduced in Sections 1.1 and 1.2. Then, the literature survey results of these two concepts are explained in Section 3. The analysis results are shown graphically and explained in this section as well. Following the analysis, current research gaps and future needs are outlined in Section 4. Conclusions based on Sections 2 through 4 are discussed in Section 5.

1.1. Origin of the "Net Energy" Concept

The concept of "net energy" has its origin in ecology and has continued to maintain a close relation to that field. In 1920, Frederick Soddy, an English chemist and Nobel prizewinner, first offered a new perspective on economics rooted in the law of thermodynamics in physics. Soddy drew attention to the importance of energy for social progress based on real wealth formation as distinct from virtual wealth and a debt accumulation process [7,8]. He suggested that detailed accounting for energy use could be a good alternative to the monetary system, as the conventional monetary system treated the economy as a perpetual motion machine, while in reality, as with any commodity, the actual wealth flow obeyed the laws of thermodynamics [9,10]. Soddy argued that real wealth was derived from the use of energy to transform materials into physical goods and services [11]. However, his theory was largely criticized and ignored in his time due to his standing as a critic—not a scholar—of orthodox economics. The contempt was mutual: in one review of his book called "Wealth, Virtual Wealth, and Debt," the Times Literary Supplement remarked that "it was sad to see a respected chemist ruin his reputation by writing on a subject about which he was quite ignorant ... " [10]. The criticism of Soddy's theory contributed to the long-term silence in associated research development between 1930 and 1970.

In the 1970s, Romanian American mathematician and economist Nicholas Georgescu-Roegen further developed ecological economics or eco-economics based on Soddy's concepts, a field of research that is transdisciplinary and interdisciplinary, encompassing ecology, economics, and physics. Georgescu-Roegen proposed the application of entropy law in the field of economics, arguing that all natural resource consumption is essentially irreversible, a concept that had a profound impact on thinking about net energy flow or the lifecycle of natural resources. He was the first economist of some standing to theorize on the premise that all of Earth's mineral resources will eventually be exhausted at some point [12]; this concept of depletion of natural resources eventually led to the movement of sustainable development.

Another important development in the 1970s was the publication of the article "Energy, Ecology, & Economics" and the book "Environment, Power, and Society" by ecologist Howard Odum, who tackled the economic issue using ecological theories based on energy fundamentals. His energy economics concept was based on the understanding that energy is the foundation for all forms of life and is transformable. He stated that "the true value of energy to society is the net energy, which is that after the costs of getting and concentrating that energy are subtracted" [13]. His view of studying ecology as large and integrative ecosystems paved the foundation for understanding how the different aspects of a whole ecosystem influence each other. In the latter part of his career, in the 1990s, he developed the concept of "emergy," which he defined as "a measure of energy used in the past and thus is different from a measure of energy now. The unit of emergy is the mjoule [14]. Emergy has attracted the attention of academic researchers and is being applied to research in the urban environment in addition to natural ecosystems [15,16]. Since the reemergence of the concept of net energy in the 1970s, energy-flow analysis and net energy simulation have been applied to many different fields beyond ecology and economics; the most visible increase of research and practice interest is in architecture, engineering, and the construction industry (AEC), and the most applicable translation of the concept is NZE building design and construction.

1.2. Origin of Concept of "Resilience"

Resilience emerged in the field of ecology at the same time that NZE studies started to catch researchers' and practitioners' attention. In the 1960s, the ecological resilience concept was introduced in studies of the stability of ecosystems. One of the pioneers in this area is C.S. Holling, who was considered by many to be the father of ecological resilience theory and who also introduced the word "resilience." Holling believed that extending the ecological framework to other fields would be useful for understanding how society, individuals, and community interact with natural ecosystems. The origin of this term has deeper roots that may be linked to the origins of ecosystem and systems ecology in the 1940s and 50s and their attempts to mathematically model dynamic ecosystems [17]. The idea that nature was composed of systems that may have properties like resilience set the stage for more formal conceptualizations of the term by Holling and colleagues throughout the 1970s and 80s [17]. Holling describes resilience as dynamic and complex in juxtaposition to other views of a stable and simple nature. The "stability" of ecological systems refers to their ability to return to an equilibrium state following a disturbance. Holling suggests that resilience "is a measure of the persistence of systems and of their ability to absorb change and disturbance and still maintain the same relationships between populations or state variables." In this early formulation by Holling, it is the instability of a system that conveys its resilience.

These ecological origins for the modern concept of resilience are in some ways at odds with notions of resilience from other disciplines. Engineering resilience applies to how a system responds to disturbances with respect to the system's stability in comparison to an equilibrium steady state [4,18,19]. Engineering resilience derives from notions of resistance to and recovery from disturbances and focuses on the ability and speed of a system to bounce back to its initial, equilibrium conditions following a disturbance event [18,19].

The ecological concept of resilience has seen an extension of its domain to include social-ecological resilience and a paradigm that is applicable to resource management [20,21]. This parallels the organization of the Resilience Alliance, a collective of institutions and researchers that implement "resilience thinking" for the study and management of systems from an interdisciplinary perspective. Here, resilience extends from a concept focused on buffering stress and maintaining function to one where the focus is on the adaptive capacity to innovate and transform a system to sustain and reorganize in the face of stress and disturbance [20] These principles can be put into practice to manage a system for resilience [22–24].

2. Research Method, Materials, and Tools (Literature Review)

The literature review of these two related topics took two forms: quantitative and qualitative research.

2.1. Quantitative Research

The screening and review entailed three steps. First, key search terms on the NZE movement and resilience concept were used to scan the Web of Science and Elsevier's Scopus databases. The screening excluded literature that was not peer reviewed, and multiple combinations of the search terms were used. Combinations of terms included (1) "resilience" with "city," "building," and "urban environment" and (2) "net zero" with "buildings" and "urban environment." Articles, conference proceedings, books, and book chapters were included. Terms appeared in titles, the abstract, and keyword lists. There were 1821 papers found from a variety of disciplines in the resilience research domain, and 592 papers were found in the NZE research domain, all from the period of 1970 through 2018. The different research fields include architecture, construction, engineering, environmental technology and science. After the initial papers were identified, the occurrence of the search terms was used as a filter to narrow down the literature: 20 occurrences were used for resilience papers and 10 were used for net zero papers to gain a comparable number of research domain clusters.

In the second step, citation analysis, co-citation analysis, and text data mining were carried out with the VOSViewer software to determine influential studies, thinkers, and concentrated research topics and their correlations. Citation analysis (CA) is the bibliometric method used to quantitatively evaluate scientific and academic literature to assess the quality of an article or the impact of study/research projects, authors, journals, or institutions. Co-citation analysis (CCA) is a bibliometric method used to measure the correlations among a variety of academic papers based on the rationale that shared references suggest an intellectual connection. This form of document coupling measures the number of documents that have cited a given pair of documents [25]; it is used in this research to trace origins and fields related to NZE and resilience studies in the urban environment. Text data mining (TM) is the process of deriving high-quality information from text-based documents (e.g., titles, keywords, and abstracts) to identify patterns and trends. Automated content analysis for text—which draws on techniques developed in natural language processing, information retrieval, and text mining—has boomed over the past several years in the social sciences and humanities fields [26,27] but is rarely used in scientific and engineering fields.

To analyze and interpret the results from CA, CCA, and TM, maps are often constructed to help visualize the data. Two types of maps are commonly used in bibliometric research: distance-based and graph-based maps [28,29]. In a distance map, the distance between two items generally indicates the relation and correlation. For representing literature review results, the degree of proximity of two research topics can be viewed as a representation of the intellectual connection of the research topics or areas. Graph-based maps indicate the distance between two items but do not need to reflect the strength of the relation between the items [30]. The relation between items is represented by a line. There is a variety of mapping techniques used in bibliometric research, such as multidimensional scaling. Van Eck's VOS mapping techniques [20,28], VXIrd, Pajek [31]. and Gephi. The most commonly used technique is multidimensional scaling.

For this project, VOSViewer was chosen for its two-dimensional distance-based map [32]. VOS, which stands for "visualization of similarities," aims to locate words in a low-dimensional space in such a way that the distance between two words reflects the similarity or relatedness of the words as accurately as possible [28]. The Pearson correlation has been the most popular indirect similarities measure in the literature review for a long time, and it is well known that the Pearson indirect similarities method is not completely satisfactory [33]. VOS is based on a more sophisticated indirect similarity measurement. More detailed technique information regarding how VOS runs the correlation analysis can be found in Van Eck et al. [28]. VOSviewer constructs a map based on a co-occurrence matrix using three steps. The first step is to obtain a similarity matrix; in the second step, a map is constructed by applying the VOS mapping technique to the similarity matrix; then, in the final step, the map is translated and reflected. The similarity or association strength between different words measured in VOSviewer depends on the total number of co-occurrences of words together and the number of occurrences of the terms separately. In a VOS-constructed map, different cluster maps represent different research foci; the sizes of the nodes indicate the relevance of the items—including research topics, authors, sources, or countries—and the distance between nodes illustrates the intellectual connections.

In this review, the following map types were created:

- Map of terms: we used all text data to generate a term map based on occurrence of texts to understand the researcher topics/clusters in one domain.
- Map of keywords: we used co-occurrence of keywords data to construct a map to understand the relation between knowledge groups and different research fields.
- Map of authors: we used citation data to construct a map to identify the influential thinkers in research domains.
- Map of countries: we used citation data to construct a map to identify the influential regions in research domains.

2.2. Qualitative Research

After creating an overview of research activities based on the map constructed with VOSviewer, the researchers identified the most active research areas, trending terms, and influential papers. Then, a qualitative review of all studies was conducted to reduce the articles to a total of 452 studies for resilience and 81 for NZE papers that were applicable to the scope of this study. After the initial VOS scan, the top 200 ranked studies for resilience and all 81 for NZE were reviewed separately by two authors to determine if they actually focused on resilience and NZE. According to the goal of this review, a paper was excluded if it (1) failed to define the term and (2) did not identify the research gaps and trends. This analysis unveiled 20 top-ranked influential papers in each area (40 papers in all). Lastly, a focused review was carried out on the 20 most influential papers in each area as a means of gleaning findings and identifying research gaps and needs. Sections 3 and 4 highlight and discuss the main results from the NZE and resilience literature review, respectively.

3. Findings: Research Clusters, Topics, Gaps, and Trends

3.1. Research Clusters on Resilience (Map of Terms)

First, in order to identify the research clusters and origin, a term map was created based on the occurrence of texts within a corpus of 1821 scientific publications. The terms (texts) appeared in titles, abstracts, and keywords. The occurrence frequencies of terms (texts) of journals were determined based on a minimum number of 20 occurrences of a term. (The default and recommended number of occurrences is 10, according to the program manual. In order to narrow down the research, we doubled the default number.).

Out of the 39,855 terms, 496 met the threshold. For each of those 496 terms, a relevance score was then calculated. Based on those scores, the most relevant terms were selected, using the program's default of the top 60% most relevant terms. Altogether, 275 terms were selected for the resilience research study; the result is shown in Figure 1. Based on the VOSviewer clustering technique, the terms in the data set were divided into four clusters, with a different color for each research cluster.

- Cluster 1 (red): technology, application, energy efficiency, performance, event (left)
- Cluster 2 (blue): factor, finding, relationship, health (right)
- Cluster 3 (green): urban resilience, governance, understanding/theory, ecosystem/eco service (middle)
- Cluster 4 (yellow): disaster, hazard, mitigation (upper)

These clusters show four separate focus areas across different disciplines, which may be referred to as techniques and application of resilient practice (cluster 1), cause and relation of resilient factors (cluster 2), the understanding of the mutual influence of urban resilience and ecosystem (cluster 3), and disaster and hazard relief (cluster 4). The clusters are closely related to the three predominant definitions of resilience in an urban environment emerging from its ecological origin: engineering resilience, ecological resilience, and social–ecological resilience [34]. Engineering resilience is defined as a system's speed of return to equilibrium following a shock, indicating that a system can have only a single stability regime [35].Ecological resilience emphasizes conditions far from any equilibrium steady state, where resilience is the measurement of the magnitude of disturbance that can be absorbed before the system changes its structure [4]. Social–ecological resilience is defined by Adger as the "Ability of communities to withstand external shocks to their social infrastructure" [36,37]. Cluster 2 is derived from ecological resilience; cluster 3 is the continuation of social–ecological resilience; and cluster 4 is related to engineering resilience. Lastly, cluster 1 is about the translation or transformation of resilience theory to practice.

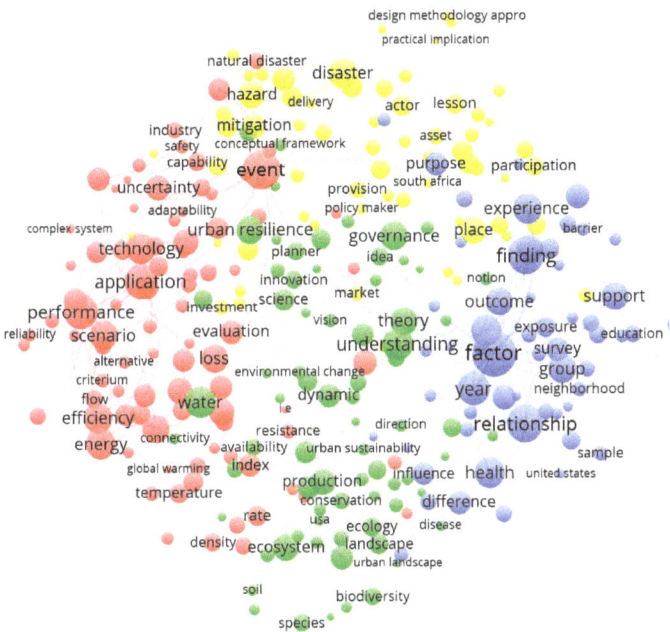

Figure 1. Term map representing the main research areas on resilience in the built environment.

The four research clusters are not of equal size and density. Cluster 1 has the most research activities and impact, while cluster 4 has the least number of studies and relevance. Clusters 2 and 3 are of very similar size. Within cluster 1, there are four sub-clusters: technology application, energy efficiency advancement, performance in different scenarios, and individual disastrous events. Among those, technology application is closely linked to energy efficiency advancement, as well as performance in different scenarios. The disastrous events sub-cluster, which is associated with damage and uncertainty, is relatively remote from the other three sub-clusters. Cluster 2 has three clear sub-clusters: finding/program, factor, and relationship. This cluster mainly represents social and community resilience, as other associated terms include family, health, culture, intervention, and education, and they can be seen as a further development of the socio–ecological resilience definition. Cluster 3 is the second-largest cluster, but unlike clusters 1 and 2 it does not have clear sub-clusters. Instead, it is interwoven into cluster 1 and cluster 2.

The connection between "event" under cluster 1 and "urban resilience" under cluster 3 appears to be very strong, even after pulling them away from the cores of their respective areas. This close connection might indicate the rising risk of shocks from natural disasters and the demand to find techniques with which to build robust urban resilience. "Understanding" under cluster 3 is connected to "finding" in cluster 2; the finding about the factors that determine the resilience of a system help to establish an understanding of what drives the transformation of an existing system into a more resilient one. There is almost no connection between cluster 4 and cluster 3, and "disaster", "natural disaster", and "hazard" are located far from "ecosystem/ecosystem service" and "green infrastructure".

This weak link could indicate a substantial gap between disaster mitigation and ecosystem service and green infrastructure development. "Understanding" (cluster 2) is situated in the center of the diagram, equally remote from both ecosystem (cluster 3) and natural disaster (cluster 4). "Understanding" also has almost no linkages to technology/application and performance/energy (cluster 1). Together, these may indicate that the current application of engineering resilience needs more scientific support, as without an in-depth understanding of the resilient factors, the performance and technical application of resilience theory cannot be improved and verified.

3.2. Research Focus/Topics and Relations on Resilience (Map of Keywords)

A keywords map was created, based on the co-occurrence of keywords, to study the underlying structure of the different research clusters explained in Section 3.1 and their relationship with knowledge groups and different research fields. In general, the closer keywords are located to each other on a map, the stronger the relation between them. Out of 2359 keywords taken from all the papers, where a minimum of 10 occurrences of a keyword was used to identify the most related knowledge and research fields, 36 keywords met the threshold. The 36 keywords came from the four research clusters identified in Section 3.1.

As can be seen in Figure 2, among all research topics, there are three linked research concentrations. The first is the linkage between "climate change" and "management" and "adaptation", which together could represent environmental science research rooted in ecology with a focus on adaptation to climate change in the urban environment. Biodiversity and the ecosystem are viewed as trending techniques for realizing climate change management and adaptation. The United Kingdom, Canada, China, and South Africa appeared as having the most research objects of interest. The second set of coupled topics is sustainability and system management, where concentrations appear around energy efficiency and buildings. The third set of coupled topics is vulnerability and urban resilience, which can be interpreted as the connection between existing urban environmental conditions and the difficulty of adapting to future conditions under climate change.

3.3. Research Gaps and Future Trends on Resilience

Overall, the term map indicates the gap between social–ecological resilience and engineering resilience. Although there are some studies that discuss future and past views of the two types of resilience, to a large extent, social–ecological resilience focuses on the understanding and interoperation of potential development (forward thinking), while engineering resilience mostly relies on what happened and responses to extreme events [4,37]. From the keywords map, despite the closely connected major topics, there is quite a large number of uncoupled research interests, such as health, information, consumption, operation, networks, heat-island, urban regeneration, and design. Those topics represent potential research gaps that require further development and study, as well as integration with existing resilience frameworks. For instance, the disconnect between ecosystem/green infrastructure and building and energy efficiency is evidence of some reluctance in the building and construction industry to consider the urban environment as part of a large ecosystem. Some fundamental thinking and knowledge could be brought into the planning and design phase. For example, the quality of ecosystem services and green infrastructure could protect urban dwellers from natural hazards. The robustness and continuity of such services under conditions of shock or stress are important.

Overall, the intellectual connection of research topics indicates potential research trends. After a focused and in-depth qualitative review of the most influential studies and articles, we categorized the resilience of urban environment research into three future research trends: urban resilience, risk management, and sustainable development.

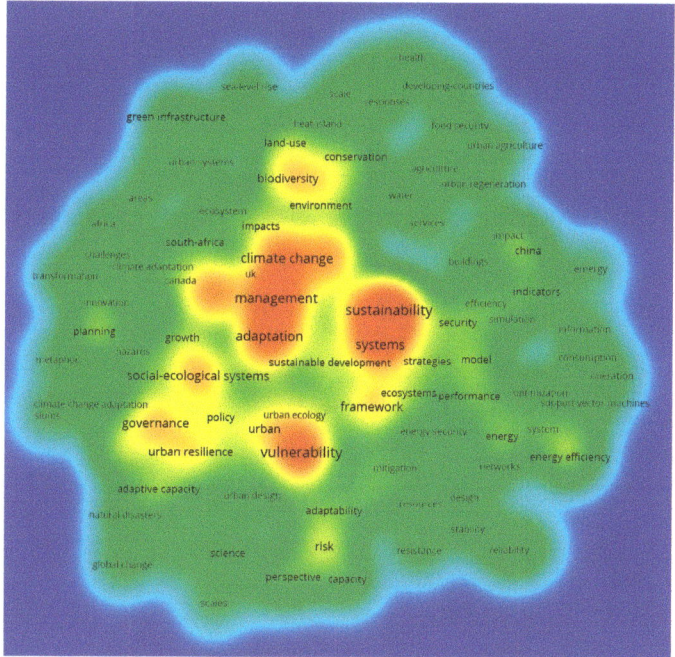

Figure 2. Keywords co-occurrence map (by linkage), showing the relationship between research areas.

Urban resilience studies are closely related to adaptation, vulnerability, and community building, and many of the research studies are written from a social–ecological perspective. There are also quite a few technical reports produced by government and not-for-profit foundations and organizations [38]. This domain represents the fusion between ecology and social science. Risk management is closely related to climate change, which has the potential to bridge the gap between disaster risk reduction and climate change adaptation, as well as re-establish the connection between green infrastructure building and urban design principles. In this domain, the research related to disaster reduction has moved away from traditional disaster risk management, which mainly focuses on specific hazards such as floods and other natural disasters. Instead, it has been expanded to a wide range of disruptive events and social disturbance and stress, both stresses and shocks that could influence governance and policy making. The third domain is sustainable development, which is mainly rooted in the engineering, building, and construction fields. It does appear to have more frequent co-occurrence with other critical terms, as the concept and practice of sustainability has a much longer history in comparison to resilience; the two terms often appear as paired concepts. Through this exercise, we could observe the evolution of the concept of sustainability.

3.4. Research Clusters on NZE (Map of Terms)

The process used in studying research domains within resilience was reapplied to the net zero research domain, first creating a term map from 519 documents (articles, book chapters, and books). This number is much lower than the quantity of research activities in the resilience domain. Therefore, the occurrence frequencies of terms (text) of journals were determined based on a minimum number of 15 occurrences of a term, which is lower than the minimum used for resilience research. Out of the 15,438 terms—only half of the number of terms associated with resilience—167 met the threshold. For each of those 167 terms, a relevance score was then calculated. Based on this score, the most relevant terms were selected using the program's default of the top 60% most relevant terms. Altogether,

100 terms were selected for NZE research; the result is shown in Figure 3. Based on the VOSviewer clustering technique, the terms in the data set were divided into four clusters, with a different color for each research cluster.

- Cluster 1 (yellow): effect, rate/period (left)
- Cluster 2 (red): project/standard, net zero energy/practice, home (right)
- Cluster 3 (green): emission, energy source, water (middle)
- Cluster 4 (blue): zero energy building, heating/cooling/temperature, ventilation (upper)

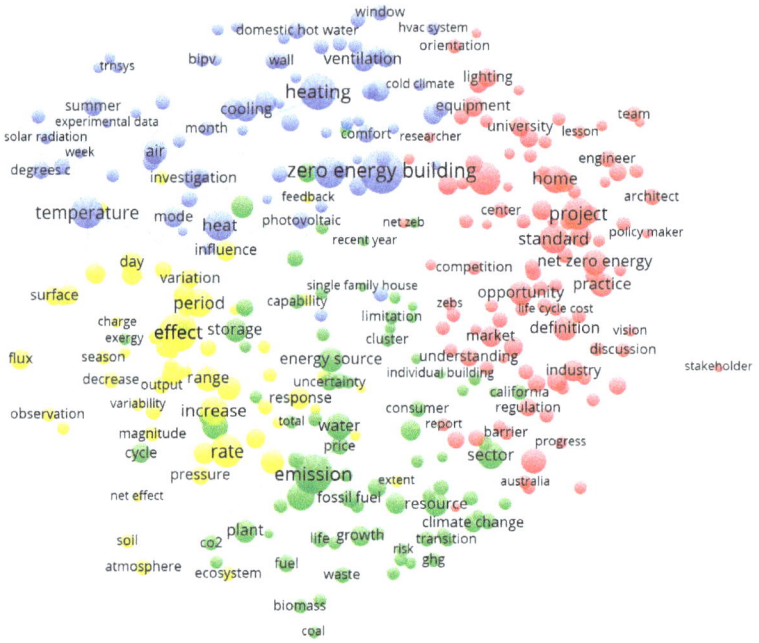

Figure 3. Term map representing the main research areas on net zero in the built environment.

These clusters show four research areas across different disciplines related to NZE research activities, which may be referred to as net zero impact (cluster 1), net zero energy practice in industry (cluster 2), net zero emission (cluster 3), and techniques to achieve net zero energy building (cluster 4). Cluster 4 has the greatest number of research activities, which center around NZE building and temperature control. It is commonly known that NZE building mainly focuses on operating energy reduction. Heating and cooling together account for more than 50% of overall operating energy consumption, and the outside ambient temperature has a large influence on building heating and cooling loads. Those facts could explain why heating, cooling, and temperature appear as critical sub-topics in cluster 4. Cluster 2 and cluster 3 are of almost equal size, with cluster 2 containing more items and cluster 3 having items aggregating toward one major focus area, "emission." Cluster 1 has the fewest research items and is connected to clusters 3 and 4, but has no connection to cluster 2. The disconnection between clusters 1 and 2 indicates the limited verification and measurement of the effectiveness of net zero development, particularly in the building industry. Such lack of verification and measurement caused some concerns about the validity of the NZE building design and approach, which presents a challenge and a gap as well as opportunities.

3.5. Research Focus/Topics and Relations on NZE (Map of Keywords)

A co-occurrence map was created, using keywords, in order to study the underlying relationship of research areas and then to predict future research trends (refer to Figure 4). Out of 2028 keywords from all the papers, where a minimum of 10 occurrences of a keyword was used to identify the most related knowledge and research fields, 43 keywords met the threshold. The 43 keywords came from the four research clusters identified. As can be seen in Figure 4, strong relationships emerged among the topics of design, system, efficiency, and performance.

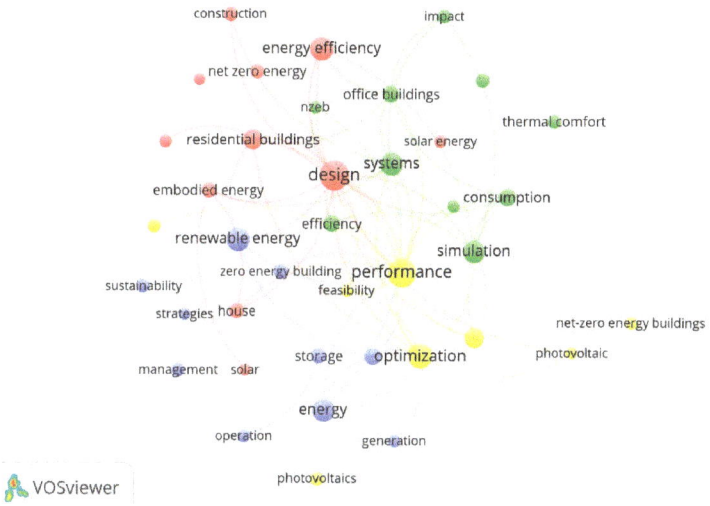

Figure 4. Keywords co-occurrence map representing the relationship between different research topics on NZE.

Among all research topics, there are two linked research concentrations. The first is the linkage between "design" and "efficiency" and "system," which reflects awareness of the important role of design and systemic consideration in energy efficiency. The second linked research concentration is "performance" and "simulation." Office buildings, residential buildings, and embodied energy are popular research topics associated with those two concentrations.

3.6. Research Gaps and Future Trends in NZE

As shown in Figure 4 (keywords map), there are two research gaps in NZE research. Firstly, current research is primarily around renewable energy sources. Only one specific renewable energy technology is present: photovoltaics. Limiting the focus to photovoltaics could hinder the search for other alternative renewable energy resources for future net zero building development. Alternative renewable energy resources such as hydro energy and biomass energy were not active in the urban environment, but are more widely studied and applied in other fields, such as industry ecology. The second research gap is identified by some important keywords that are missing, such as insulation, standards and regulations Most of those missing keywords are associated with the practicality and feasibility of NEZ. This might be explained by most studies found in the Web of Science database and Elsevier's Scopus database being academic research focusing on system, method, and framework studies. But it does show that more future studies and research could be undertaken in the areas of technology, techniques, and practical standards of NEZ. The lack of technique keywords might also indicate a disconnection between academia and practice. The lack of research indicates gaps in both knowledge and opportunities for future work.

Based on the quantitative and qualitative research from previous sections, we then categorized future research on NZE into four directions: office building; residential building; sustainability management/strategies; and design. NZE research activities in office building have largely been related to operating energy efficiency and environmental impact. With respect to residential building, NZE research activities are mainly focused on embodied energy reduction and achieving an overall net zero energy goal. Another difference between the first and second trends is the fact that in residential building, design is a focal point, and the active research and practice activities involve design practitioners such as architects, planners, and engineers. In contrast, in the office building field the focus is around system and simulation, due to the large scale and complexity of office building. Also, the inputs from numerous stakeholders demand a more systematic and high-level approach in achieving a net zero goal in office building. The third and fourth trends are not associated with any particular building types; instead, the focus is around the optimization of energy performance and research on renewable energy production.

3.7. Most Influential Studies and Active Regions

The most influential studies of resilience, based on the citations and connections to others, are Pickett's "Resilient Cities: Meaning, Models, and Metaphor for Integrating the Ecological, Socio-economic, and Planning Realms" [39] (ecology), which has been cited 490 times since 2003, and Head's "Suburban Life and the Boundaries of Nature: Resilience and Rupture in Australian Backyard Gardens" [40], which has been cited 103 times since 2006 (refer to Figure 5). These studies cover diverse fields that include ecology, economics, environmental science, geology, landscape architecture and regional planning. Generally, they could be categorized as social-ecological research and studies. To date, there are no influential studies about engineering resilience.

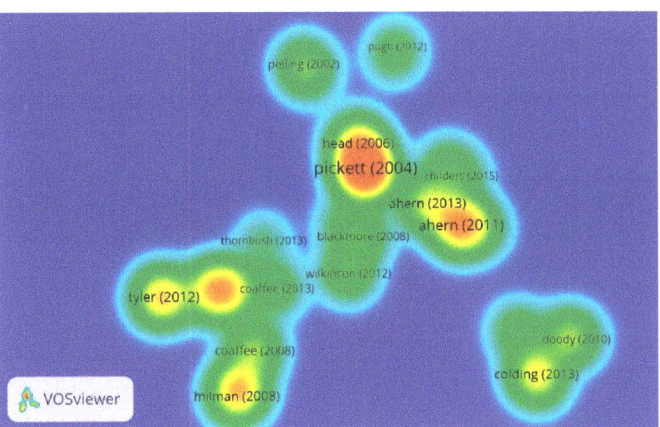

Figure 5. Citation map representing influential studies (resilience).

The most influential studies of NZE are Hernandez and Kenny's "From Net Energy to Zero Energy Buildings: Defining Life Cycle Zero Energy Buildings" [10] (engineering), which has been cited 332 times since 2010, Marszal and team's "Zero Energy Building: A Review of Definitions and Calculation Methodologies" [41] (engineering), which has been cited 598 times since 2011, and Sartori and team's "Net Zero Energy Buildings: A Consistent Definition Framework" [42] (architectural engineering), which has been cited 375 times since 2012. These studies are mainly from the architectural engineering field; research activities accelerated from 2011 and reached a peak in 2015 (refer to Figure 6).

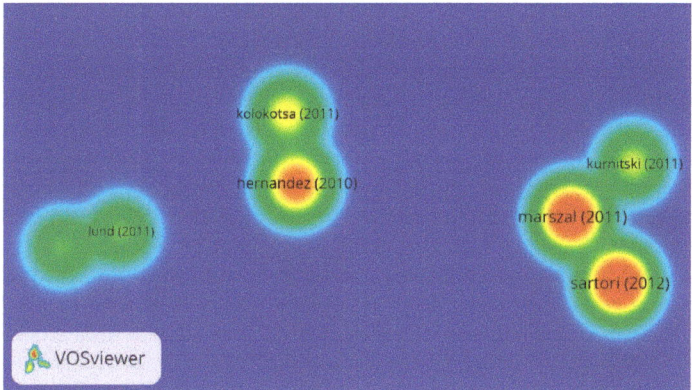

Figure 6. Citation map representing influential studies (net zero).

The most influential countries in the research domain of NZE in the urban environment are: the United States, with 156 research documents and 814 total citations; Italy, with 49 documents and 891 total citations; and Norway, with 24 documents and 615 total citations. China is the only developing country on the list and ranks fifth, after Canada, with 44 publications and 320 citations. This is likely due to China's rapid urbanization and the demand for the control of pollution caused by excessive emissions (refer to Figure 7).

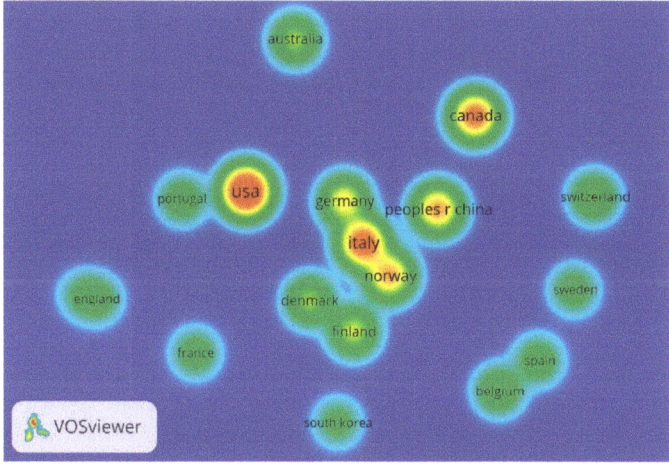

Figure 7. Citation map representing influential countries (net zero).

4. Discussion: Difference and Divergence of Research Activities

The concepts of net zero and resilience are both rooted in ecology, originally looking at the built environment from a systematic perspective [4,7,8,15–19]. However, the two disciplines have taken quite different approaches and directions in recent years. Resilience studies were promoted due to the urgency of severe climate change events such as flooding, wildfire, and extreme temperatures. The three research focus areas of urban resilience [38], risk management [43], and sustainable development recognize the vulnerability of the urban ecosystem [39,44]. and the necessity to adapt to climate change [37]. In general, resilience studies in the urban environment take a more holistic and comprehensive approach and take into account long-term strategies. In comparison to resilience, the research around

NZE has limited scope. Unlike resilience studies, the research activities on NZE have moved away from its ecological origin, to focus on particular building types [42] and design applications [41]. The lifecycle approach of the original net energy concept, which developed from ecology, was replaced by a performance-driven, building-types-driven engineering approach [10]. This relatively narrow focus might be seen as a deviation from the original ecological roots, which used energy flow to measure the sustainability, accessibility, and efficiency of a system. The narrowed focus was driven by multiple market needs, industry preference, and political incentives. Among many factors, the search for a consistent and common definition of net zero building has been a singular focus over the past five years due to strong interest and the quantity of new construction projects around the world [42].

Unlike research related to resilience, the three most influential papers for NZE focus on net zero building measurement and application, and the most influential thinkers, researchers, and institutions are primarily in the building and construction industry. Those net zero studies only address CO_2 emissions from buildings that affect climate change, and do not consider the influence of climate change on urban environments at a higher level. They pay only limited attention to the effects of climate change or ecosystem change on net zero building design strategies.

Another main difference between net zero research and resilience research is coverage. It appears that resilience not only covers a wide range of fields, but also has research contributions from both developed nations (such as the US, the Netherlands, and Sweden) [4,35,36] and developing countries (such as China and India) [18]. This might be because resilience is in more emergent need in developing countries, since they are more vulnerable to natural disasters due to their lack of robust infrastructure and prevention methods. On the other hand, net zero research is mostly undertaken in developed countries, including the United States and some of the western European countries.

Resilience and net zero are two primary research focus areas in the urban environment related to sustainable development, and bridging them could potentially yield rich results. Recently, there have been some studies that integrated the holistic and multi-disciplinary resilient approach in energy development. Sharifi and Yamagata [45] used a resilience framework to measure the sustainability of an energy system. They suggest that the energy system could be evaluated based on availability, accessibility, affordability, and acceptability, which is more holistic than the current efficiency-centric criteria. Torcellini et al. [46] identified metrics useful for implementing guidance for energy-related planning, design, investment, and operation. The metrics include energy reduction, energy resource diversity, energy storage, preventive maintenance on energy systems, and other metrics in physical, information, and human interaction categories. Beheshtian et al. [47] created a model to measure the long-term resilience of transportation energy infrastructure in Manhattan, NY.

5. Conclusions

This paper presents findings from a literature review of two topics with shared ecological intellectual roots. The literature review was conducted with both quantitative and qualitative approaches. It scanned 1821 papers in the resilience research domain and 592 in the NZE research domain, all from the period of 1970 through 2018. Citation analysis, co-citation analysis, and text data mining were carried out using new data mining software, VOSviewer, to determine influential studies, thinkers, and concentrated research topics and their correlations. Based on the findings, conclusions can be drawn as follows. This paper provides an initial step in understanding the research activities of the past five decades in these two areas and their connection to their ecological roots.

'Net zero' and 'resilience' have overlapping origins in ecosystem and systems ecology. 'New zero' stems from Georgescu-Roegen's application of entropy to ecological economics and Odum's systems ecology and development of the emergy concept. 'Resilience' stems from Holling being influenced by developments in systems ecology that drew together a more formal mathematical approach to systems analysis of ecological cases. Developments in systems ecology in the 1960s largely forms the connection between the two research fields through Holling and Odum. However, despite the richness of research on both NZE and resilience in the urban environment, and despite the shared ecological roots of NEZ and resilience, the literature on both energy resilience and comparisons between NZE and resilience is very limited, almost non-existent. Unlike resilience studies, NZE research has relatively narrow coverage and has moved away from its ecological origin. The transformation of current NZE practice and concept will be crucial for climate change stabilization. Applying the existing resilience ecological framework to NEZ research could shift the focus on application and individual buildings to broader systematic consideration.

Author Contributions: Conceptualization, M.H. and M.P.-Z.; methodology, M.H. and M.P.-Z.; software, M.H.; validation, M.H. and M.P.-Z.; formal analysis, M.H.; investigation, M.H. and M.P.-Z.; data curation, M.H.; writing—original draft preparation, M.H.; writing—review and editing, M.H. and M.P.-Z.; visualization, M.H.

Funding: This research received no external funding.

Acknowledgments: We would like to acknowledge the support given from School of Architecture, Planning and Preservation, and Department of Environmental Science and Technology in the University of Maryland.

Conflicts of Interest: The authors declare no conflict of interest.

References

1. Seto, K.C.; Dhakal, S.; Bigio, A.; Blanco, H.; Delgado, G.C.; Dewar, D.; Huang, L.; Inaba, A.; Kansal, A.; Lwasa, S.; et al. *Human Settlements, Infrastructure and Spatial Planning*; International Institute for Applied System Analysis: Laxenburg, Austria, 2014.
2. Lucon, O.; Ürge-Vorsatz, D.; Ahmed, A.Z.; Akbari, H.; Bertoldi, P.; Cabeza, L.F.; Eyre, N.; Gadgil, A.; Harvey, L.D.; Jiang, Y.; Liphoto, E. *Buildings*; Cambridge University Press: Cambridge, UK, 2014.
3. Allen, M.R.; Barros, V.R.; Broome, J.; Cramer, W.; Christ, R.; Church, J.A.; Clarke, L.; Dahe, Q.; Dasgupta, P.; Dubash, N.K.; et al. *IPCC Fifth Assessment Synthesis Report-Climate Change 2014 Synthesis Report*; IPCC: Geneva, Switzerland, 2014.
4. Holling, C.S. Engineering resilience versus ecological resilience. In *Engineering within Ecological Constraints*; National Academies Press: Washington, DC, USA, 1996; p. 32.
5. Yi, H.; Srinivasan, R.S.; Braham, W.W.; Tilley, D.R. An ecological understanding of net-zero energy building: Evaluation of sustainability based on emergy theory. *J. Clean. Prod.* **2017**, *143*, 654–671. [CrossRef]
6. New Building Institute. Getting to Zero Status Update and Zero Energy Building List, 2018. Available online: https://newbuildings.org/resource/2018-getting-zero-status-update/ (accessed on 5 January 2018).
7. Rapoport, A. (Ed.) *The Mutual Interaction of People and Their Built Environment*; Walter de Gruyter: Berlin, Germany, 1976.
8. Spash, C.L. (Ed.) *Routledge Handbook of Ecological Economics: Nature and Society*; Taylor & Francis: Milton Park, UK, 2017.
9. Hassler, U.; Kohler, N. *Resilience in the Built Environment*; Taylor & Francis: Milton Park, UK, 2014.
10. Hernandez, P.; Kenny, P. From net energy to zero energy buildings: Defining life cycle zero energy buildings (LC-ZEB). *Energy Build.* **2010**, *42*, 815–821. [CrossRef]
11. Soddy, F. *Wealth, Virtual Wealth and Debt*; George Allen and Unwin Ltd.: London, UK, 1933.
12. Boulding, K.E. *Ecodynamics: A New Theory of Societal Evolution*; SAGE Publications, Incorporated: Thousand Oaks, CA, USA, 1978.
13. Odum, H.T. Energy, ecology, and economics. *Ambio* **1973**, *2*, 220–227.
14. Odum, H.T. *Environmental Accounting: Emergy and Environmental Decision Making*; Wiley: Hoboken, NJ, USA, 1996.
15. Pulselli, R.M.; Simoncini, E.; Pulselli, F.M.; Bastianoni, S. Emergy analysis of building manufacturing, maintenance and use: Em-building indices to evaluate housing sustainability. *Energy Build.* **2007**, *39*, 620–628. [CrossRef]
16. Pulselli, R.M.; Simoncini, E.; Marchettini, N. Energy and emergy based cost–benefit evaluation of building envelopes relative to geographical location and climate. *Build. Environ.* **2009**, *44*, 920–928. [CrossRef]
17. Lindseth, B. The pre-history of resilience in ecological research. *Limn* **2011**, *1*, 1.
18. Liao, K.H. A theory on urban resilience to floods—A basis for alternative planning practices. *Ecol. Soc.* **2012**, *17*, 48. [CrossRef]

19. Wang, C.H.; Blackmore, J.M. Resilience concepts for water resource systems. *J. Water Resour. Plan. Manag.* **2009**, *135*, 528–536. [CrossRef]
20. Berkes, F.; Folke, C. Linking social and ecological systems for resilience and sustainability. In *Linking Social and Ecological Systems: Management Practices and Social Mechanisms for Building Resilience*; Cambridge University Press: Cambridge, UK, 1998.
21. Folke, C. Resilience: The emergence of a perspective for social–ecological systems analyses. *Glob. Environ. Chang.* **2006**, *16*, 253–267. [CrossRef]
22. Adger, W.N.; Hughes, T.P.; Folke, C.; Carpenter, S.R.; Rockström, J. Social-ecological resilience to coastal disasters. *Science* **2005**, *309*, 1036–1039. [CrossRef] [PubMed]
23. Biggs, R.; Schlüter, M.; Biggs, D.; Bohensky, E.L.; BurnSilver, S.; Cundill, G.; Dakos, V.; Daw, T.M.; Evans, L.S.; Kotschy, K.; et al. Toward principles for enhancing the resilience of ecosystem services. *Annu. Rev. Environ. Resour.* **2012**, *37*, 421–448. [CrossRef]
24. Walker, B.; Salt, D.; Reid, W. *Resilience Thinking: Sustaining People and Ecosystems in a Changing World*; Island Press: Washington, DC, USA, 2006.
25. Small, H. Visualizing science by citation mapping. *J. Am. Soc. Inf. Sci.* **1999**, *50*, 799–813. [CrossRef]
26. O'Connor, B.; Bamman, D.; Smith, N.A. Computational Text Analysis for Social Science: Model Assumptions and Complexity. Available online: https://people.cs.umass.edu/~{}wallach/workshops/nips2011css/papers/OConnor.pdf (accessed on 28 March 2019).
27. Zhai, C.; Massung, S. *Text Data Management and Analysis: A Practical Introduction to Information Retrieval and Text Mining*; Morgan & Claypool: Rafael, CA, USA, 2016.
28. Van Eck, N.J.; Waltman, L.; Dekker, R.; van den Berg, J. A comparison of two techniques for bibliometric mapping: Multidimensional scaling and VOS. *J. Assoc. Inf. Sci. Technol.* **2010**, *61*, 2405–2416. [CrossRef]
29. Garfield, E. From the science of science to Scientometrics visualizing the history of science with HistCite software. *J. Informetr.* **2009**, *3*, 173–179. [CrossRef]
30. Van Eck, N.J.; Waltman, L. Software survey: VOSviewer, a computer program for bibliometric mapping. *Scientometrics* **2010**, *84*, 523–538. [CrossRef] [PubMed]
31. Davila, C.C.; Reinhart, C. Urban energy lifecycle: An analytical framework to evaluate the embodied energy use of urban developments. In Proceedings of the BS2013: 13th Conference of International Building Performance Simulation Association, Chambéry, France, 26–28 August 2013; pp. 26–28.
32. Moed, H.F. *Citation Analysis in Research Evaluation*; Springer Science & Business Media: Berlin, Germany, 2006; Volume 9.
33. McCain, K.W. Mapping authors in intellectual space: A technical overview. *J. Am. Soc. Inf. Sci.* **1990**, *41*, 433–443. [CrossRef]
34. Quinlan, A.E.; Berbés-Blázquez, M.; Haider, L.J.; Peterson, G.D. Measuring and assessing resilience: Broadening understanding through multiple disciplinary perspectives. *J. Appl. Ecol.* **2016**, *53*, 677–687. [CrossRef]
35. Borg, I.; Groenen, P.J. *Modern Multidimensional Scaling: Theory and Applications*; Springer Science & Business Media: Berlin, Germany, 2005.
36. Adger, W.N. Social and ecological resilience: Are they related? *Prog. Hum. Geogr.* **2000**, *24*, 347–364. [CrossRef]
37. Colding, J.; Barthel, S. The potential of 'Urban Green Commons' in the resilience building of cities. *Ecol. Econ.* **2013**, *86*, 156–166. [CrossRef]
38. Schipper, E.L.F.; Langston, L. A Comparative Overview of Resilience Measurement Frameworks. Available online: https://www.odi.org/sites/odi.org.uk/files/odi-assets/publications-opinion-files/9754.pdf (accessed on 28 March 2019).
39. Pickett, S.T.; Cadenasso, M.L.; Grove, J.M. Resilient cities: Meaning, models, and metaphor for integrating the ecological, socio-economic, and planning realms. *Landsc. Urb. Plan.* **2004**, *69*, 369–384. [CrossRef]
40. Head, L.; Muir, P. Suburban life and the boundaries of nature: Resilience and rupture in Australian backyard gardens. *Transact. Inst. Br. Geogr.* **2006**, *31*, 505–524. [CrossRef]
41. Marszal, A.J.; Heiselberg, P.; Bourrelle, J.S.; Musall, E.; Voss, K.; Sartori, I.; Napolitano, A. Zero energy building—A review of definitions and calculation methodologies. *Energy Build.* **2011**, *43*, 971–979. [CrossRef]
42. Sartori, I.; Napolitano, A.; Voss, K. Net zero energy buildings: A consistent definition framework. *Energy Build.* **2012**, *48*, 220–232. [CrossRef]

43. Tyler, S.; Moench, M. A framework for urban climate resilience. *Clim. Dev.* **2012**, *4*, 311–326. [CrossRef]
44. Ahern, J. From fail-safe to safe-to-fail: Sustainability and resilience in the new urban world. *Lands. Urban Plan.* **2011**, *100*, 341–343. [CrossRef]
45. Sharifi, A.; Yamagata, Y. A conceptual framework for assessment of urban energy resilience. *Energy Procedia* **2015**, *75*, 2904–2909. [CrossRef]
46. Torcellini, P.; Pless, S.; Leach, M. A pathway for net-zero energy buildings: creating a case for zero cost increase. *Build. Res. Inf.* **2015**, *43*, 25–33. [CrossRef]
47. Beheshtian, A.; Donaghy, K.P.; Geddes, R.R.; Gao, H.O. Climate-adaptive planning for the long-term resilience of transportation energy infrastructure. *Transp. Res. Part E Logist. Transp. Rev.* **2018**, *113*, 99–122. [CrossRef]

© 2019 by the authors. Licensee MDPI, Basel, Switzerland. This article is an open access article distributed under the terms and conditions of the Creative Commons Attribution (CC BY) license (http://creativecommons.org/licenses/by/4.0/).

Article

Environmental and Efficiency Analysis of Simulated Application of the Solid Oxide Fuel Cell Co-Generation System in a Dormitory Building

Han Chang * and In-Hee Lee *

Department of Architecture, Pusan National University, Busan 46241, Korea
* Correspondence: changhan@pusan.ac.kr (H.C.); samlih@pusan.ac.kr (I.-H.L.)

Received: 23 August 2019; Accepted: 11 October 2019; Published: 15 October 2019

Abstract: The problem of air pollution in Korea has become progressively more serious in recent years. Since electricity is advertised as clean energy, some newly developed buildings in Korea are using only electricity for all energy needs. In this research, the annual amount of air pollution attributable to energy under the traditional method in a dormitory building, which is supplying both natural gas and electricity to the building, was compared with the annual amount of air pollution attributable to supplying only electricity. The results showed that the building using only electricity emits much more air pollution than the building using electricity and natural gas together. Under the traditional method of energy supply, a residential solid oxide fuel cell cogeneration system (SOFC–CGS) for minimizing environmental pollution of the building was simulated. Furthermore, as a high load factor could lead to high efficiency of the SOFC–CGS, sharing of the SOFC–CGS by multi-households could increase its efficiency. Finally, the environmental pollution from using one system in one household was compared with that from sharing one system by multi-households. The results showed that the environmental pollution from sharing the system was relatively higher but still similar to that when using one system in one household.

Keywords: environmental impact; device efficiency; air pollutant; multi-households; solid oxide fuel cell cogeneration system

1. Introduction

1.1. Research Background

Electricity as non-combusted energy is advertised to the public as clean energy. Since the price of natural gas and the price of electricity for residential use are similar [1], in the areas of architecture and construction, some real estate developers have been developing new buildings using only electricity for power. However, as a secondary energy resource, electricity is either clean or not clean depending on the electricity generation procedure. In Korea, 39% of electricity is generated from burning coal and only 4% is generated from renewable resources as seen in Figure 1 [2].

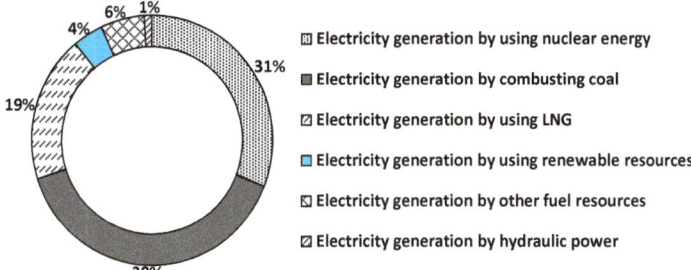

Figure 1. Electricity generation system in Korea.

The traditional method of supplying energy to a building in Korea is mainly using natural gas for heating and hot water and using electricity for lighting, cooling, and equipment. A recent method of supplying energy to a building is using only electricity for all energy needs in the building. This research compared the annual amount of air pollution attributable to energy under the traditional method in a dormitory building, which is supplying both natural gas and electricity to the building and the annual amount of air pollution attributable to supplying only electricity.

A solid oxide fuel cell (SOFC) is a clean energy device for generating electricity by consuming hydrogen or natural gas. By utilizing chemical interaction in the cell stack during the process of electricity generation in SOFC, there is no environmental pollution. However, there is a large amount of heat dissipation during SOFC operation. In order to improve the efficiency of SOFC, the SOFC cogeneration system (SOFC–CGS) was developed to collect and use the heat loss. The heat dissipation generated from SOFC is collected by residential SOFC–CGS and supplied to a hot water tank for hot water demand of a household. A fuel cell as a clean energy system is widely studied for residential use. In recent years, various researches were done about optimizing the operation efficiency of a fuel cell. Some scholars proposed efficiency optimization methods of a fuel cell through improving working principles of electronic components. Wang et al. proposed a novel stack and converter interpreted design scheme to improve the converter efficiency [3]. A micro tri-generation system could increase the system efficiency to over 90% [4]. An energy management system could also achieve cost reduction, CO_2 mitigation, and energy consumption [5]. Some scholars aim to find a way to improve the efficiency of a fuel cell from analysis of energy demand, energy distribution, and energy supplication of a fuel cell. Adam et al. presents a multi-objective modeling approach that allows an optimized design of micro-combined heat and power (CHP) systems considering the source, distribution and emission requirements in unison to achieve more efficient whole systems [6]. Coordination between utilization factor of SOFC and features of a DC/AC (Direct current/Alternating current) inverter is also an efficient control strategy for SOFC operation [7]. Some researchers have combined a photovoltaic, battery, or heat pump with a fuel cell to satisfy the energy needs by a household while reducing cost and CO_2 emission [8–10]. From a wider energy supply perspective, SOFC was considered to supply energy for both households and EV (Electric vehicle) to improve the efficiency and energy supply capacity of SOFC. Using a hydrogen supply system for a residential fuel cell to provide energy for fuel cell vehicles is also a method to solve the problem of lacking hydrogen stations for fuel cell vehicles [11,12]. Field experimental studies demonstrated that a residential fuel cell could significantly reduce the CO_2 emissions and primary energy consumption [13]. Accurate mathematical models were developed to analyze the performance of SOFC. A 1D dynamic model was built for studying system integration and developing adequate energy management and control strategies of SOFC systems [14]. Gallo et al. developed a model of a non-conventional SOFC system by applying a lumped energy balance to each component. This model could efficiently increase the heat exchanges inside the system. This model is verified by experiments and applicable to numerous layouts [15]. According to reference [16], a fuel cell is mainly used in America, Europe, South Korea and Japan worldwide. From environmental aspect, analysis of CO_2 reduction by a fuel cell was also shown in recent researches from Europe, Japan and South Korea. Fuel cell for EV is mainly applied and investigated in Korea. Environmental impact of CO_2 reduction and emission by fuel cell EV is conducted [17,18]. In order to encourage the utilization residential fuel cell in households in Korea, Lim et al. did a study in the view of CO_2 emissions reduction by applying residential fuel cell [19]. European researchers did several analyses of applying SOFC in the industrial area to reduce CO_2 emission [20,21]. On the other hand, a residential fuel cell is mainly studied in Japan in recent years. The CO_2 emission density of a public power plant was used to calculate the amount of CO_2 emission and CO_2 reduction can be calculated by a simulating application of a fuel cell in a household [22–24]. The same calculation method was used for environmental analysis in this research. From the energy demand oriented research, Ozawa et al. indicates that SOFC–CHP systems can drastically reduce greenhouse gas (GHG) emissions from a particularly small-sized household [24]. Another aspect studied in this research is the efficient operation of SOFC–CGS in small-sized households in a dormitory building by using a exist model of SOFC–CGS from Japan [25–27]. This study analyzed efficiency improvement and environmental impact of operating SOFC–CGS in a dormitory building in Korea by simulating SOFC–CGS operation in multi-households. Under the traditional energy supply method, solid oxide fuel cell cogeneration system (SOFC–CGS) was attempted to simulate applying in the target building of this research. The target building was a dormitory building near a university in Korea. In Korea, the dormitories inside the university campus usually cannot accommodate all of the students registered in the university. Thus, most of the students have to rent a small room for living off campus. Therefore, there are a lot of dormitory buildings near universities in Korea. The dormitory building usually contains several households in each floor and area of one household is around 20 to 30 m^2, which can only accommodate one or two students. The target building is near a university campus in Busan. Specifications of the target building can be checked in Table 1 in Section 2.1. SOFC–CGS is a clean energy system that does not release any air pollutants when it generates electricity [6,14,24,28]. The device efficiency and environmental impact were analyzed under the conditions of applying one system to one household and applying one system to multi-households to identify the most appropriate way to operate SOFC–CGS in the building. The model of SOFC–CGS is introduced in Section 4. Four different types of energy supply were considered in this research. The first type was supplying only electricity to the building, and the second type was supplying natural gas and electricity together. According to the analysis result, the second energy supply type emits a relative smaller amount of air pollutants than the first type and the efficiency of using natural gas directly by combustion is much higher than using natural gas by SOFC for heating. Therefore, the third type was applying one SOFC–CGS to one household under the second energy supply type, and the fourth type was applying one SOFC–CGS system to multi-households under the second energy supply type.

Table 1. Building specifications.

Parameter	Specific Characteristic
Area	172.3 m^2
Structure	Steel–concrete
Envelop	Concrete
Aboveground Stories	4
Underground Stories	0
Story height	2.6 m
Height	10.4 m
Width	12.6 m
Length	14.5 m
Service life	50 years

1.2. Research Methodology

This study mainly used the commercial software Design Builder (Version 4.5, DesignBuilder Software Ltd, London, UK) to predict energy consumption of the building. In the second part of this research, to simulate operating the SOFC–CGS in the building, a calibrated SOFC–CGS mathematical model was programmed using Visual Basic.NET. The research target building was modeled using Design Builder, and the one-year hourly energy consumption of the building was also predicted using this software. Design Builder as a calibrated software is made in England. This software was mainly used for energy analysis of the building. The calculation model of this software is EnergyPlus (Version 8.2.0, Department of Energy, Washington, USA), which is a calculation engine invented in America for analyzing the heating, cooling, lighting, ventilation, etc., of a building. There are several calculation engines for energy consumption prediction from which DOE-2 (Version 2.1E, Department of Energy, Washington, USA), DeST (Version 2.0, Tsinghua University, Beijing, China), and EnergyPlus are mainly used for the analysis of the energy consumption of buildings. Xin et al. compared the simulation performance among these three different calculation engines. Compared with EnergyPlus and DeST, DOE-2 is limited in the basic assumption of the heating, ventilation and air conditioning (HVAC) system calculations [29]. Therefore, the commercial software Design Builder with the calculation engine EnergyPlus was chosen for doing this research. The air pollutants of the research target building were calculated and compared under the conditions of using the traditional method of supplying energy to the building and new method of supplying energy to the building by referring to the air pollutant emission densities of electricity and natural gas in Korea. Finally, the environmental impact and efficiency analyses of operating the SOFC–CGS in the building were conducted.

2. Energy and Environmental Analysis of the Dormitory Building

2.1. Modeling of the Dormitory Building

There are a lot of dormitories for accommodating students near the campus of universities in Korea. Each room in the buildings typically can accommodate one or two students. Most of the dormitory buildings are constructed so that there is a supply of natural gas and electricity to the building. However, recently, some newly developed dormitory buildings have been constructed to use only electricity for all of the energy needs. One of these newly developed buildings was selected for this research. To predict the energy consumption of the building, a model of the building was made using Design Builder, and the 3D model and configuration are shown in Figure 2.

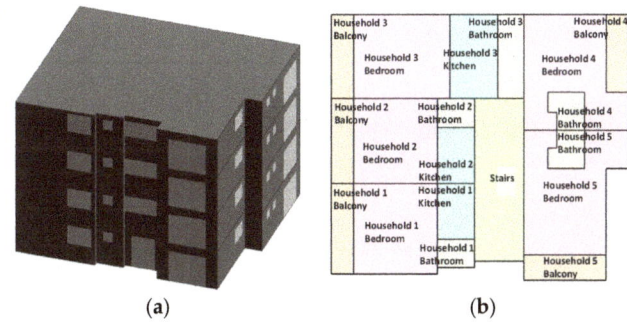

Figure 2. 3D model and configuration of the dormitory building. (**a**) 3D model; (**b**) Configuration.

According to the energy and environmental analysis of the dormitory building in a previous study [30], to predict the energy consumption of the building precisely, weather data of Busan, Korea, were acquired, and the occupancy schedule of different partitions of the building were referred from ASHRAE standard 90.2, 2010 [31]. Basic weather data were listed in Table 2. Moreover, the energy consumption standard of a building in Korea was referred from the Korean Energy Agency to build the model. According to this standard, heating, cooling, resident density, lighting density, and equipment density were set as in Table 3 [32]. According to Korean Construction Law, the construction thermal specifications of external wall, internal wall, roof, and floor were set, as listed Table 3 [32–34]. Since this building uses only electricity, cooling and heating were both set from electricity. Table 3 briefly summarizes the building specifications of the dormitory building.

Table 2. Climate specifications of Busan.

Climate Specification	Value
Latitude	35°32′ N
Longitude	129°19′ E
Average annual temperature	14.05 °C
Average max annual temperature	35.4 °C
Average min annual temperature	−8.8 °C
Average annual relative humidity	63.40%
Average solar radiation	106.15 Wh/m^2
Average annual wind speed	1.93 m/s
Elevation above sea level	33 m

Table 3. Construction specifications.

Parameter	Specific Characteristic
Orientation	West to East
External wall	0.58 W/m^{2}K
Internal wall	2.92 W/m^{2}K
Roof	0.27 W/m^{2}K
Glazing	3.12 W/m^{2}K
Infiltration rate	0.3 ac/h
Ventilation rate	3 ac/h
Equipment	5.38 W/m^2
Lighting	3.88 W/m^2
Occupancy	0.054 people/m^2
HVAC/Heating	Dedicated hot water boiler (electricity)
HVAC/Cooling	Electricity from grid
Set point temperature	Summer 26 °C, Winter 22 °C

2.2. Energy and Environmental Analyses

According to the building model made using Design Builder in Section 2.1, the one-year energy consumption of the building was predicted. The energy analysis result showed that the one-year energy consumption of the building was 87,830.78 kWh when only electricity was used in the building. On the other hand, when natural gas and electricity were used together, the one-year electricity consumption was 18,013.7 kWh and the one-year natural gas consumption was 69,817.1 kWh.

To calculate the annual air pollutants emission of the building, the air pollutant emission densities of electricity and natural gas were referred from the Korea Environmental Industry and Technology Institute, as listed in Table 4 [32,35]. The one-year air pollutants of the building when only electricity was supplied and when natural gas and electricity were supplied together were calculated and compared. The environmental analysis result showed that the amount of CO_2 emission was much larger when the building used only electricity than when the building used electricity and natural gas together, as shown in Figure 3. On the other hand, emissions of other air pollutants, such as CH_4, N_2O, SO_2, CO, NO_X, and NMVOC, were much lower than CO_2 emission. This means that the building using natural gas and electricity together was much cleaner than the building using only electricity for all energy needs.

Table 4. Air pollutant emission densities of electricity and natural gas.

Air Pollutant	Electricity (kg/kWh)	Natural Gas (kg/kWh)
CO_2	4.87×10^{-1}	3.94×10^{-2}
CH_4	3.50×10^{-4}	2.01×10^{-4}
N_2O	1.35×10^{-6}	1.72×10^{-7}
SO_2	0.00×10^{0}	4.22×10^{-4}
CO	5.00×10^{-5}	9.91×10^{-5}
NO_X	1.20×10^{-4}	5.67×10^{-4}
NMVOC	2.00×10^{-5}	1.04×10^{-6}

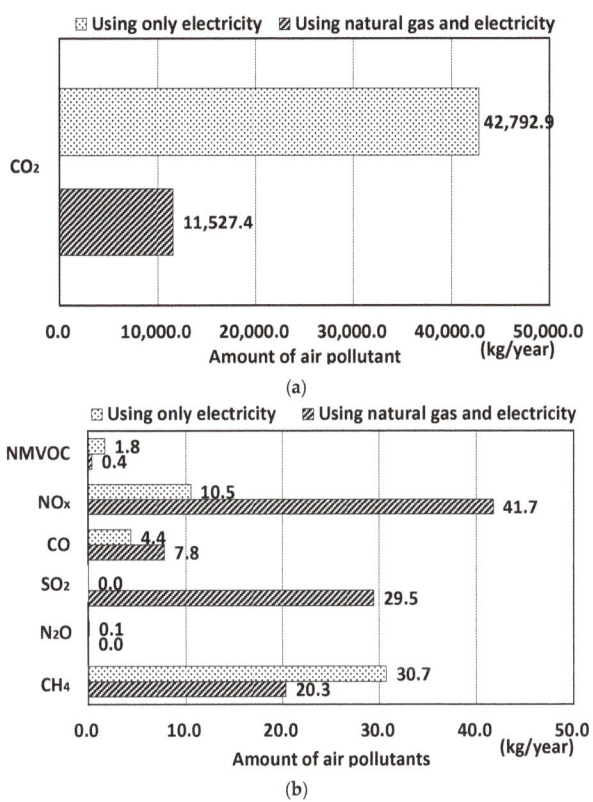

Figure 3. Comparison of air pollutant amounts between the building using only electricity and the building using electricity and natural gas together. (a) Comparison of amount of CO_2 emission (b) Comparison of amount of other air pollutants emission.

3. Modeling of SOFC–CGS

The residential fuel cell as a clean energy device is used widely in Japan. Since environmental problems including air pollution have been getting worse recently in Korea, implementation of the residential fuel cell as a clean energy device in buildings in Korea is necessary. Therefore, this research analyzed the efficiency and environmental impact of operating SOFC–CGS in a dormitory building in Korea, as mentioned in previous sections.

Typical approaches of fuel cell modeling in recent research were summarized in Table 5. In order to evaluate and optimize the performance of fuel cell, electrochemical models were developed to simulate the specific chemical and energy interaction inside the cell stack [36–38]. On the other hand, in order to evaluate and improve the device operation efficiency and CO_2 reduction by using fuel cell for transportation or residential, fuel cell modeling by mainly using device efficiency was developed to simulate operating fuel cell in household or for EV [11,22,24]. In this research, the modeling of SOFC–CGS by mainly using the efficiency of the device was developed to simulate the application of SOFC–CGS in a dormitory building in Korea.

Table 5. Typical approaches of fuel cell modeling.

System	System Application	Model Application	Main Design Variables
SOFC [36]	Cell stack	Electrochemical model	Volumetric rate generated thermal energy, rate of production of species, area specific resistance
SOFC–CGS [37]	Stationary	Electrochemical model	Partial pressure of methane, reaction rate of steam reforming, cell temperature
SOFC [38]	Cell Stack	Electrochemical model	Partial pressure of species, anodic and cathodic charge-transfer coefficients, exchange current density
SOFC–CGS [22]	Stationary	Device efficiency	Efficiency of electricity generation, efficiency of heat collection, water temperature
SOFC–CGS [11]	Stationary/Transportation	Device efficiency	Electric demand of cafeteria, required electricity for EV charging, electric power generated by SOFC–CGS
SOFC–CHP/PEMFC–CHP [24]	Stationary	Device efficiency	Power generation efficiency of the fuel cell, heat recovery efficiency of the fuel cell, heat loss rate of the tank

3.1. Explanation of SOFC–CGS

Residential SOFC, as a clean energy supply device, has relatively low efficiency. Due to heat losses, the highest electricity generation efficiency of SOFC is about 46.4% [24,27]. To improve the efficiency of SOFC, SOFC–CGS was developed [24,25,27]. Heat losses from SOFC are collected to heat water, and a backup boiler is used for reheating the water to meet the demand temperature. A 28 L water storage tank was used in this system, as illustrated in Figure 4. SOFC uses natural gas to generate electricity to supply the electricity demand of a household. The heat losses of SOFC are collected and transferred to the water storage tank by a water tube. The limit temperature control of the storage tank is 65 °C, and the temperature of the storage tank is adjusted by controlling the water flow rate in the tube that collects heat from the SOFC. If the input water temperature through the radiator is higher than 34 °C, the radiator reduces the water temperature to 34 °C, then the water can enter the SOFC to collect heat. Hot water outputs from the storage tank to the household when hot water is needed. In this system, the city water temperature was set as 15 °C, and the hot water temperature demand was set as 40 °C. The output water temperature from the water mix device was set to be equal to or less than 33 °C. If the water output from the storage tank was equal to or less than 33 °C, hot water directly entered the backup boiler for reheating of the water to 40 °C and then the hot water at 40 °C was provided to the household. If the output water temperature from the storage tank was higher than 33 °C, the water mix device mixed city water and water from the storage tank to attain the temperature of 33 °C, and then, through the backup boiler, water was heated to 40 °C to supply the household. This system is illustrated as Figure 4. The specifications of the SOFC–CGS in this study are shown in Table 6, and the efficiency of electricity generation and heat collection of the SOFC–CGS are shown in Figure 5. The mathematical equation of the electricity generation efficiency is written as Equation (1). The efficiency of heat collection of the SOFC–CGS was constantly 31%.

$$EFF_{ElSup} = 0.6868LF^3 - 1.6829LF^2 + 1.4601LF. \tag{1}$$

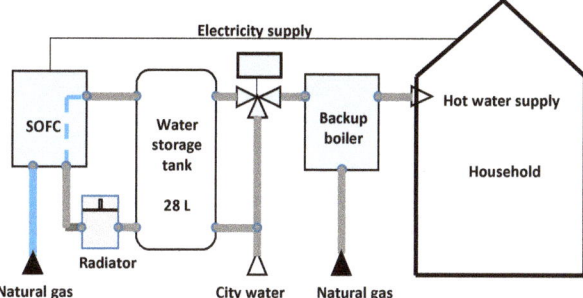

Figure 4. Working principle of the solid oxide fuel cell cogeneration system (SOFC–CGS).

Table 6. Specifications of the SOFC–CGS.

Type of Cell Stack	SOFC (Solid Oxide Fuel Cell)
Rated electrical output	700 W
Max running time	26 days
Standby time for start	1 day
Water tank volume	28 L
Reaction to overheating	Heat dissipation by radiator
Efficiency of heat collection	31%
Water temperature from tank to backup boiler	40 °C
Efficiency of electricity generation	46.4%

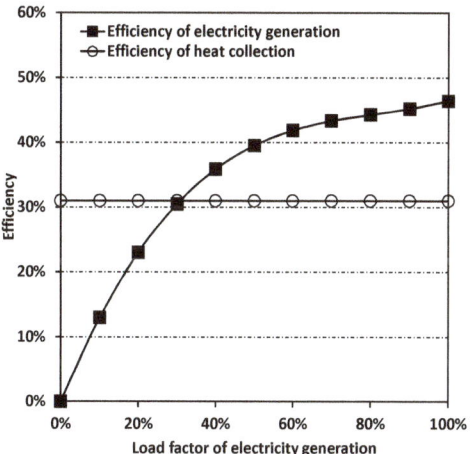

Figure 5. Efficiency of the SOFC–CGS.

In the equation, EFF_{ElSup} is the efficiency of electricity generation of the SOFC and LF is the load factor of electricity generation.

3.2. Mathemetical Model of SOFC–CGS

In this study, the mathematical model of SOFC–CGS was referred from recent research from Kyushu University [25–27]. The electricity generation calculation follows the steps below. Firstly, depending on the electricity generation tracking speed, the maximum electricity supply capability can be determined by Equation (2):

$$P_{Supmax} = -P_{ElSup} + V_{ElSupSp}I_{tint}, \quad (2)$$

where P_{Supmax} is the maximum electricity supply capability, P_{ElSup} is the amount of electricity generation, $V_{ElSupSp}$ is the electricity generation tracking speed, and I_{tint} is the calculation time interval. Then, according to the household electricity demand, the electricity supply from SOFC can be determined. Following Equations (3) and (4), the amount of natural gas consumption by SOFC can be calculated:

$$LF = \frac{P_{ElSup}}{P_{ElMax}} \quad (3)$$

$$P_{EhCell} = \frac{P_{ElSup}}{EFF_{ElSup}}, \quad (4)$$

where LF is the load factor of electricity generation, P_{ElMax} is maximum electricity generation capacity of SOFC, P_{EhCell} is the amount of natural gas consumption of SOFC, and EFF_{ElSup} is the efficiency of electricity generation. The amount of heat output from SOFC can be calculated by Equation (5):

$$P_{HtSup} = P_{EhCell}EFF_{HtSup}, \quad (5)$$

where P_{HtSup} is the amount of heat output from SOFC and EFF_{HtSup} is the heat output efficiency of SOFC, which is 31%.

The water flow rate in the water tube that collects heat can be calculated using either Equation (6) or Equation (7). If the input water temperature through the radiator unit is higher than or equal to 34 °C, Equation (6) is used to calculate the water flow rate in water tube; otherwise, Equation (7) is used to calculate it.

$$V_{wCell} = \frac{P_{HtSup}}{C_w(T_{out}-34)} \quad (6)$$

$$V_{wCell} = \frac{P_{HtSup}}{C_w(T_{out}-T_{in})}. \quad (7)$$

In the equations, V_{wCell} is the water flow rate in the water tube; P_{HtSup} is the amount of heat supply from SOFC; T_{out} is the target output water temperature from the SOFC, which was set as 65 °C in this study; T_{in} is the input water temperature to the radiator unit; and C_w is the specific heat of water.

If the water temperature is higher than 34 °C, the heat dissipation of the radiator unit is determined by Equation (8); otherwise, the heat dissipation of the radiator unit is zero:

$$P_{Rad} = V_{wCell}C_w(T_{out} - T_{in}), \quad (8)$$

where P_{Rad} is the heat dissipation from the radiator unit. The amount of heat output from the backup boiler is determined by Equation (9), and the amount of gas used in the backup boiler is calculated from Equation (10):

$$P_{BBSup} = Vol_{HwDmd}C_w(T_{Dmd}-T_{BBin}), \quad (9)$$

$$P_{BBUse} = \frac{P_{BBSup}}{EFF_{BB}}, \quad (10)$$

where P_{BBSup} is the amount of heat output from the backup boiler, P_{BBUse} is the amount of gas used in the backup boiler, T_{Dmd} is the water temperature demand from the household, T_{BBin} is the input water temperature to the backup boiler, Vol_{HwDmd} is the amount of hot water demand from the household, P_{BBUse} is the amount of gas used in the backup boiler; and EFF_{BB} is the working efficiency of the backup boiler, which was set as 95% in this study.

The calculation flow of the program and the input data are shown as Figure 6 and Table 7.

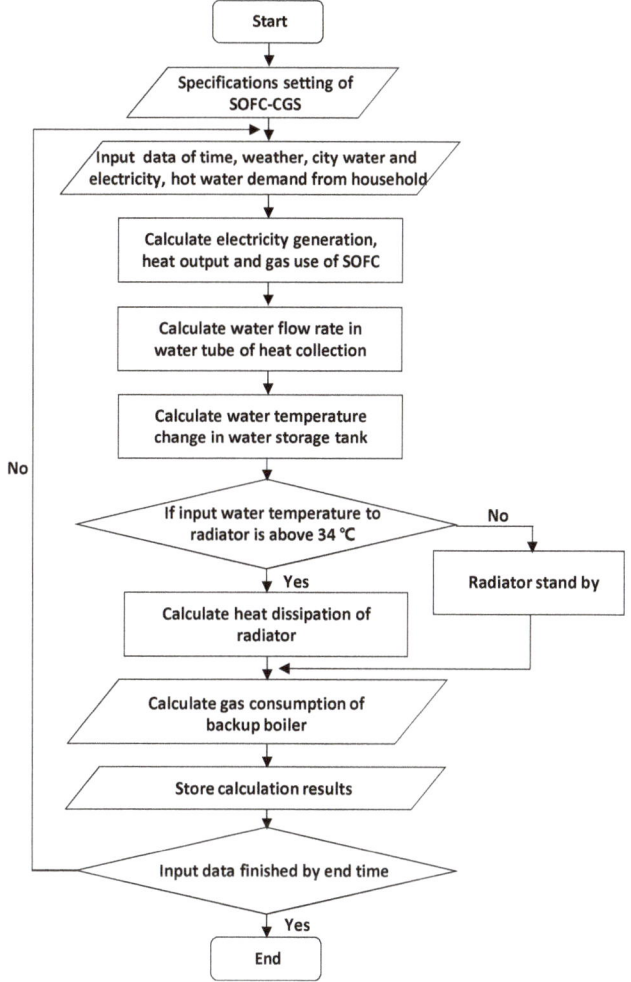

Figure 6. Calculation diagram of the program.

Table 7. Input data of the program.

Input Data	Unit/Value
Electricity demand	W
Hot water demand	L/Time unit
Setting temperature of backup boiler	40 °C
Outside temperature	Data from ASHRAE
Temperature of city water	15 °C
Temperature of hot water demand	40 °C

4. Simulation of Operating SOFC–CGS in the Dormitory Building

As mentioned in Section 2, the energy consumption for heating the research target building in winter was significant. Since using electricity from the public power plant results in much more air pollution than using natural gas for heating and because the efficiency of directly using natural gas for heating was much higher than that of using natural gas to generate electricity for heating by the SOFC, in this research, energy for heating was not considered as being supplied from the SOFC. Aside from heating, the SOFC–CGS provides electricity and hot water to the households.

The research target building had four floors, and each floor had the same structure with five households. As the energy consumption of each household was predicted by the calibrated commercial software Design Builder, the energy consumption pattern and amount were the same for each floor. Thus, the first floor was chosen for this research. As mentioned previously, there were five households in first floor, and the energy consumption pattern was similar among the five households in the first floor. According to schedule of ASHRAE standard for residential house [31], the electricity and hot water consumption patterns of representative dates of each season of household 1 in the first floor are shown as Figure 7. As heating energy was provided directly by natural gas, it was not considered in simulating the operating SOFC–CGS. The yearly energy consumption of each household in the first floor is shown as Figure 8.

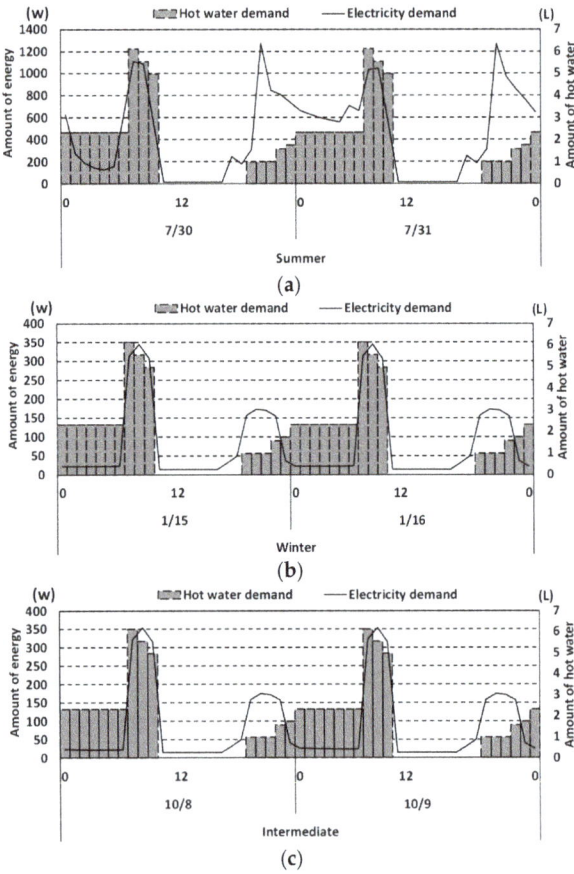

Figure 7. Representative daily data of electricity demand and hot water demand of household 1 in each season. (**a**) Data in summer; (**b**) Data in winter; (**c**) Data in intermediate seasons.

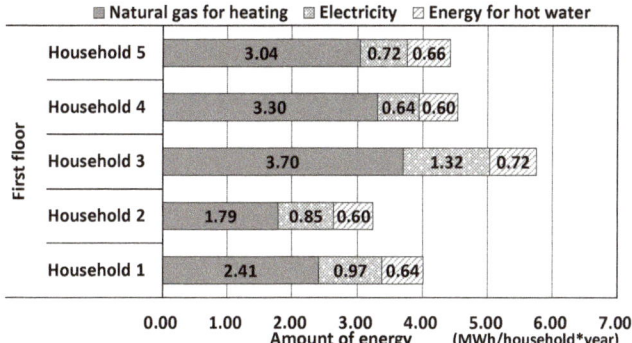

Figure 8. Yearly energy consumption of each household in the first floor.

To improve and analyze the efficiency of SOFC–CGS, one SOFC–CGS was simulated to operate in either only one household or in multi-households (two or three households). According to the yearly energy consumption of each household, as shown in Figure 8, household 2 ranked the lowest and household 3 ranked the highest. Therefore, in order to form a balance, households 2 and 3 were simulated as sharing one SOFC–CGS, and households 1, 4, and 5 were simulated as sharing one SOFC–CGS. Hourly data of electricity demand, electricity generation, and heat output from the SOFC and the heat reduction amount from the radiator at representative days in the year in summer are shown as Figures 9–11 below.

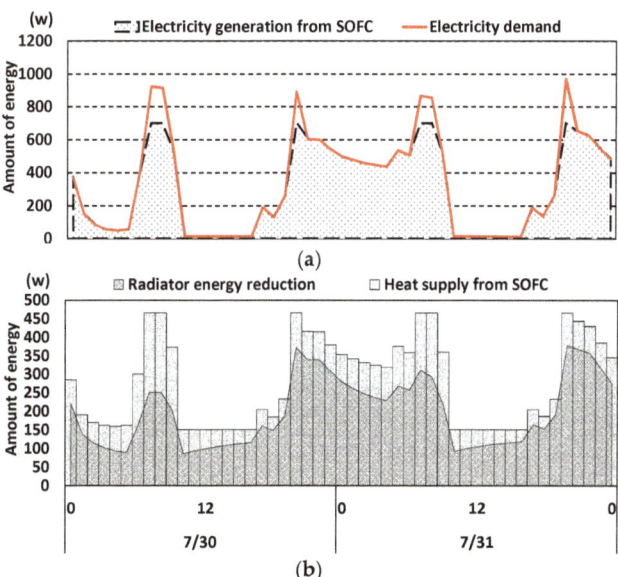

Figure 9. Hourly simulation data of household 2. (**a**) Data of electricity generation and electricity demand; (**b**) Data of heat reduction and heat supply.

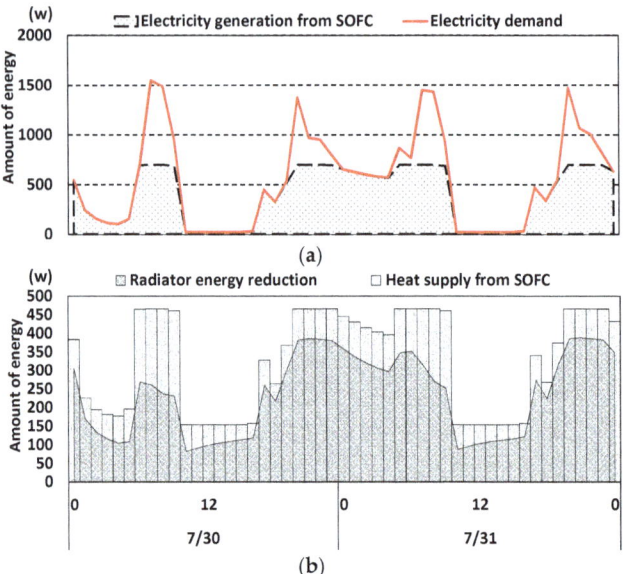

Figure 10. Hourly simulation data of household 3. (**a**) Data of electricity generation and electricity demand; (**b**) Data of heat reduction and heat supply.

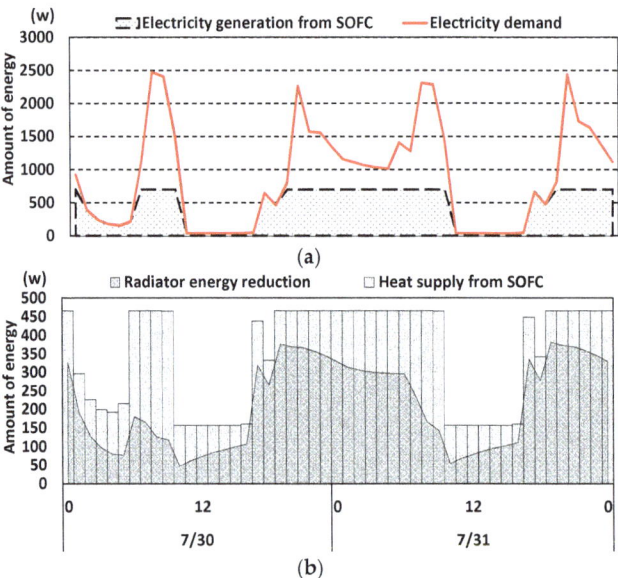

Figure 11. Hourly simulation data of households 2 and 3. (**a**) Data of electricity generation and electricity demand; (**b**) Data of heat reduction and heat supply.

As the data of the simulation results were large, the simulation results of households 2 and 3 were selected for presentation in this research. Figures 9 and 10 show the simulation results when operating one SOFC–CGS for one household. Figure 11 shows the simulation results when sharing one SOFC–CGS by multi-households. The load factor of SOFC–CGS was increased when multi-households share one SOFC–CGS, as in Figure 11. This means that the electricity generation efficiency of SOFC–CGS was improved. Additionally, the ratio of heat reduction amount by the radiator was reduced when SOFC–CGS was shared by multi-households, as shown in Figure 11, which means that the efficiency of heat usage for hot water was increased.

5. Analysis of Efficiency and Environmental Impact of SOFC–CGS

5.1. Analysis of SOFC–CGS Efficiency

The yearly energy consumptions of simulating the operation of one SOFC–CGS to one household and multi-households are shown as Figure 12. The natural gas consumption by SOFC, electricity generation by SOFC, purchased electricity, and natural gas consumption by the backup boiler were higher when simulating the operation of one SOFC–CGS for multi-households than when simulating for one household. The electricity consumption by SOFC was constant because electricity was only used when the SOFC started or restarted. Increasing electricity generation led to increasing efficiency of electricity generation by SOFC. The reason for this was that sharing one SOFC–CGS to multi-households led to increasing electricity and hot water demand for the one SOFC–CGS.

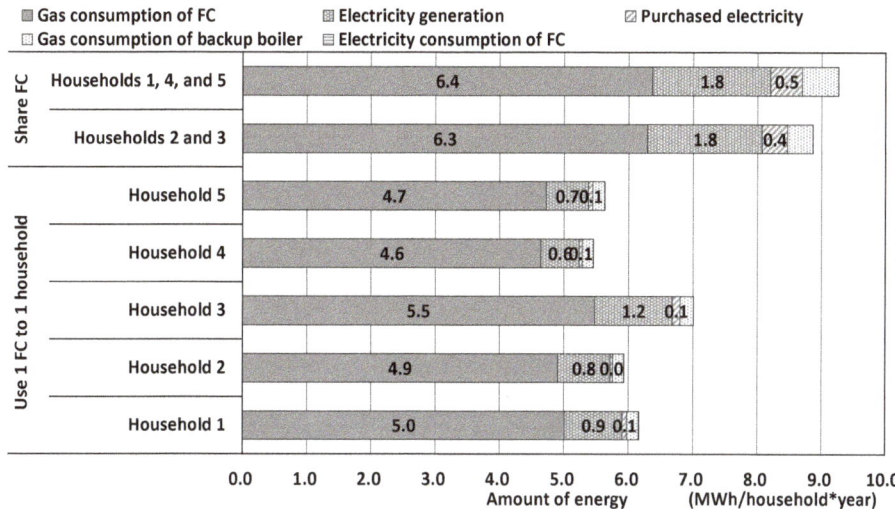

Figure 12. Yearly energy consumptions of simulating one SOFC–CGS for one household and multi-households.

According to Figure 13, the electricity generation efficiency of SOFC ranged from 13% to 22% and the heat usage efficiency ranged from 22% to 23% when simulating the operation of one SOFC–CGS for one household. The efficiency of electricity generation of SOFC was increased to range from 28% to 29% and the heat usage efficiency was increased to range from 24% to 25% when multi-households share one SOFC–CGS. The percentages of losses from different components of SOFC–CGS of simulating operation of one SOFC–CGS to one household and sharing one SOFC–CGS by multi-households are compared in Figure 14. The losses in fuel cell and radiator were decreased when multi-households share one SOFC–CGS. In another aspect, sharing one SOFC–CGS by multi-households increased the electricity demand from the public power plant, which leads to more air pollution. The specific environmental analysis is shown in Section 5.2.

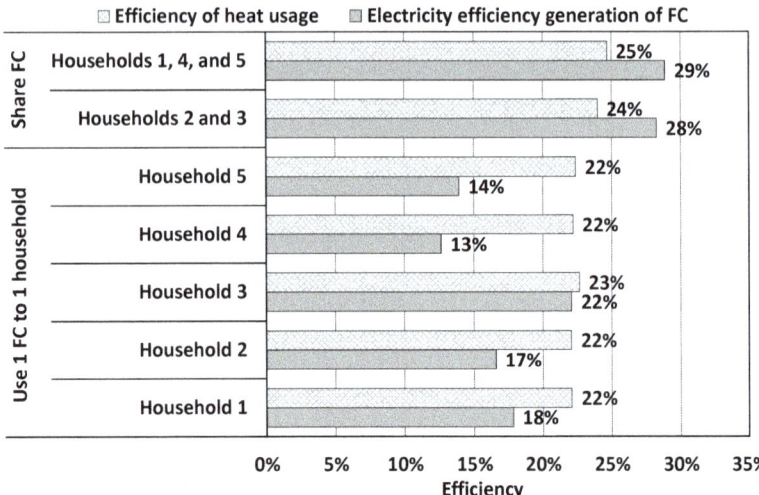

Figure 13. Yearly electricity generation and heat output efficiencies of simulating using one SOFC cogeneration system for one household and multi-households.

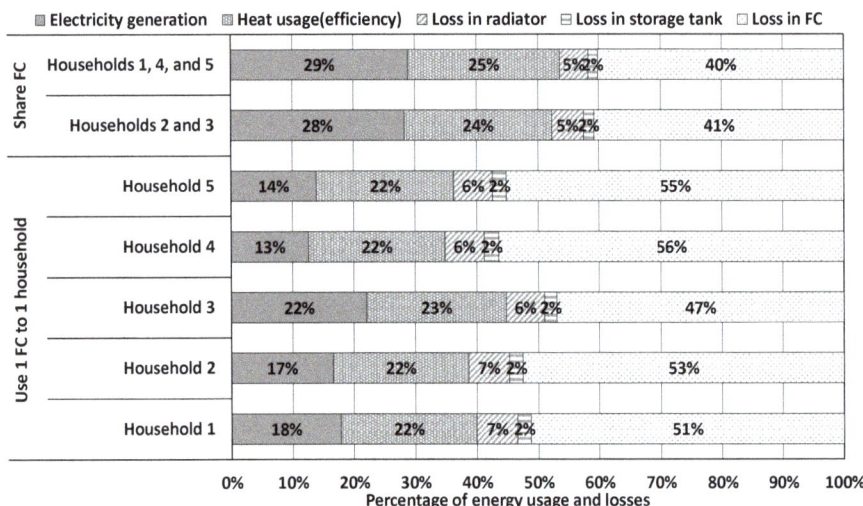

Figure 14. Yearly energy losses of simulating using one SOFC–CGS to one household and multi-households.

5.2. Analysis of Environmental Impact of SOFC–CGS

Including the two previous cases of energy supply types in the building in Section 2, four different cases of energy supply types are explained and summarized in Table 8. In the first case, only electricity was supplied to the building, while in the second case, natural gas and electricity together were supplied. According to the analysis result, the second energy supply type emitted a relatively smaller amount of air pollutants than the first type and the efficiency of using natural gas directly by combustion was much higher than using natural gas by SOFC for heating. Therefore, in the third case, one SOFC–CGS was applied to one household under the second energy supply type, and in the fourth case, one SOFC–CGS was applied to multi-households under the second energy supply type. The environmental analysis for the entire building (four stories) under the four different cases of energy application types is shown as Figure 15. According to the result of the environmental analysis, the amount of CO_2 emission was most significant in the case of using only electricity in the building. The CO_2 emission of the second case, using natural gas and electricity together in the building, was much lower than in the case of using only electricity in the building. Under the energy supply type of the second case, when SOFC–CGS was simulated as operating in the building, the CO_2 emission was significantly reduce to range from 3057.2 to 4120.6 kg/year. The case of operating one SOFC–CGS for one household gave the lowest CO_2 emission. Due to the purchased electricity from the public power plant was increased from the third case to the fourth case as shown in the energy analysis in Figure 12, the fourth case resulted in a higher air pollutants emission than the third case. However, the case of sharing one SOFC–CGS among multi-households had a higher system efficiency than the case of operating one SOFC–CGS for one household, as indicated by the analysis result in Section 5.1. In another aspect, the difference of CO_2 emission between cases three and four was not significant, as in Figure 15; therefore, the case of sharing one SOFC–CGS among multi-households was recommended. According to Figure 15, the emission amounts of the pollutants other than CO_2 of the four different cases were much lower than the CO_2 emissions of the four different cases.

Table 8. Explanation of cases for environmental analysis.

Case Number	Case Name	Energy for Heating/Hot Water	Energy for Other Needs in Household
1	Using only electricity	Purchased electricity from power plant	Purchased electricity from power plant
2	Using natural gas and electricity	Direct natural gas (heating and hot water)	Purchased electricity from power plant
3	One SOFC–CGS for one household	Direct natural gas (heating)	Individual SOFC–CGS for one household and purchased electricity from power plant
4	One SOFC–CGS to multi-households	Direct natural gas (heating)	Shared SOFC–CGS for multi-households and purchased electricity from power plant

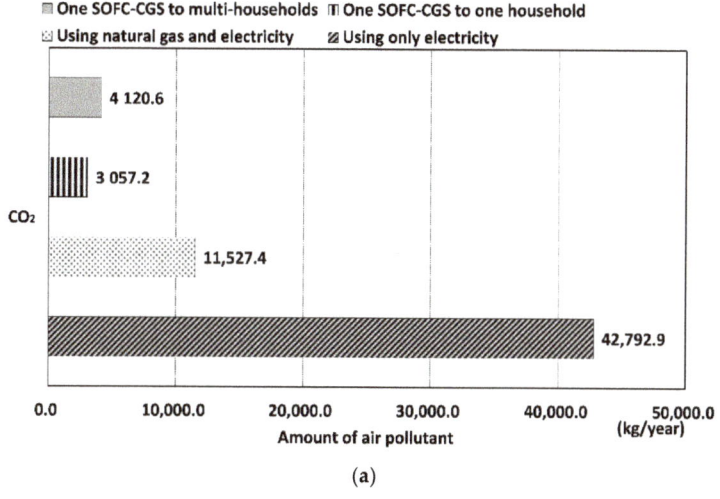

(a)

Figure 15. Cont.

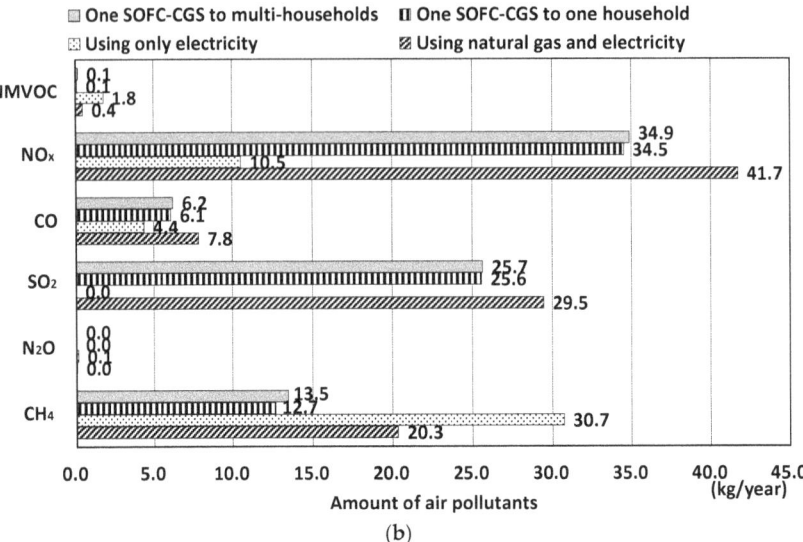

Figure 15. Comparison of air pollutant amounts of the building under the four different energy supply types. (a) Comparison of amount of CO$_2$ emission (b) Comparison of amount of other air pollutants emission.

6. Conclusions

Residential SOFC, as a clean energy device, continues to be researched by researchers in the field of architecture. However, as the efficiency of SOFC is relatively low, SOFC–CGS has been studied recently by many researchers. In this research, the environmental and efficiency impacts of operating SOFC–CGS in a dormitory building in Korea were analyzed. As a newly developed building, the research target building was constructed with electricity as the only energy resource. Analysis of this strategy was required. Firstly, in this study, the environmental impact was compared between the cases of using only electricity (new way) and using electricity and natural gas (traditional way) in the building. The analysis result showed that the traditional way of energy supply type in the building emits less air pollution than the new way. Secondly, based on the traditional energy supply type, operating SOFC–CGS in the building was simulated. The efficiency and environmental impact were analyzed, and these were compared between the cases of simulating one SOFC–CGS in one household and sharing one SOFC–CGS among multi-households. The specific conclusions are listed below.

(1) The comparison analysis between supplying only electricity to the building and supplying electricity and natural gas together showed that supplying electricity and natural gas together results in much less air pollution than supplying only electricity.

(2) Under the energy supply type of supplying electricity and natural gas together to the building, SOFC–CGS was simulated in the building under the conditions of operating one SOFC–CGS in one household and sharing one SOFC–CGS among multi-households. The electricity generation efficiency of SOFC ranged from 13% to 22% and the heat usage efficiency ranged from 22% to 23% when simulating operation of one SOFC–CGS in one household. The electricity generation efficiency of SOFC increased to range from 28% to 29% and the heat usage efficiency increased to range from 24% to 25% when multi-households shared one SOFC–CGS.

(3) The four different cases of energy supply type for environmental analysis were summarized in Table 8. According to the results of the environmental analysis, the CO$_2$ emission was significantly reduced from 11,527.4 kg/year to 3057.2 kg/year when simulating the operation of one SOFC–CGS in one household under the energy supply type of case two. On the other hand, the CO$_2$ emission

was significantly reduced from 11,527.4 kg/year to 4120.6 kg/year when simulating the sharing of one SOFC–CGS among multi-households. The emission amounts of pollutants other than CO_2 of the four different cases were much lower than the CO_2 emissions of the four different cases.

(4) The difference of CO_2 emission between a simulating operation of one SOFC–CGS in one household and the sharing of one SOFC–CGS among multi-households was not significant. Meanwhile, simulating the sharing one SOFC–CGS among multi-households led to higher efficiency than simulating the operation of one SOFC–CGS in one household. Therefore, sharing one SOFC–CGS among multi-households in a dormitory building is recommended.

Author Contributions: H.C. conceived and completed this paper. I.-H.L. supervised and reviewed this paper.

Funding: This work was supported by a National Research Foundation of Korea grant funded by the Korean Government (NRF-2018R1D1A1B07051106).

Conflicts of Interest: The authors declare no conflict of interest.

Abbreviations

SOFC	Solid oxide fuel cell
SOFC–CGS	Solid oxide fuel cell cogeneration system
LNG	Liquefied Natural Gas
CHP	Combined heat and power
DC	Direct current
AC	Alternating current
EV	Electric vehicle
SOFC–CHP	Solid oxide fuel cell combined heat and power
GHG	Greenhouse gas
HVAC	Heating, Ventilation and Air Conditioning
ASHRAE	American Society of Heating, Refrigerating and Air-Conditioning Engineers
PEMFC–CHP	Proton exchange membrane fuel cell combined heat and power

Nomenclatures

EFF_{ElSup}	Efficiency of electricity generation
LF	Load factor of electricity generation
P_{Supmax}	Maximum electricity supply capability
P_{ElSup}	Amount of electricity supplication
$V_{ElSupSp}$	Electricity generation tracking speed
I_{tint}	Time interval
P_{ElMax}	Maximum electricity generation capacity of SOFC
P_{EhCell}	Amount of natural gas consumption of SOFC
P_{HtSup}	Amount of heat output from SOFC
EFF_{HtSup}	Heat output efficiency of SOFC
V_{wCell}	Water flow rate in the water tube
T_{out}	Target output water temperature from the SOFC
T_{in}	Input water temperature to the radiator unit
C_w	Specific heat of water
P_{Rad}	Heat dissipation from the radiator unit
P_{BBSup}	Amount of heat output from the backup boiler
P_{BBUse}	Amount of gas used in the backup boiler
T_{Dmd}	Water temperature demand from the household
T_{BBin}	Input water temperature to the backup boiler
Vol_{HwDmd}	Amount of hot water demand
EFF_{BB}	Working efficiency of the backup boiler

References

1. Doosoon, L. An Analysis of the Effect of Relative Prices between Electricity and Gas on the Sectoral Demand of City Gas. Master's Thesis, Public Administration of Seoul National University, Seoul, Korea, 2014.

2. Korea Electric Power Corporation. Sustainable Report. 2016. Available online: www.kepco.co.kr (accessed on 6 August 2018).
3. Wang, Y.; Zhao, Q.; Borup, U.; Choi, S. Stack and converter-integrated design for efficient residential fuel cell system. *Int. J. Hydrogen Energy* **2009**, *34*, 7316–7322. [CrossRef]
4. Yuqing, W.; Yixiang, S.; Meng, N.; Ningsheng, C. A micro tri-generation system based on direct flame fuel cells for residential applications. *Int. J. Hydrogen Energy* **2014**, *39*, 5996–6005.
5. Hirohisa, A.; Tetsuya, W.; Ryohei, Y. Development of an energy management system for optimal operation of fuel cell based residential energy systems. *Int. J. Hydrogen Energy* **2016**, *41*, 20314–20325.
6. Alexandros, A.; Eric, S.F.; Dan, J.L.B. Options for residential building services design using fuel cell based micro-CHP and the potential for heat integration. *Appl. Energy* **2015**, *138*, 685–694.
7. Mohammad, H.-M. Transient performance improvement of a stand-alone fuel cell for residential applications. *Sustain. Cities Soc.* **2019**, *50*, 101650.
8. Hong, B.; Qiong, W.; Weijun, G.; Weisheng, Z. Optimal operation of a grid-connected hybrid PV/fuel cell/battery energy system for residential applications. *Energy* **2016**, *113*, 702–712.
9. Yasuhiro, H.; Kiyotaka, T.; Ryuichirao, G.; Hideki, K. Hybrid utilization of renewable energy and fuel cells for residential energy systems. *Energy Build.* **2011**, *43*, 3680–3684.
10. Marco, S.; Marta, G.; Massimo, S. Modeling and techno-economic analysis of the integration of a FC based micro-CHP system for residential application with a heat pump. *Energy* **2017**, *120*, 262–275.
11. Ramadhani, F.; Hussain, M.A.; Mokhlis, H.; Fazly, M.; Ali, J.M. Evaluation of solid oxide fuel cell based polygeneration system in residential areas integrating with electric charging and hydrogen fueling stations for vehicles. *Appl. Energy* **2019**, *238*, 1373–1388. [CrossRef]
12. Takahide, H.; Yusuke, O.; Takashi, I.; Atsushi, A. Technological assessment of residential fuel cells using hydrogen supply systems for fuel cell vehicles. *Int. J. Hydrogen Energy* **2017**, *42*, 26377–26388.
13. Kazunori, N.; Joshua, D.; Michael, E.W. Assessment of primary energy consumption, carbon dioxide emissions, and peak electric load for a residential fuel cell using empirical natural gas and electricity use profiles. *Energy Build.* **2018**, *178*, 242–253.
14. Van Biert, L.; Godjevac, M.; Visser, K.; Aravind, P.V. Dynamic modelling of a direct internal reforming solid oxide fuel cell stack based on single cell experiments. *Appl. Energy* **2019**, *250*, 976–990. [CrossRef]
15. Marco, G.; Dario, M.; Marco, S.; Cesare, P.; Siu, F.A. A versatile computational tool for model-based design, control and diagnosis of a generic Solid Oxide Fuel Cell Integrated Stack Module. *Energy Convers. Manag.* **2018**, *171*, 1514–1528.
16. Weidner, E.; Ortiz, C.R.; Davies, J. *Global Development of Large Capacity Stationary Fuel Cells*; JCR Technical Reports; Publications Office of the European Union: Brussels, Belgium, 2019.
17. Kim, J.H.; Kim, H.J.; Yoo, S.H. Willingness to pay fuel-cell electric vehicles in South Korea. *Energy* **2019**, *174*, 497–502. [CrossRef]
18. Eunji, Y.; Myoung, K.; Han, H.S. Well-to-wheel analysis of hydrogen fuel-cell electric vehicle in Korea. *Int. J. Hydrogen Energy* **2018**, *43*, 9267–9278.
19. Lim, S.Y.; Kim, H.J.; Yoo, S.H. Household willingness to pay for expanding fuel cell power generation in Korea: A view from CO_2 emissions reduction. *Renew. Sustain. Energy Rev.* **2018**, *81*, 242–249. [CrossRef]
20. Giuseppe, D.; Piero, B.; Erasmo, M.; Francesco, P.; Fabio, M.; Dawid, P.H.; Vasilije, M. Feasibility of CaO/CuO/NiO sorption-enhanced steam methane reforming integrated with solid-oxide fuel cell for near-zero-CO2 emissions cogeneration system. *Appl. Energy* **2018**, *230*, 241–256.
21. Rao, M.; Fernandes, A.; Pronk, P.; Aravind, P.V. Design, modelling and techno-economic analysis of a solid oxide fuel cell-gas turbine system with CO2 capture fueled by gases from steel industry. *Appl. Therm. Eng.* **2019**, *148*, 1258–1270. [CrossRef]
22. Koki, Y.; Akihito, O. Study on Performance Evaluation of Residential SOFC Co-Generation Systems Part 2 Equipment characteristics and installation effect. *Soc. Heat. Air-Cond. Sanit. Eng. Jpn.* **2013**, *70*, 281–284.
23. Koki, Y.; Akihito, O.; Myonghyang, L. Study on Performance Evaluation of Residential SOFC Co-Generation Systems. *Soc. Heat. Air-Cond. Sanit. Eng. Jpn.* **2014**, *203*, 25–33.
24. Akito, O.; Yuki, K. Performance of residential fuel-cell-combined heat and power systems for various household types in Japan. *Int. J. Hydrogen Energy* **2018**, *43*, 15412–15422.
25. Sumiyoshi, D. The examination for suitable spec of solid oxide fuel cell co-generation system and the proposal of tank minimization by using Bas-Tab in collective housing. *J. Environ. Eng.* **2015**, *80*, 441–450. [CrossRef]

26. Takahiro, Y.; Koshiro, A.; Daisuke, S. Estimation energy saving effects of sharing fuel cell with two households by simulation program: Study on effective usage of the home fuel cell in the apartment housing. *J. Environ. Eng.* **2018**, *83*, 365–374.
27. Takahiro, Y. Analysis of Impact to Applying Residential Fuel Cell by Different Lifestyle. Ph.D. Thesis, Department of Architecture of Kyushu University, Fukuoka, Japan, 2018.
28. Taehee, L.; JinHyeok, C.; TaeSung, P. Development and Performance Test of SOFC Co-generation System for RPG. In Proceedings of the Korean Society for New and Renewable Energy, Seoul, Korea, 25 June 2009; pp. 361–364.
29. Zhou, X.; Hong, T.; Yan, D. *Comparison of Building Energy Modeling Programs: HVAC Systems*; Lawrence Berkeley National Lab. (LBNL): Berkeley, CA, USA, 2013.
30. Han, C.; Inhee, L. A case study of decreasing environment pollution caused by energy consumption of a dormitory building which only using electricity by efficiently simulating applying residential SOFC (solid oxide fuel cell). *J. Archit. Res. Archit. Inst. Korea* **2019**, *21*, 21–29.
31. American Society of Heating, Refrigeration, and Air-Conditioning Engineers. *Energy Efficient Design for Low-Rise Residential Buildings*; ASHRAE: Atlanta, GA, USA, 2010.
32. Jehean, S. The Analysis of Environmental Impact Index Using Life Cycle Assessment for Korean Traditional Building. Ph.D. Thesis, Department of Architecture of the Pusan National University, Busan, Korea, 2017.
33. Cheolyong, J.; Byunglip, A.; Chihoon, K.; Wonhwa, H. A Study on the insulation performance of the super window considering the evaluation of building energy rating. *J. Korean Sol. Energy Soc.* **2009**, *29*, 1598–6411.
34. Kim, C.; Ahn, B.; Kim, J.; Jang, C. A Study on the Evaluation of Building Energy Rating Considering the Insulation Performance of the Window and Wall in Apartment Houses. In Proceedings of the Conference of The Society of Air-conditioning and Refrigerating Engineers of Korea, Seoul, Korea, 25 June 2009.
35. Korea Environmental Industry & Technology Institute. Available online: www.epd.or.kr (accessed on 6 August 2018).
36. Arata, N.; Zacharie, W.; Patrick, M.; Stefan, D.; Gunter, S.; Jan, V.; Daniel, F. Electrochemical model of solid oxide fuel cell for simulation at the stack scale. Part I: Calibration procedure on experimental data. *J. Electrochem. Soc.* **2011**, *158*, B1102–B1118.
37. Marcin, M.; Katarzyna, B.; Shinji, K.; Janusz, S.S.; Grzegorz, B. A Multiscale Approach to the Numerical Simulation of the Solid Oxide Fuel Cell. *Catalysts* **2019**, *253*, 1–28.
38. Kwang, H.L.; Richard, K.S.; Shinji, K. SOFC cogeneration system for building applications, part 1: Development of SOFC system-level model and the parametric study. *Renew. Energy* **2009**, *34*, 2831–2838.

© 2019 by the authors. Licensee MDPI, Basel, Switzerland. This article is an open access article distributed under the terms and conditions of the Creative Commons Attribution (CC BY) license (http://creativecommons.org/licenses/by/4.0/).

MDPI
St. Alban-Anlage 66
4052 Basel
Switzerland
Tel. +41 61 683 77 34
Fax +41 61 302 89 18
www.mdpi.com

Energies Editorial Office
E-mail: energies@mdpi.com
www.mdpi.com/journal/energies

www.ingramcontent.com/pod-product-compliance
Lightning Source LLC
LaVergne TN
LVHW071944080526
838202LV00064B/6675